MODERN LOGIC

THE UNIVERSITY
of LIVERPOOL

**University
Library**

MODERN LOGIC

A Text in Elementary Symbolic Logic

GRAEME FORBES

New York   Oxford
OXFORD UNIVERSITY PRESS
1994

Oxford University Press

Oxford   New York   Toronto
Delhi   Bombay   Calcutta   Madras   Karachi
Kuala Lumpur   Singapore   Hong Kong   Tokyo
Nairobi   Dar es Salaam   Cape Town
Melbourne   Auckland   Madrid

and associated companies in

Berlin   Ibadan

Published by Oxford University Press, Inc.

198 Madison Avenue, New York, New York 10016-4314

Library of Congress Cataloging-in-Publication Data
Forbes, Graeme
Modern logic: a text in elementary symbolic logic/Graeme Forbes
p. cm.
Includes bibliographical references and index.
ISBN 0-19-508028-9— ISBN 0-19-508029-7 (pbk.)
1. Logic, Symbolic and mathematical.   2. Logic, Modern—20th century.
I. Title
BC 135.F57 1994   160—dc20   93-17282 CIP

5 7 9 8 6

Typeset in 9 on 11 Lucida Bright by Graeme Forbes

Printed in the United States of America
on acid-free paper

Preface

I first taught logic in 1976, at the behest of Gareth Evans and John Mackie, for University College, Oxford. The text then in use was E. J. Lemmon's *Beginning Logic*, and though I have subsequently used a variety of books, Lemmon's remains one of my two favorites. The other is Gustason and Ulrich's *Elementary Symbolic Logic*. The strength of Lemmon's book is its system of natural deduction, and the strength of Gustason and Ulrich's, the variety and interest of their material on the symbolization of English. But I have never been completely satisfied with any symbolic logic textbook, and so I have added my own to the groaning shelves.

The inspiration for *Modern Logic* is not dissatisfaction, however—I have lived with that quite comfortably for years—but rather the appearance of the logic program *MacLogic* from the Machine-Assisted Logic Teaching project at St Andrews University. I wrote *Modern Logic* because I wanted to teach from a text with the same system of natural deduction as *MacLogic;* hence the systems in this book are essentially those of that program. But I recognize that there are many students who will resist, or will not have access to the equipment for, computer-aided instruction. So this textbook is quite orthodox: it is self-contained and *MacLogic* is not required for using it. However, for those who may be interested in a description of how the program works, I have included a brief appendix.

Modern Logic was written with a specific audience in mind. It is intended in the first instance to provide an adequate grounding in logic for undergraduates pursuing courses of study with a substantial philosophy component—in the United States, philosophy majors, joint majors, minors, and students in interdisciplinary programs such as cognitive science, and in the United Kingdom and elsewhere, students in equivalent courses of study. So the book is definitely not a 'critical thinking' textbook. In the second instance, *Modern Logic* is intended for courses which a mathematics, computer science or linguistics department would be happy to give credit for at the beginning level in their undergraduate programs. At various points in the book I have indicated developments of the subject matter with specific mathematical, computer science or linguistics interest, but in the space available, of course, I have not been able to pursue such topics in any detail. In the third instance, the book is intended for courses which satisfy the mathematics or 'formal thought' requirement that some universities have. Its intellectual content, I hope, is at least as substantial as that of a good introductory calculus or computer science course.

The book has four parts, the material in Part IV being of a more advanced nature. I think that Parts I–III could be exhaustively covered in one semester only by going at a rather fast pace with a class of rather good students, though I have tried to write so that a considerable amount of self-teaching is possible. Typically, in a one-semester course, I have been able to cover most of Parts I and II and some of Part III, choosing in the latter between the semantics and proofs involving identity. For those who have the constraint of a one quarter course, or the luxury of a two quarter sequence, I have separated out monadic predicate logic in Part II from full first-order logic with identity in Part III, so that a coherent first course can be given based on Parts I and II alone. However, in order to avoid cognitive overload on students, the whole story about the semantic treatment of individual constants, though strictly a part of monadic predicate logic, is postponed until Part III. Parts III and IV together make a substantial second course, and I have found that Part IV by itself, supplemented with some extra reading, makes a good supervised independent study.

By inspecting the table of contents, instructors will easily see how to tailor the book to their own requirements, by omitting sections on which later material does not depend. Chapters 1, 2, 3.1–3.4, 4.1–4.8, 5, 6.1–6.5, 7.1–7.3, and 8.1–8.4, constitute the core of the book. For those who are going on to Part IV, I recommend Chapter 4.10 of the earlier material. And Chapter 8.5 is a prerequisite for Chapter 10.

There are some conventions which I have tried to follow consistently throughout the text. After Chapter 1, most sections finish with a number of exercises, arranged in roughly ascending order of difficulty; solutions to starred problems can be found at the end of the book. References for further reading are given at various points, though these are usually to convenient sources rather than original ones. Cross-references to other parts of the book are in two styles, verbose and terse, and in both styles may be context-sensitive. 'Example 4 of §5 in Chapter 3' is a verbose cross-reference, while its terse counterpart would be 'Example 3.5.4' (in terse style, the numbers are given in order of largest domain, chapter, to smallest, object-token identifier). But if these cross-references occurred in Chapter 3, they would be abbreviated to 'Example 4 of §5' and 'Example 5.4' respectively.

For those who are interested in such matters, camera-ready copy for *Modern Logic* was produced by the author on a Macintosh Quadra 700 and output to a LaserJet 4M. The text was composed in Nisus and transferred into FrameMaker for page layout; some graphical elements were created in CA-Cricket Draw III, and Chapter 11's giraffes are based on a ClickArt Studio Series image. The display font for the chapter numbers is Vag Rounded Bold and the text fonts are from the various Lucida families.[1] I am grateful to Louis Vosloo of

---

[1] Macintosh is a registered trademark, and Quadra a trademark, of Apple® Computer. Laserjet is a registered trademark of Hewlett-Packard®. Nisus is a registered trademark of Nisus® Software. FrameMaker is a registered trademark of Frame® Technology. CA-Cricket® Draw III™ is a trademark of Computer Associates®. ClickArt® Studio Series™ is a trademark of T/Maker®. Vag Rounded Bold is copyright by Adobe™ Systems. Lucida is a registered trademark of Bigelow & Holmes.

Y&Y Software for technical assistance with fonts and to the staff of Oxford University Press for advice on page layout.

Probably the hardest part of writing a logic textbook is coming up with a sufficiently large and sufficiently varied set of exercises. There are over 900 problems in *Modern Logic*, some of them borrowed. I thank Simon and Schuster for permission to reprint the problems for Chapter 3.1 from *What is the Name of This Book?* by Raymond Smullyan. I have used about twenty problems from old Oxford exams; the bulk of these are in the symbolization exercises of Chapter 2.4 and in the symbolization and proof exercises of Chapter 8.3–4. I have also used problems from old tests of my own and I suspect that some of these may have been taken from other sources whose identity I do not recall. If so, I apologize in advance to their originators for failing to credit them.

I thank my beta-testers, Robert Berman, Anne Hunt, Norton Nelkin, Michael Ochoa, and Tracy Tucker, who taught from drafts of the book; David Boerwinkle for his proofreading help; and David Bostock, Kit Fine, Steve Kuhn, Xiaoping Wang and Timothy Williamson for suggesting a number of improvements. For reading the whole manuscript carefully, catching many errors, and making many invaluable suggestions, I am particularly indebted to Roy Dyckhoff, Stephen Read and two sets of readers for Oxford University Press, one of whom did it *twice!*

Finally, I thank my wife, Marilyn Brown, for a number of ideas, one of which, that the book be called *Postmodern Logic*, I rejected with some regret.

New Orleans   G. F.
June 1993

Contents

Part I
CLASSICAL SENTENTIAL LOGIC

Part II
MONADIC PREDICATE LOGIC

Part III
FIRST-ORDER LOGIC WITH IDENTITY

Part IV
EXTENSIONS AND ALTERNATIVES TO CLASSICAL LOGIC

PART I

CLASSICAL SENTENTIAL LOGIC

1   What Is Logic?

1   Arguments

Symbolic logic is usually described as the study of the difference between *valid* and *invalid* arguments, so we begin with an explanation of this terminology. An argument is a piece of discourse which contains some *premises*, a *conclusion*, and perhaps also some reasoning in which an attempt is made to *derive* the conclusion *from* the premises. The premises are those statements which, for the purposes of the argument, are being accepted as true. The conclusion is the statement whose truth the argument aims to demonstrate, given acceptance of the premises. Hence this 'logical' sense of 'argument' is quite different from its sense in 'Smith and Jones are having an argument', though an argument in the logical sense might be reconstructible from what Smith or Jones says.

    For present purposes, we take an argument to consist just in a listing of the premises, and then the conclusion. A *valid* argument is an argument whose conclusion *follows* from its premises, and correspondingly, an invalid argument is an argument whose conclusion *does not follow* from its premises. Here are two very simple arguments, one valid, the other invalid, which illustrate the difference between the conclusion following from the premises and its not following:

> A: (1) If our currency loses value then our trade deficit will narrow.
> (2) Our currency will lose value.
> (3) ∴ Our trade deficit will narrow.[1]

> B: (1) If our currency loses value then our trade deficit will narrow.
> (2) Our trade deficit will narrow.
> (3) ∴ Our currency will lose value.

In argument A, the truth of the conclusion (3) is guaranteed by the truth of the two premises (1) and (2): *if* (1) and (2) are true, then (3) will be true as well. So (3) follows from (1) and (2). But it is not difficult to see that in argument B, the truth of (1) and (2) does not guarantee the truth of (3). (1) of B says that a depre-

---

[1] The symbol '∴' abbreviates 'therefore'.

ciating currency will lead to a narrowing trade deficit but leaves it open whether there are other conditions which would also lead to a narrowing trade deficit,
for example, the institution of protectionist trade policies. So the truth of (2) in
B does not justify concluding that our currency will lose value: if there *are* other
conditions which would lead to a narrowing trade deficit, it may be one of them
which accounts for the truth of (2), without our currency having lost value at
all. This means that in B, (3) does not follow from (1) and (2).

This discussion of A and B brings out two important points:

- In discerning the difference between A and B we did not rely on any
  special knowledge about economics. The fact that the conclusion of A
  follows from its two premises is something that should be evident
  even to the reader who has no views about how the trade deficit could
  be narrowed: all that is required is an understanding of the English
  words 'if...then...' and an ability to recognize the recurrence of the sentences 'our currency loses value' and 'our trade deficit will narrow'.[2]
  Equally, even if you cannot think of any alternative method of producing a narrowing of the trade deficit, you should still be able to see that
  nothing in the premises of B rules out the existence of such a method,
  and that is enough to tell you that the conclusion of B does not follow
  from its premises.

- In assessing A and B we did not try to settle whether the premises are
  in fact true, or whether the conclusion is in fact true. Whether or not
  the conclusion of an argument follows from its premises is a conditional matter, having to do with whether, *if* the premises are true, that
  by itself guarantees the truth of the conclusion. If the premises of an
  argument are in fact true and the conclusion in fact false, then of
  course the premises do not guarantee the truth of the conclusion. But
  if, say, the premises and the conclusion are all in fact false, the argument may still be valid, and if they are all true, the argument may still
  be an invalid one. For example, A is a valid argument, but it is quite
  conceivable that its premises and conclusion are all false, and B is an
  invalid argument, though it is equally conceivable that its premises
  and conclusion are all true.

We can label these two points *topic neutrality* and *independence of actual
truth or falsity*. The techniques of logical evaluation which we apply to arguments do not depend on an argument's topic or on whether its premises and
conclusion are actually true or false. Their focus lies elsewhere, because the
property of an argument which determines whether it is valid or invalid has
nothing to do with its topic or with whether its premises and conclusion are

[2] In A and B the sentence 'our currency loses value' does not strictly speaking itself recur, since for
grammatical reasons we have to change the tense from future to present. Tense differences are one
kind of difference between sentences which are often unimportant from the point of view of sentential logic (others will emerge as we go along). In such cases, we count sentences which differ only as
regards tense as having the same content, and we use the same sentence letter for them.

actually true or false. To see what the relevant property of an argument is, compare A and B with the following two arguments:

C: (1) If the Continuum Hypothesis is true, the Zermelo-Frankel universe is constructible.
(2) The Continuum Hypothesis is true.
(3) ∴ The Zermelo-Frankel universe is constructible.

D: (1) If the Continuum Hypothesis is true, the Zermelo-Frankel universe is constructible.
(2) The Zermelo-Frankel universe is constructible.
(3) ∴ The Continuum Hypothesis is true.

In C the conclusion follows from the premises (even though (3) in C is false) while in D it does not. And a comparison with A and B shows that the reason C is valid is the same as the reason A is, and the reason D is invalid is the same as the reason B is. The reader is not expected to know so much as what the premises and conclusion of C and D mean. But in considering D, it should nevertheless be evident that there may be some other condition which would confer constructibility (whatever that is) on the Zermelo-Frankel universe (whatever that is) besides the truth of the Continuum Hypothesis (whatever that is).[3] Hence (1) and (2) do not warrant drawing the conclusion (3).

2 Logical form and validity

If A and C are valid for the same reason, and B and D are invalid for the same reason, this suggests that what makes A and C valid is something they have in common, and what makes B and D invalid is something they have in common. What is common to A and C is their pattern. Each argument has as its first premise an 'if...then...' sentence; in each argument, the sentence following the 'if' in premise 1 is itself premise 2 and the sentence following the 'then' in premise 1 is itself the conclusion. Similarly, B and D have the same pattern, importantly different from that of A and C: the first premise of each is an 'if...then...' sentence, the second premise is the sentence following the 'then' in the first premise, and the conclusion is the sentence following the 'if' in the first premise. We can exhibit the pattern common to A and C by abstracting from the particular English sentences which occur in them. In place of sentences, we use letters 'P', 'Q', 'R' and so on. Then the common pattern of A and C can be written out as:

---

[3] An excellent source for information on the Continuum Hypothesis and the Zermelo-Frankel universe (and on many other topics not discussed in much detail in this book) is Partee *et al;* see Part A for set theory.

E: If P then Q
    P
    ∴ Q

Similarly, the common pattern of B and D is given by

F: If P then Q
    Q
    ∴ P

When an argument instantiates a certain pattern, that pattern is called the *logical form* of the argument, and it is to its logical form that we look to settle whether an argument is valid or invalid. However, there is a complication. In this book we shall successively present three systems of logic, each of which extends the previous system, beginning with *sentential* logic, the system of Part I. In sentential logic, the basic units are letters like 'P' and 'Q' above, which stand for complete sentences, like the sentences 'the Continuum Hypothesis is true' and 'our trade deficit will narrow' from which the arguments of the previous section were constructed (hence the name 'sentential' logic). Note that we do not allow the use of letters to stand for complete sentences which contain other complete sentences as constituents, connected with expressions like 'if...then...'; for example, we would never represent (1) in D by a single letter.

With a more powerful system of logic, in which the basic units are smaller than complete sentences, we can often give a more discerning description of the form of an English argument than is possible in sentential logic. Thus it is a misnomer to speak of 'the' logical form of an argument; the form is relative to the system of logic under discussion. Hence E and F give what we call the *sentential logical forms* of A through D. Notions like validity and invalidity for English arguments are also relativized to the system of logic, and so at the moment we are discussing just validity in sentential logic, or *sentential validity*. For an argument to be sententially valid is for it to have a valid sentential form. A sentential form is a form whose basic units are letters like the 'P' and 'Q' of E and F. A *valid* sentential form is one such that in any English argument with that form, the conclusion follows from the premises. For the moment, we will assume that with each English argument we can associate just one sentential form, so that it is permissible to speak of 'the' sentential logical form of an English argument; later, we shall see that this is debatable. In addition, we shall often omit the qualifier 'sentential' in Part I of this book, on the understanding that by the logical form of an English argument we mean, in Part I, the form ascribed to the argument in sentential logic by the translation procedures we are going to develop.

A little reflection tells us that any English argument which fits the pattern E, that is, which has the logical form E, will be a sententially valid argument, regardless of its subject matter and regardless of whether its premises and conclusion are true or false. Similarly, any English argument which has the logical form F will be a sententially invalid argument: the first premise is consistent with there being other conditions besides P which would bring about the truth of Q, so despite the second premise informing us of the truth of Q, there is no warrant for inferring the truth of P.[4]

In making these judgements, we use a relativized notion of validity and a relativized notion of invalidity. However, there is an asymmetry between these notions. If an English argument is sententially valid, then it is valid absolutely; even if we can construct a more finely grained form for it in the logical systems of Parts II and III of this book, those forms will still pronounce the argument valid. This is because the methods of sentential logic remain available in more powerful systems, so if an argument's conclusion has been shown by these methods to follow from its premises, it will still follow in more powerful systems. Thus a positive verdict on an English argument at the level of sentential logic cannot be overturned later. But a negative verdict *can* be overturned: if an argument is sententially *in*valid, it can still turn out to have a valid form in a more advanced system, with more powerful methods for demonstrating that the argument's conclusion follows from its premises (think of telescopes with different magnifications—what can be seen with the less powerful telescope can be seen with the more powerful, but not necessarily vice-versa). The curious reader who would like examples of sententially invalid arguments which are valid in stronger systems will find some at the start of Part II. However, the examples of sententially invalid arguments which are to be found in Part I will not be like those later examples; as well as being sententially invalid, they will be invalid in stronger systems of logic as well, and thus invalid absolutely. In particular, there will be no future redemption for arguments B and D.

Here are two more examples of English arguments, one valid, the other invalid, and so simple as hardly to merit the designation 'argument'.

> G: (1) Nixon was president and Agnew was vice president.
>   (2) ∴ Nixon was president.

> H: (1) Either Nixon was president or Agnew was.
>   (2) ∴ Nixon was president.

In G the conclusion follows from the premise since the truth of the premise guarantees the truth of the conclusion. The sentential logical form of G is

> I: P and Q
>   ∴ P

and it is clear that any English argument with this form, be it about politics, economics, mathematics or sport, will be a sententially valid argument and therefore valid absolutely. In H the conclusion does not follow from the premise: if our only information is that one or the other of Nixon or Agnew was president, then we have no warrant for concluding that it was Nixon. The form of H is

---

[4] A note to the instructor: no argument whose conclusion is a tautology could have the form F since we do not allow the use of a sentence-letter for an English sentence which contains sentential connectives.

> J: Either P or Q
>   ∴ P

and again it is clear that any English argument will be sententially invalid if our procedures for determining logical form ascribe the form J to it: if our only information is that one or the other of two basic alternatives is the case, we are not warranted in drawing any conclusion about which it is. Of course, it would be simple to turn H into a valid argument by adding an extra premise:

> K: (1) Either Nixon was president or Agnew was.
>   (2) It was not Agnew.
>   (3) ∴ Nixon was president.

The sentential form of K is

> L: Either P or Q
>   It is not the case that Q
>   ∴ P

and any English argument with this form will be an absolutely valid argument, since if one of two alternatives holds (premise 1) and we know which one does not hold (premise 2), it must be the other which does hold.

We said above that logic is the study of the difference between valid and invalid arguments. We have now seen that whether or not an argument is valid depends on its form, so we can revise our definition of logic to say that it is the study of the difference between valid and invalid argument-forms. An important point to note is that we are interested in the difference between forms that are *in fact* valid and forms that are *in fact* invalid. This is not quite the same as the difference between what *seems* valid to us and what *seems* invalid, even for the simplest sentential forms. The latter distinction is investigated in cognitive psychology, and there is plenty of experimental evidence that the judgments people make about validity and invalidity do not always reflect the facts. For example, experimental subjects sometimes resist arguments with the form I. It seems that in judging the 'acceptability' of a piece of reasoning, we often use criteria in addition to validity; for instance, we prefer not to draw conclusions which involve a loss of information from the premises, such as occurs in G. This preference has a certain practical utility, but it is irrelevant to the logical issue of whether the conclusion of the argument follows from the premises, and it is only this issue with which we are here concerned. (For more information about psychological aspects of deductive reasoning, see Johnson-Laird and Byrne.)

3 Complications

The examples we have looked at so far have been very simple, and their intuitive statuses as valid or invalid more or less evident to inspection. When we look at more realistic examples of English arguments, there are at least two ways in which they may be harder to assess than any of the foregoing. One problem is that it may be difficult to determine what the argument *is*, for in ordinary discourse, we rarely make our premises completely explicit, list them in order and clearly demarcate our conclusions. Here is an example of a real argument which leaves the reader a lot of work to do. It occurred at an American Bar Association debate on a motion calling for legislation barring discrimination against homosexuals in employment and housing. The *New York Times* reported one delegate speaking against the motion as making the following argument:

> This is not to me a question of unjust discrimination, because it's not to punish someone for what they are, but for what they do. They don't have to be homosexuals.

Perhaps we can reconstruct the intended argument as follows:

M: (1) If a type of behavior is freely chosen then it is permissible to discriminate against people who behave in that way.
(2) Homosexual behavior is freely chosen.
(3) ∴ It is permissible to discriminate against homosexuals.

Reconstructed in this manner, the argument is clearly valid, so if its premises are true then its conclusion must be true as well. But it is not certain that M captures the speaker's intention—does 'They don't have to be homosexuals' mean that sexual orientation is a matter of choice or just that *manifestation* of sexual orientation is under voluntary control?—and second-guessing what someone may have intended is not really a part of logic: logic is about what makes arguments in general valid or invalid, and psychological speculation about which particular argument a speaker may have been expressing on a given occasion is out of place. So we will not be concerned with this kind of interpretative difficulty in what follows but instead will confine ourselves to arguments which are expressed much more explicitly. The example, however, does help to emphasize the distinction between the questions (a) does the conclusion of the argument follow from its premises? and (b) are the premises true? If the answer to (a) is 'yes', then the argument is valid. If the answer to (b) is also 'yes', then the conclusion must be true as well. Of course, the answer to (b) in the case of M is highly controversial.[5]

It would be useful to have a term for arguments to which the answer to *both* (a) and (b) is 'yes'. Such an argument, one whose form is valid *and* whose premises are true, is said to be *sound*. So we have the following definitions:

> A *valid* argument is an argument with a valid form; concomitantly, an invalid argument is one that does not have a valid form.

---

[5] Premise 1 could justify religious discrimination. Premise 2 appears to ignore the likely contribution to an individual's sexual orientation of factors beyond his or her control. Aspiring lawyers concerned that law school may be too intellectually demanding should take heart from this example.

A *valid argument-form* is a form which guarantees that the conclusion of an argument with that form is true if its premises are (refer to E, I and L in §2).

A *sound* argument is a valid argument with true premises.

Hence the conclusion of a sound argument is guaranteed to be true. These definitions are not limited to sentential logic. They capture the intuitive idea of validity which all our systems of logic aim to articulate.

The second way in which arguments may be harder to assess than were our earlier examples is that the complexity of the argument may be much greater. Here is an example which we will meet again:

> If the safe was opened, it must have been opened by Smith, with the assistance of Brown or Robinson. None of these three could have been involved unless he was absent from the meeting. But we know that either Smith or Brown was present at the meeting. So since the safe was opened, it must have been Robinson who helped open it.

Some readers may be able to discern that this is a valid argument, though they may feel unsure of their judgement. Is there any methodical procedure we could follow to determine without question that the argument is valid? Since validity is fundamentally a property of logical forms, we could begin by writing out the sentential form of the argument (this is an exercise in Chapter 2). However, the result will be much more complicated than such forms as I or J or L above. What do we do with a complicated form once we have it?

There are in fact various tests which can be applied to an argument-form to find out if it is valid or invalid, and in later chapters we will develop the techniques on which tests of this sort are based. However, the result of a test on the argument-form gives us information about the English argument whose form we say it is only if we are *right* that the form in question is that of the argument. To begin with, therefore, we need to acquire facility in the representation of arguments, and of English sentences in general, in the kind of symbolic notation to which the argument-forms we have already exhibited approximate.

4 A note to the reader

What can the reader expect to learn from the study of logic? What 'use' is it? Grasp of logic plays a central role in reasoning capacities, and humans seem to be endowed with these to a greater degree than other animals, at least on the average. On the other hand, a sense of humor is also fairly characteristic of humans—again, on the average—so the rationale for studying logic is not just that logic is a distinguishing feature of human nature. The main reason is that the distinction between valid and invalid arguments gives rise to a rich theoretical structure in a way that the distinction between funny and unfunny jokes does not. And rich theoretical structures are worth studying for their own sake.

There is in addition a more practical aspect of the subject, though one that has only emerged in the last few decades, concerning the role of logic in computation. It is not for nothing that the main component board of the modern personal computer is called a 'logic board'. On the software side, the most prominent application of the principles developed and studied in this book is in a kind of program called an 'expert system'. An expert system contains a database of facts about some domain linked to an inference engine which allows the program to draw conclusions if it is given information about a situation within its domain as premises. Expert systems are used, for example, in medical diagnosis, oil and gas exploration, and financial decision making, in all of which they are generally more reliable than people. The deductive systems studied in this book are abstract inference engines of the sort implemented by such programs.

A natural question to ask about the subjective impact of studying logic is whether working through a book like this will make the reader a more logical person. Some systematic research has been done in this area, with results that are not entirely comforting to logic teachers. On the anecdotal side, however, a student once told the author that after a month of his symbolic logic course, she had found herself able to rout her boyfriend every time they had an argument. She added that she had then lost the boyfriend. Fortunately, she seemed quite pleased with this outcome.

5 Summary

- Logic is the study of the distinction between valid and invalid arguments.
- A valid argument is one whose conclusion follows from its premises.
- Whether or not an argument's conclusion follows from its premises depends on the form of the argument.
- The form we discern in an argument depends on which logical system we have at our disposal.
- We begin with the simplest system of logic, known as sentential logic.
- In sentential logic the forms we discern are called sentential logical forms.
- To test an English argument for sentential validity, we first write out its sentential form and then test that form for validity.
- An argument which is sententially valid is valid absolutely, but a sententially invalid argument may prove to be valid in the framework of a more powerful system of logic.
- While studying this book, readers should be careful not to intimidate their friends unintentionally.

2 First Steps in Symbolization

1 The sentential connectives

The premises and the conclusions of the arguments of Chapter One are complete sentences of English, but some of these sentences themselves consist in shorter complete sentences combined into a single sentence using 'and', 'if... then...' and so on. For example, premise 1 of A on page 3, 'If our currency loses value then our trade deficit will narrow', is a *compound* sentence whose *constituent* sentences are 'our currency loses value' and 'our trade deficit will narrow', connected together using 'if...then...'. A phrase like 'if...then...', which can join shorter sentences to make a longer one, is called a *sentential connective*, sentential logic being that system of logic concerned with the logic of sentential connectives. To be more specific, in our presentation it is concerned with the logic of exactly five sentential connectives:

- it is not the case that...
- either...or...
- both...and...
- if...then...
- ...if and only if...

In this list, each occurrence of an ellipsis '...' indicates a position which may be filled by a complete sentence (which may itself contain further connectives). A connective which attaches to one sentence to form a compound sentence is called a *one-place* connective, and a connective which combines with two sentences to form a compound sentence is called *two-place*. So in our list, there is a single one-place connective, 'it is not the case that...', and the others are all two-place.

Here are some technical terms which we will often use in talking about sentences and sentential connectives:

- A sentence which contains no occurrence of any sentential connective is called a *simple* or *atomic* sentence.
- A sentence which contains at least one occurrence of at least one sentential connective is called a *complex* or *compound* sentence.

Writing out the connectives in English whenever they occur rapidly becomes tiresome: a shorthand notation saves space and time. We will therefore settle on the following abbreviations:

- For 'it is not the case that...' write '~...'
- For 'both...and...' write '... & ...'
- For 'either...or...' write '... ∨ ...'
- For 'if...then...' write '... → ...'
- For '...if and only if...' write '... ↔ ...'

There is important terminology which attaches to these abbreviations, and which should be memorized:

- A sentence beginning with '~' is a *negation*, and the subsentence to which the '~' is prefixed is the *negated sentence*.
- A sentence consisting in two shorter sentences connected by '&' is a *conjunction*, and the connected sentences are the *first* and *second conjuncts* respectively.
- A sentence consisting in two shorter sentences connected by '∨' is a *disjunction*, and the connected sentences are the *first* and *second disjuncts* respectively.
- A sentence consisting in two shorter sentences connected by '→' is a *conditional;* the shorter sentence before the '→' is the *antecedent* and the shorter sentence after the '→' is the *consequent*.
- A sentence consisting in two shorter sentences connected by '↔' is a *biconditional;* the shorter sentence before the '↔' is the *left-hand side* and the shorter sentence after the '↔' is the *right-hand side*.

The symbols themselves also have names, though these are less important. '~' is the familiar *tilde* symbol from typewriter or computer keyboards; '∨' is called the *wedge,* or sometimes the *vel* (after the Latin word 'vel' for 'or'); '&' is the *ampersand* symbol, again familiar from typewriter or computer keyboards; '→' is called the *arrow;* and '↔' is called the *double-arrow.* In other logic textbooks one sees different symbols for these five English sentential connectives, but we will use the ones just listed exclusively.

There are of course many other sentential connectives besides the five listed. For example, there is 'neither...nor...' as in 'neither will our currency lose value nor will our trade deficit narrow' (or more colloquially, 'our currency will not lose value, nor will our trade deficit narrow'), and there is '...after...', as in 'our currency will lose value after our trade deficit narrows'. These sentential connectives and others like them fall into two groups: in the first group there are connectives like 'neither...nor...' which are *definable* in terms of the five connectives we have already introduced, and in the second group there are connectives like 'after' which are, as we might put it, beyond the scope of classical sentential logic (what this means will be fully explained in §6 and §7 of Chapter 3). So in Part I of this book our five connectives '~', '&', '∨', '→' and '↔' are all we need.

2 Negations, conjunctions and disjunctions

The simplest sentences to formalize are those which contain no occurrences of connectives, that is, atomic sentences. To symbolize an atomic sentence one simply uses a single letter, known as a *sentence-letter*. Thus

> (1) Our currency will lose value

is symbolized as, say,

> (1.s) C.

Sentences with one explicitly employed connective are equally straightforward, as in

> (2) Our currency will lose value and our trade deficit will narrow

which is

> (2.s) C & N

where 'N' stands for 'our trade deficit will narrow'.
   However, when two or more connectives occur in a sentence, problems of interpretation can arise. Consider

> (3) Either our currency will lose value, or exports will decrease and inflation will rise.

Using 'E' for 'exports will decrease' and 'I' for 'inflation will rise', we could symbolize (3) as

> (3*) C ∨ E & I.

But (3*) can be read two ways, as a conjunction whose first conjunct is 'C ∨ E' or as a disjunction whose second disjunct is 'E & I'. These correspond to two different ways of reading the English, but in English the correct interpretation is settled by the position of the comma: everything that follows the comma is grouped together. So (3) is a disjunction whose second disjunct is 'exports will decrease and inflation will rise'. In symbolizations we do not use commas, but we can get the same effect by using parentheses to group the appropriate sentence-letters together. The correct symbolization of (3) is written:

> (3.s) C ∨ (E & I).

(3.s) is to be distinguished from (4), which is the incorrect reading of (3):

> (4) (C ∨ E) & I.

The difference between (3.s) and (4) can be brought out by comparing the following two English sentences which instantiate them, 'either I will keep to the speed limit, or I will break the speed limit and I will be killed', versus 'either I will keep to the speed limit or I will break the speed limit, and I will be killed'. Only the second of these says that death is inevitable.

The difference between (3.s) and (4) introduces the idea of the *scope* of a connective in a sentence. We will later give a precise definition of the notion of scope, but for the moment let us say that the scope of a two-place connective in a sentence, or *formula*, is that part of the formula enclosed in the closest pair of matching parentheses within which the connective lies, if there is such a pair, or else the entire formula, if there is not. So the scope of '&' in (3.s) is 'E & I' while the scope of '&' in (4) is the entire formula. The function of the parentheses in (3.s) and (4) is to tell us what reading of (3*) is intended, but we can also describe this function as that of *fixing the scope* of the connectives in (3*), since the different ways of reading (3*) differ only in assigning different scopes to the two connectives which occur in each: in (3.s), '∨' has '&' within its scope, while in (4), '&' has '∨' within its scope. The *main* connective of a formula is that connective in it which is not within the scope of any other. So '∨' is the main connective of (3.s) and '&' is the main connective of (4). Again, we will give a more adequate definition later.

Parentheses are only really required when a string of symbols can be read in two or more ways *and* the different readings mean different things. So there is no need to use parentheses on the string of symbols 'C & E & I'; though we can group the conjuncts either as '(C & E) & I' or as 'C & (E & I)', there is no difference in meaning between these two formulae. However, we will later adopt some simple rules about how to use parentheses, and these rules will indeed require parentheses in 'C & E & I'. The reason for this is that following a universal rule mechanically is easier than qualifying it with a list of exceptions. It is also useful to settle on a particular occurrence of '&' in 'C & E & I' as the main connective.

The use of parentheses could be minimized by introducing a collection of conventions about how a formula is to be read if it lacks parentheses, in other words, by introducing a collection of conventions about the scope of connectives in otherwise ambiguous strings of symbols like (3*). For example, we could have a scope convention that '&' takes the smallest scope possible, which would force the reading 'C ∨ (E & I)' on (3*). But conventions for two-place connectives can be difficult to remember, so we will only have one scope convention, (SC~), for the one-place connective '~':

(SC~)   '~' takes the smallest possible scope.

Hence although the string of symbols '~P & Q' can be read in two different ways, as '(~P) & Q' or as '~(P & Q)', (SC~) requires that we interpret it in the first of these two ways. If the second is what we want, then we must put the parentheses in explicitly.

In many cases, the exact symbolization of a sentence of English is not immediately obvious. For example, although

(5)   Our budget deficit and our trade deficit will both rise

contains the connective 'and', the connective does not join two complete sentences, since the phrase 'Our budget deficit' on its left is not a complete sentence. But (5) means the same as a sentence in which 'and' does join two complete sentences, namely

(6)   Our budget deficit will rise and our trade deficit will rise.

(5) is merely a more economical syntactic variant of (6); in English grammar, (5) is said to result from applying *conjunction reduction* to (6). In symbolizing sentences which contain sentential connectives that apparently do not join complete sentences, but which are simply economical variants of sentences where the connectives do join complete sentences, the first step is to recover the longer sentence in which the connectives join only complete sentences, such as (6); the second step is to choose sentence-letters for the atomic sentences which appear in the longer sentence; and then the last step is to write out the formula. Let us call the longer sentence in which the connectives join complete sentences the *full sentential form* of the original sentence. We have expanded (5) to its full sentential form (6), so if we use the sentence-letter 'B' for the atomic sentence 'our budget deficit will rise' and the sentence-letter 'T' for the atomic sentence 'our trade deficit will rise', we can symbolize (5) as

(5.s)   B & T.

Notice, however, that there are examples rather like (5) in appearance but where expansion in the style of (6) would be wrong. For instance, 'Bill and Ted are friends' does not mean 'Bill is a friend and Ted is a friend'. The 'and' in this example is said to have a 'collective' sense, which is not expressed by '&'. In symbolizing a sentence as a conjunction, therefore, we must make sure that its 'and' really does join two complete sentences. 'Bill and Ted are friends' could only be symbolized by a single sentence-letter in sentential logic if justice is to be done to the 'and'.

Another common complication in symbolizing an English sentence is that it may contain a connective which is not one of the five connectives listed at the start of this chapter. However, the new connective may be expressible by one of our five or by some combination of them. The simplest example involves 'but'. Using 'J' for 'John will attend the meeting' and 'M' for 'Mary will attend the meeting', we can symbolize

(7)   John will attend the meeting but Mary will not

as

(7.s)   J & ~M

since for logical purposes there is no difference between 'and' and 'but': so far as who is at the meeting is concerned, (7) may as well say 'and' instead of 'but'.[1]

Negation is sometimes expressed not by 'not', but rather by some prefix attached to a word. For instance,

(8) Insider trading is unethical and illegal

contains two occurrences of 'not', one as 'un' and the other as 'il'. So using 'E' for 'insider trading is ethical' and 'L' for 'insider trading is legal', and expanding (8) to its full sentential form 'Insider trading is unethical and insider trading is illegal', we would symbolize it as

(8.s) ~E & ~L.

Notice, however, that not every use of a prefix like 'un' can be treated as the sentential negation 'it is not the case that'. For example, 'an unintelligible book often sells well' does not mean the same as 'it is not the case that an intelligible book often sells well'. Each example must be considered on its own merits.

Just as negation need not be expressed by 'not', conjunction need not be expressed by 'and'. Consider the following:

(9) At least one of our two deficits, budget and trade, will rise.

Though (9) contains 'and', the crucial phrase is 'at least one', which has the effect of disjunction. Thus the full sentential form of (9), paraphrasing to make the occurrence of disjunction explicit, is simply

(10) Either our budget deficit will rise or our trade deficit will rise.

Using the same symbols as for (6), the symbolization would therefore be

(10.s) B ∨ T.

However, the claim that (9) and (10) mean the same, that is, that (10) is a correct paraphrase of (9), may be challenged. Suppose that both our deficits rise. Then (9) is certainly true, but is (10)? Or does (10) carry the implication that one deficit or the other will rise, *but not both?* If it does, then (10) would be false if both deficits rise, and so cannot mean the same as (9). Sometimes the word 'or' seems to carry the implication 'but not both', for example in the prediction 'tonight I will either watch television or read a good book'; and sometimes it does not carry this implication, as in 'you can take an advanced logic course if either you have taken an introductory course or you have taken an advanced mathematics course'—obviously, one would not exclude a student

---

[1] Of course, there is *a* difference between 'and' and 'but'. (7) conventionally carries the suggestion that there is some antecedent expectation of Mary's accompanying John, whereas 'and' would not suggest this. But that difference does not affect what can be validly deduced from (7) (or what (7) can be validly deduced from), so it is not a difference of any *logical* import. It is not a difference in what would make the statement *true*, but only in what would make 'but' rather than 'and' *appropriate*.

from an advanced logic course on the grounds that she has taken *both* an introductory logic course and an advanced mathematics course! What are we to
make of this apparently shifting nature of the meaning of 'or'?

One explanation of the shift is that the English connective 'either...or...' is
ambiguous between two senses, an *inclusive* sense, meaning at least one, and
an *exclusive* sense, meaning one or the other but not both, that is, exactly one.
The sense 'either...or...' has in any particular occurrence would then depend on
context. In the exclusive sense of 'or', (10) would be false if both deficits rise,
while in the inclusive sense, it would be true. However, there is another possible
account of the disagreement. Perhaps those who hear (10) as false in the case
where both deficits rise are confusing the question of whether or not (10) is
true in this situation with whether or not it would be appropriate to assert (10)
if one knew this was the situation. On this view, (10) would be *inappropriate* if
both deficits were known to be rising—one should assert the conjunction
instead—but it is still literally *true*.

If the first explanation of the disagreement over (10) is correct, then we
should simply *decide* which of the two senses of 'or' we will use our symbol '∨'
to express. If the second explanation is correct, then the inclusive sense is the
only sense of 'or'. Since using '∨' to express the inclusive sense is unobjectionable by both accounts, that is the use we will make. *In this book, then, 'either...-
or...' is always understood inclusively;* when we want to express the exclusive
sense, we will explicitly use the words 'and not both'.

Finally, there are English connectives which require a combination of our
five to express. 'Neither...nor...' is a familiar example, as in

(11)   Neither John nor Mary will attend the meeting.

To say that neither of the two will attend is to say that John will not attend *and*
Mary will not attend (this unpacks the meaning of 'neither...nor...' and also
gives a full sentential form). So using 'J' for 'John will attend the meeting' and
'M' for 'Mary will attend the meeting' as before, (11) comes out as

(11.s)   ~J & ~M.

(11) should be distinguished carefully from

(12)   John and Mary won't both attend the meeting

which does not mean that they will *both* be absent, but rather, means that *at
least one* will be absent (that is, at most one will attend). In other words, there
is a significant difference between 'won't both attend' and 'both won't attend'.
(12) can be formalized equally naturally in two ways (still using 'J' and 'M' as
above). If we follow the structure of the English, we have a formula in which '~'
has a conjunction within its scope, reflecting the occurrence of "won't" in (12)
preceding 'both'. This gives

(12.s)   ~(J & M).

On the other hand, if we use our paraphrase of (12) as 'at least one will be absent', where 'absent' is 'not in attendance', and expand the paraphrase to its full sentential form 'one will not attend or the other will not attend', the symbolization we get is

(12.s′)   ~J ∨ ~M.

We will later explain what it is for two formulae to be *logically equivalent*, and (12.s) and (12.s′) will be an example of a pair of equivalent formulae. But (12.s) is our preferred symbolization of (12), since it follows (12)'s syntax more closely. It is by following the English syntax as closely as possible that we get a single preferred symbolization of an English sentence (*modulo* choice of sentence-letters). This procedure is also more likely to lead to a *correct* symbolization. Nevertheless, since (12.s) and (12.s′) are equivalent, there is no logical objection to using one rather than the other in symbolizing (12).

❏ Exercises

Symbolize the following sentences of English, saying what atomic sentences your letters stand for. Write out these sentences so that they are grammatically complete, and also explicit (replace words like 'they' and 'it' with the phrases which they stand for in the context). Be sure that your simple sentences do not contain any connective-words, and that you do not use two sentence-letters for what is essentially the same English sentence. Use the suggested sentence-letters where given, and use no connectives other than '~', '&' and '∨'.

(1) Van Gogh's pictures are the world's most valuable, yet they are not the most profound. (Use 'E', 'P')

(2) Van Gogh's pictures are not the world's most valuable, but they are the most profound.

(3) Van Gogh's pictures are neither the world's most valuable nor the most profound.

(4) Van Gogh's pictures aren't both the world's most valuable and the most profound.

(5) Neither digital computers nor neural networks can simulate every aspect of human intelligence, though each can simulate some. (D, N, E, O)

*(6) Even though inflation is falling, the government can't guide the economy wisely and at the same time regain its popularity.

(7) Smith, Brown and Robinson are politicians, yet all three of them are honest. (6 letters)

(8) Smith, Brown and Robinson are persuasive, though none of them is honest.

*(9) Smith, Brown and Robinson are lawyers—still, at least two of them are honest.

(10) Smith, Brown and Robinson are ethicists, but just one of them is honest.

(11) The favorite won't win, but not for want of trying. (W, T)

(12) It never rains but it pours. (R, P) [Hint: (12) is a saying to the effect that troubles don't come just in ones and twos but in large numbers.]

3 Conditionals and biconditionals

English has many ways of expressing the indicative conditional. So far we have met only the most straightforward 'if...then...' construction, as in

>   (1) If Smith bribes the instructor, then Smith will get an A

for which, using 'R' for 'Smith bribes the instructor' and 'G' for 'Smith will get an A', we have the formula

>   (1.s) R → G.

In a conditional formula such as (1.s), the part or *subformula* on the left of the arrow is the antecedent and the subformula on the right, the consequent. So the antecedent is the first subformula we meet. But in English, the antecedent does not have to come first. (1) means exactly the same as

>   (2) Smith will get an A if Smith bribes the instructor

and has exactly the same symbolization (1.s). In (1) and (2), the antecedent is the subsentence which follows the 'if', which in an inversion like (2) is not the subsentence which comes first.
　　Another way of expressing the same idea as (1) and (2) is

>   (3) Smith's bribing the instructor is a *sufficient condition* for Smith to get an A.

One condition is sufficient for another when the first's obtaining is *all* that is required for the second's obtaining, and since (1) and (2) both say that all that is required for getting an A is bribing the instructor, (1) and (2) mean the same as (3). But we do not symbolize (3) using sentence-letters for 'Smith's bribing the instructor' and 'Smith to get an A', since these phrases are not complete sentences. The complete sentence which corresponds to the condition of Smith bribing the instructor is just 'Smith bribes the instructor', and similarly, the sentence which corresponds to the condition of Smith getting an A is 'Smith gets an A'. So we can use 'R' and 'G' again and (3) is symbolized by (1.s), just like (1) and (2).
　　(1), (2) and (3) contrast with

>   (4) Smith will get an A *only* if Smith bribes the instructor

and

>   (5) Smith's bribing the instructor is a *necessary* condition for Smith to get an A.

One condition is said to be *necessary* for another when the obtaining of the first is *required* for the obtaining of the second. Thus (5) says that a bribe is required for getting an A, which is what (4) says as well. This means that there are two ways in which (4) and (5) differ in meaning from (1), (2) and (3). First, whereas (1), (2) and (3) say that *all* Smith has to do to get an A is to bribe the instructor, (4) and (5) do not say this: they both leave open the possibility that there is *more* Smith must do to get an A besides give the bribe. Hence (1), (2) and (3) would all be false in a situation where Smith's instructor demands both a bribe and regular class attendance before he gives anyone an A, while (4) and (5) would be true in such a situation.

On the other hand, there is something (1), (2) and (3) leave open that (4) and (5) do not, and that is whether there is a way of getting an A *without* bribing the instructor, say by doing excellent work. (4) and (5) exclude this possibility, but (1), (2) and (3) do not. So how should (4) and (5) be symbolized? There is a natural way of paraphrasing (4) and (5) which answers this question, for both say the same as

    (6) If Smith does not bribe the instructor he will not get an A,

that is,

    (6.s) ~R → ~G.

Another approach is suggested by the thought that (4) seems to say that if Smith gets an A then he has bribed the instructor. Though 'bribes' changes to 'has bribed' in this paraphrase, we can abstract from this. Thus as an alternative to (6.s) we could also have

    (6.s') G → R.

(6.s') is sometimes said to be the *converse* of (1.s), and (6.s) is said to be the *contrapositive* of (6.s'), though we will not make much use of these terms. Since (6.s) and (6.s') are correct symbolizations of English sentences with the same meaning, those formulae have the same meaning as well, and since (1.s) symbolizes a sentence with a different meaning, it should turn out to mean something different from (6.s) and (6.s').

For the sake of having a definite policy, we should decide between (6.s') and (6.s) for dealing with 'only if'. In some ways (6.s) is a more natural means of expressing the effect of 'only if', but when we come to manipulate formulae it will be an advantage to work with the simplest ones possible. Since (6.s') is simpler than (6.s), we will use it as the pattern for 'only if', as stated in the following rule. Let p and q be any two sentences of English. Then

- 'if p then q' and 'q if p' are both symbolized '$p → q$'
- 'q only if p' is symbolized '$q → p$'

The rule for 'only if' must be carefully memorized. Probably the most common error in symbolizing sentences of English in sentential logic is to symbolize 'only if' sentences the wrong way around. Notice, by the way, how the letters '*p*' and '*q*' are used in stating this rule. We are using them to make a generalization which is correct for all English sentences, in the same way that mathematicians use the letters '*x*' and '*y*' to make generalizations about numbers, as in 'for any numbers *x* and *y*, if *x* and *y* are even so is *x* + *y*'. '*x*' and '*y*' are sometimes called *numerical variables*, because they stand for numbers, but not any particular numbers fixed in advance; so they contrast with numerals like '0' and '5', which do stand for particular numbers. Similarly, '*p*' and '*q*' are *sentential variables*, because they do not stand for fixed sentences, by contrast with sentence-letters like 'A' and 'B'.

There are a number of other locutions used in English to express indicative conditionals, such as 'provided' and 'whenever', and to determine exactly what conditional is being expressed it is often useful to ask what is being said to be a *necessary condition* for what or what is being said to be a *sufficient condition* for what. Thus

(7)   Smith gets an A whenever he bribes the instructor

says that Smith's bribing the instructor is a sufficient condition for getting an A, since according to (7), that is all he has to do to get an A. Consequently, (7) amounts to the same as (1), and so has the symbolization (1.s).

(8)   Smith will get an A provided he bribes the instructor

also says that a bribe is sufficient for an A and again is symbolized as (1.s). In general, when a new way of expressing a conditional construction is encountered, we ask what is being said to be sufficient for what, or necessary for what, and then give the appropriate translation into symbols using the symbolizations of (3) and (5) as a guide.

The difference between necessary and sufficient conditions is encountered often enough in real life. For example,

(9)   Your car will start only if there is fuel in its tank

says, correctly, that having fuel in its tank is necessary for your car to start. If we use 'S' for the atomic sentence 'your car will start' and 'F' for the atomic sentence 'your car has fuel in its tank' we can symbolize (9) following the pattern of (4) and (5) as

(9.s)   S → F.

Again, it helps to see why this is right if we note that (9.s) could be read back into English as 'if your car starts then (it follows that) it has fuel in its tank'. For its having fuel in its tank to be a consequence of its starting, its having fuel in its tank must be necessary for it to start.

While (9) is in most circumstances true, and will be until electric cars become more common,

(10) Your car will start if it has fuel in its tank

is very often false. It is not *sufficient* for a car to start that it have fuel in its tank—usually when one's car refuses to start, it does have fuel in its tank (however, a mechanic who has checked that everything *else* is working would be in a position to assert (10)). (10)'s symbolization, on the model of (2), is

(10.s) F → S.

One other common English connective which can be captured by the conditional is 'unless', as in

(11) Smith will fail unless he bribes the instructor

and

(12) Your car will not start unless it has fuel in its tank.

It is not too difficult to see that (11) and (12) can be accurately paraphrased respectively by the following:

(11.a) If Smith does not bribe the instructor then he will fail

and

(12.a) If your car does not have fuel in its tank, it will not start.

In these paraphrases, the *negation* of the sentence immediately following the 'unless' becomes the antecedent of the conditional, while the other constituent subsentence becomes the consequent (in an 'unless' sentence, this subsentence does not have to precede the 'unless', in view of the grammatical acceptability of, for instance, 'unless he bribes the instructor, Smith will fail'). So using 'L' for 'Smith fails', the symbolizations of (11) and (12) at which we arrive are:

(11.s) ~R → L

and

(12.s) ~F → ~S.

It is important to note that (11) does not say that bribing the instructor is sufficient for not failing, so 'R → ~L' would be wrong. Similarly, (12) does not say that having fuel in its tank is sufficient for your car to start, so 'F → S' would be wrong. The rule we extract from this discussion is therefore:

- for any sentences p and q, 'p unless q' and 'unless q, p' are symbolized '$\sim q \to p$'

The last of our five basic connectives is the biconditional, '...if and only if...', for which the symbol is '↔'. Suppose Smith's instructor will only give an A to a student who bribes him but requires nothing more of such a student. Then (2) and (4) are both true, and we could express this simply by writing their conjunction

> (13)   Smith will get an A if he bribes the instructor and Smith will get an A only if he bribes the instructor

for which we have the symbolization

> (13.s) $(R \rightarrow G) \& (G \rightarrow R)$.

But (13) is rather long-winded, and by using '...if and only if...' we can avoid its repetitious aspect:

> (14)   Smith will get an A if and only if he bribes the instructor.

(13) and (14) both say that a bribe is necessary *and* sufficient for getting an A; the 'if' captures sufficiency while the 'only if' captures necessity. (14) could be symbolized by (13.s), but instead we use the special symbol '↔', which allows us to contract (13.s) into

> (14.s) $G \leftrightarrow R$.

We think of the symbol '↔' in a slightly different way from the others. The others were introduced as symbols for connectives having meaning in their own right, but we treat '...if and only if...' as merely as an abbreviation for '...if...and ...only if...'. In other words, (14) abbreviates (13) and so (14.s) is an abbreviation of (13.s). This different attitude toward '↔' will have certain consequences in later chapters.

Like the conditional, the biconditional can be expressed in English in more than one way. A common variant of '...if and only if...', at least in mathematical writing, is '...just in case...', as in

> (15)   Smith will get an A just in case he bribes the instructor

which also receives the symbolization (14.s). Other variants of '...if and only if...' will be encountered in the exercises.

Our collection of connectives allows for the symbolization of sentences of some complexity, for example

> (16)   John will study hard and also bribe the instructor, and if he does both then he'll get an A, provided the instructor likes him.

In approaching the symbolization of a complicated compound sentence, the first step is to classify the whole sentence as a conjunction or conditional or whatever, and then to identify the atomic sentences which occur in it. Reading through (16) we see that it is a conjunction whose first conjunct is another conjunction, 'John will study hard and also bribe the instructor', and whose second conjunct is some kind of complicated conditional construction. As for the atomic sentences which occur in it, we have

      S: John will study hard
      R: John will bribe the instructor
      G: John will get an A
      L: The instructor likes John

Here we list the relevant atomic sentences alongside the sentence-letter we have chosen for each. Such a list is called a *dictionary*, and every solution to a symbolization problem should begin with one. Note also that the English phrases are *complete* sentences: we do not write 'R: bribe the instructor', since 'bribe the instructor' is not a complete sentence. In addition, no words whose meaning or reference is settled by linguistic context appear in the dictionary: we do not write 'L: The instructor likes him' since the reference of 'him' is clear from context: looking at (16), we see that 'him' refers to John, so we make this explicit in the dictionary.

Once the dictionary has been given, the best way to proceed is to go through the sentence to be translated and substitute letters for atomic sentences, leaving the connectives in English. This intermediate step makes it more likely that the final formalization we arrive at will be correct. So using the dictionary to substitute in (16), we obtain

(16.a)   S and R, and if S and R then G, provided L.

Note that we handle 'he does both' by expanding it to full sentential form: 'he does both' is simply an economical way of repeating 'he studies hard and bribes the instructor'.

The remaining problem is to place 'provided L' properly in the formula. The effect of 'provided L' is to say that the joint truth of S and R is sufficient for the truth of G *if* L is also the case. In other words, if L is the case, the truth of S and R suffices for the truth of G. So the most natural formalization is produced by rewriting (16.a) as

(16.b)   S and R, and if L, then if S and R then G.

(16.b) is a conjunction. Its first conjunct is 'S and R', its second, 'if L, then if S and R then G'. Symbolizing 'S and R' is trivial. 'If L, then if S and R then G' is a conditional with antecedent 'L' and consequent 'if S and R then G'. So as another intermediate step on the way to a final answer, we can rewrite (16.b) as

(16.c)   (S & R) & (L → (if S and R then G)).

'If S and R then G' is symbolized '(S & R) → G', so substituting this in (16.c) we get the following full symbolization of (16),

(16.s) (S & R) & (L → ((S & R) → G)).

This procedure illustrates the best way of approaching complicated symbolization problems, which is to use a step-by-step method. If one attempts to write down the formula straight off, without any intermediate steps, it is very likely that something will go wrong. As a final check, we should inspect for unmatched parentheses: we count the number of left parentheses and make sure there is an equal number of right parentheses. The step-by-step procedure by which we arrived at (16.s) guarantees that the parentheses are in the correct positions.

❏ Exercises

Symbolize the following sentences of English, giving a dictionary for each example. Write down the simple English sentences so that they are grammatically complete and also explicit (apart from (1) and (2), replace words like 'they' and 'it' with the phrases which they stand for in the context of the compound sentence). Be sure that your simple sentences do not contain any connective-words, and that you do not use two sentence-letters for what is essentially the same English sentence.

(1)   If it's Tuesday then it must be Belgium. (T, B)
(2)   If it's Tuesday and it's not Belgium, I'm lost. (T, B, L)
(3)   If the military relinquishes power, then the new government will release all political prisoners if it has any respect for human rights.
(4)   If the economy declines, then unless there is a change of leadership there will be a recession. (D, C, R)
*(5)   Grades may be posted provided students so request and are not identified by name. (P, R, S)
(6)   The applicants may examine their dossiers only if they have not already waived their right to do so and their referees approve.
(7)   If enough people show up we will take a vote, otherwise we will reschedule the meeting.
*(8)   One has the right to run for president if and only if one is an American citizen.
(9)   The unrestricted comprehension principle is true just in case arithmetic is inconsistent.
(10)   Not only will John and Mary have to live on campus if they aren't local residents, but neither will be allowed to park on campus—whether or not they are local residents.
(11)   The award of the license is conditional upon passing the exam. (A: The license is awarded)
*(12) All that's required to pass the exam is that one makes an effort. (E: One makes an effort)
(13)   The exam can't be passed without making an effort. (P: One passes the exam)

(14) If, but only if, they have made no commitment to the contrary, may reporters reveal their sources, but they always make such a commitment and they ought to respect it.

*(15) An increase in government funding is a necessary condition of an improvement in the quality of education, but not a sufficient condition. (F: There is an increase in government funding)

(16) A right to smoke in public is contingent upon such activity's not significantly affecting the health of others, but nowadays we know that the idea that it doesn't is wrong.

4 Symbolizing entire arguments

So far we have symbolized only single sentences, but the problems which arise in symbolizing entire arguments are similar. In this section we discuss four examples.

> A: If Smith bribes the instructor then he'll get an A. And if he gets an A potential employers will be impressed and will make him offers. But Smith will receive no offers. Therefore Smith will not bribe the instructor.

The first step in symbolizing an argument is to identify its conclusion, which is often, but not always, the last complete sentence in the argument. Conclusions are usually indicated by the presence of words or phrases like 'therefore', 'so it follows that', 'hence', or as in Example C on the next page, 'so'. All these *conclusion indicators* may be abbreviated by '∴'. The conclusion in A is 'Smith will not bribe the instructor'.

Next, we need to identify the atomic sentences in the argument and construct a dictionary, *taking care to avoid duplication*. That is, we should not introduce a new sentence-letter for what is essentially a recurrence of the *same* atomic sentence. Reading through A closely indicates that the following atomic sentences occur in it, which we list in dictionary format:

> R: Smith bribes the instructor
> G: Smith gets an A
> E: Potential employers will be impressed
> O: Potential employers will make Smith offers

We then go through the argument symbolizing each premise and the conclusion in turn. The first premise is (3.1) ((1) from §3), and is symbolized as (3.1.s), 'R → G'. The second premise with letters substituted for its constituent atomic sentences is

> (1) If G, E and O

and so the formula is

(1.s) G → (E & O).

We do not symbolize the initial 'And' of the second premise, since it is present in the English only for stylistic reasons, to aid the flow of the passage. The third premise does not get its own sentence-letter, 'C', for 'Smith will receive offers', since in the context of this argument, 'Smith will receive offers' is simply another way of saying 'potential employers will make Smith offers'. Hence the third premise says that these employers will not make him offers, and so is symbolized as '~O'. The conclusion is '~R', and we therefore have the whole argument symbolized as

B: R → G
    G → (E & O)
    ~O
    ∴ ~R

Our second example is a little more complex:

C: If logic is difficult, then few students will take logic courses unless such courses are obligatory. If logic is not difficult then logic courses will not be obligatory. So if a student can choose whether or not to take logic, then either logic is not difficult or few students take logic courses.

The conclusion in C is the complex conditional 'If a student can choose whether or not to take logic, then either logic is not difficult or few students choose to take logic courses'. The following atomic sentences occur in C, which we list in dictionary format:

T: Logic is difficult
F: Few students take logic courses
O: Logic courses are obligatory

We do not use a separate sentence-letter for 'such courses are obligatory' since 'such courses' simply means 'logic courses'; and we do not use a separate sentence-letter for 'a student can choose whether or not to take logic' since the meaning of this sentence does not differ significantly from that of 'Logic courses are not obligatory', for which we have '~O'.

Turning to the symbolization of the premises and conclusion, we see that the first premise contains 'unless'. Applying our rule on page 23 for this connective, where *p* is 'F' and *q* is 'O', we obtain

(2) If T then (~O → F)

and hence for the whole premise,

(2.s) T → (~O → F).

The second premise is 'if not-T then not-O' and the conclusion is 'if not-O then either not-T or F'. So we can give the final symbolization of the whole argument as follows:

D: T → (~O → F)
    ~T → ~O
    ∴ ~O → (~T ∨ F)

Now an example with still more complexity:

E: If God exists, there will be no evil in the world unless God is unjust, or not omnipotent, or not omniscient. But if God exists then He is none of these, and there is evil in the world. So we have to conclude that God does not exist.

As before, we begin by identifying the conclusion of E, which is indicated by the phrase 'so we have to conclude that'; 'so we have to conclude that' is simply a wordier version of 'so', and thus E's conclusion is the negative sentence 'God does not exist'.

Next, we identify the atomic sentences in the argument and construct a dictionary, again taking care to avoid duplication: we do not want to introduce a new sentence-letter for a statement whose meaning can be expressed, perhaps using connectives, by sentence-letters already in the dictionary. Reading through E closely suggests the following dictionary:

X: God exists
V: There is evil in the world
J: God is just
M: God is omnipotent
S: God is omniscient

Are there any other atomic sentences in E? 'God is none of these' is presumably not atomic because 'none' contains the connective 'not'; 'God is none of these' means 'it is not the case that God is some of these'. So should 'God is some of these' be in our dictionary? Again the answer is no, for 'God is some of these', like 'he does both' in (3.16), is simply a way of getting the effect of repeating sentences without actually repeating their very words. In this case, 'God is some of these' is an economical version of 'God is unjust, or God is not omnipotent or God is not omniscient', which is a disjunction of negations of atomic sentences already listed in the dictionary (we will group the last two disjuncts together). It would therefore be a mistake to introduce a new sentence-letter for 'God is some of these'. Note also that we use 'or' rather than 'and' in the full sentential form of 'God is some of these'. Using 'and' would result instead in the full sentential form of 'God is *all* of these'.

We can now symbolize the argument premise by premise. We start with the first premise and substitute the sentence-letters of our dictionary into it. This gives us

(3) If X then (~V unless (~J or (~M or ~S)))

and symbolizing the 'if...then...' yields

(3.a) X → (~V unless (~J or (~M or ~S))).

Next, we recall our rule for 'unless', that '*p* unless *q*' becomes 'if not-*q* then *p*'; here the '*p* unless *q*' component is '~V unless (~J or (~M or ~S))', and so *p* is '~V' and *q* is '~J or (~M or ~S)' (note again the similarity between our use of the sentential variables '*p*' and '*q*' to stand for any sentences and the mathematician's use of '*x*' and '*y*' to stand for any numbers). Therefore we can rewrite (3.a) as the intermediate form

(3.b) X → (if not-[~J or (~M or ~S)] then ~V).

Putting the remaining connectives into symbols produces

(3.s) X → (~[~J ∨ (~M ∨ ~S)] → ~V).

(3.b) nicely illustrates the use of brackets rather than parentheses as an aid to readability, since it makes it easier to see which left marker belongs with which right one.

The second premise of E is 'if God exists then he is none of these, and there is evil in the world', which is a conjunction with 'there is evil in the world' as second conjunct. To say God is none of these, in the context of our argument, is to say that he is not unjust, not not-omnipotent and not not-omniscient. Although the repeated 'not's cancel each other, in what is known as a *double-negative*, we will retain them in our symbolization for the sake of being faithful to the English, which also contains double-negatives (in the combination of 'none' with '*un*just', 'not omnipotent', etc.). So when the second premise is expanded to full sentential form and substituted into from the dictionary, we obtain

(4) If X then not not-J and not not-M and not not-S, and V

which fully symbolized becomes

(4.s) [X → (~~J & (~~M & ~~S))] & V.

Note the role of the brackets in ensuring that (4.s) is a conjunction; without them it would be possible to read the formula as a conditional with V as the second conjunct of its consequent.

We can now write out the whole argument in symbols:

F: X → [~(~J ∨ (~M ∨ ~S)) → ~V]
    [X → (~~J & (~~M & ~~S))] & V
    ∴ ~X

F gives the sentential logical form of E, and with the techniques to be developed in the next chapter we will be able to determine whether or not this argument-form is valid, and hence whether or not E is sententially valid.

Our last example is quite tricky:

> G: We can be sure that Jackson will agree to the proposal. For otherwise the coalition will break down, and it is precisely in these circumstances that there would be an election; but the latter can certainly be ruled out.

G illustrates a common phenomenon in ordinary argumentative discourse, which is that the conclusion need not always be the *last* proposition to be stated. A careful reading of G indicates that it is 'Jackson will agree to the proposal' and *not* 'the latter can be ruled out' which is the conclusion of the argument: what follows the opening sentence of G is a list of the *reasons why* Jackson will agree, in other words, what follows is a list of the premises from which 'Jackson will agree' is supposed to follow. Phrases like 'for' and 'after all' are often used to introduce the premises for a conclusion which has already been stated. Another clue that the conclusion of G occurs at its start is the use of 'we can be sure that', which is a conclusion indicator like 'we have to conclude that' in E.

What atomic sentences occur in G? We have already identified 'Jackson will agree to the proposal'. Evidently, 'the coalition will break down' and 'there will be an election' also occur. 'The latter can be ruled out' is not a new atomic sentence, since 'the latter' refers to 'there will be an election', so that 'The latter can be ruled out' is just another way of saying that there will not be an election. Hence we need the following dictionary:

> J: Jackson will agree to the proposal
> C: The coalition will break down
> E: There will be an election.

The first conjunct of the first premise is 'otherwise the coalition will break down', where the 'otherwise' refers to what has already been stated, that Jackson will agree. To say that *otherwise* the coalition will break down is to say that if he does *not* agree, the coalition will break down. Substituting into the first conjunct from the dictionary and symbolizing the connectives then gives

> (5.s) $\sim J \to C$

as the formula.

The second conjunct of the first premise is 'it is precisely in these circumstances that there would be an election'. Rereading G we see that by 'these circumstances' is meant the circumstances of the coalition breaking down, so that the second conjunct, in full sentential form, says that it is precisely if the coalition breaks down that there will be an election. What is the import of 'precisely if'? Since the best way to assess the meaning of an unfamiliar connective which embodies some kind of conditional construction is to express it in terms of necessary and sufficient conditions, we ask what it is that the second premise says is necessary, or sufficient, for what. According to the second conjunct of premise 1, there *would* be an election in the circumstances of the coalition breaking down, so a breakdown is being said to be *sufficient* for the occurrence of an election. Part of the meaning of the second conjunct is therefore 'if the coalition breaks down there will be an election'. But that is not the whole meaning, since the import of 'precisely' is that there are no *other* circumstances in which an election would occur; hence, a breakdown of the coalition is necessary for an election as well. Consequently, we need the biconditional to get the full effect of 'precisely if'. The second conjunct of the first premise is:

(6) There will be an election if and only if the coalition breaks down

or in symbols, 'E ↔ C'. As we already remarked, the conclusion of G is 'Jackson will agree to the proposal', so the logical form at which we arrive for the whole argument is

H: (~J → C) & (E ↔ C)
    ~E
    ∴ J

Two questions about this argument are worth pursuing. First, would it have made any difference if instead of formalizing the whole argument with two premises, the first of which is a conjunction, we had formalized it with three premises, breaking the first premise up into its conjuncts and treating each as a separate premise? This would have given us

I: (~J → C)
    (E ↔ C)
    ~E
    ∴ J

Intuitively, the answer should be that as far as validity is concerned, there is no difference between H and I, since a conjunction contains exactly the same information as its two conjuncts. This is just to say that a conjunction is true if and only if its two conjuncts are true. However, as we shall see in Chapter 3, the equivalence between H and I is a special feature of conjunction.

Secondly, though (6) seems the natural way of expressing the second conjunct of premise 1, would it have made any difference if we had given the full sentential form as

(7) The coalition will break down if and only if there is an election

instead? According to (6), the coalition's breaking down is necessary and sufficient for there being an election, and what makes (6) natural is that it expresses the relationship in the direction of causality: it is a breakdown which would cause an election and not vice versa. However, 'if and only if' is not tied to expressing causal relationships: it simply abbreviates the conjunction of two conditionals, and the same two are conjoined in (6) and (7), though in different orders. But since the order of conjuncts does not affect what a conjunction says, (6) says the same as (7).

Another way of seeing this is to note that if a breakdown is sufficient for an election, as (6) says, then an election is necessary for a breakdown, as (7) says, in the sense that a breakdown cannot occur unless an election does too (because a breakdown would cause an election). Equally, if a breakdown is necessary for an election, as (6) says, an election is sufficient for a breakdown, as (7) says, in the sense that if an election occurs, that is all the information we need to conclude that a breakdown has also occurred. So from (6) it follows that an election's occurring is sufficient and necessary for a breakdown to have occurred, which is exactly what (7) says. Consequently we could just as well have 'C ↔ E' in place of 'E ↔ C'.

These last two examples illustrate our earlier warning that there is a certain license in speaking of 'the' sentential form of an argument. Keeping as close as possible to English structure will very often produce a 'preferred' form, but sometimes there will still be degrees of freedom, as, for example, in G, over whether to use 'C ↔ E' or 'E ↔ C', and in F, over how to parenthesize its three-disjunct disjunctions and three-conjunct conjunctions. This is why our official definition of 'sententially valid English argument' requires only that the argument have *a* sententially valid form. Fortunately, there is no harm in continuing to avail ourselves of the convenient fiction that there is always a unique preferred form for an argument, since it is never the case that of two equally acceptable symbolizations of an argument in sentential logic, one is valid and the other invalid.

❏ Exercises

Symbolize the following arguments, giving a complete dictionary for each. Be sure not to use different sentence-letters for what is essentially the same simple sentence.

(1) If the government rigs the election there will be riots. However, the government will play fair only if it is guaranteed victory, and isn't guaranteed victory unless it rigs the election. So there will be riots. (G, R, V)

(2) If the price of oil rises then there will be a recession, and if there is a recession, interest rates will fall. But interest rates will fall only if the government allows them to, and the government will keep them high if the price of oil rises. So either the price of oil won't rise or interest rates will fall.

(3) The players will go back to work if agreement is reached about their salaries, but this will be achieved, if at all, only if some of them take early retirement. So the players will not go back to work unless some retire early.

*(4) If Homer did not exist, it follows that the *Odyssey* was written by a committee or, if Butler was right, that it was written by a woman. But it was not written by a woman. So it was written by a committee.

(5) If I know I exist then I exist. I know I exist if I know I think, and I know I think if I think. But I do think. Therefore I exist.

(6) If God knows today what I will do tomorrow, then what I do tomorrow is foreordained. And if it's foreordained, then either I have no freedom of choice, or else I will freely choose to do what's foreordained. However, the latter is impossible. Therefore I have no freedom of choice unless God doesn't know today what I will do tomorrow.

(7) If internment is ended, the IRA will negotiate. But they will negotiate if either they lose Catholic support or the UDA disbands, and not otherwise. And the UDA will disband only on condition that internment is not ended. So if it isn't ended, then provided the IRA negotiate, they will keep Catholic support.

(8) John will not get a seat at the front of the class unless he arrives on time. The instructor will call roll right at the beginning of class provided she herself is not late. If she's not on time and John does get a seat at the front, he'll be able to finish his homework. So, if John isn't late in getting to class, then if the instructor does not call role right at the beginning of class, then he will get his homework done.

*(9) If parking is prohibited in the center of town then people will buy bicycles, in which case the general inconvenience is offset by the fact that the city shops will pick up some business. However, shops go bankrupt in town centers with restrictive parking policies. So it looks as though bicycles don't attract people. (P, B, O, U)

(10) Suppose no two contestants enter; then there will be no contest. No contest means no winner. Suppose all contestants perform equally well. Still no winner. There won't be a winner unless there's a loser. And conversely. Therefore, there will be a loser only if at least two contestants enter and not all contestants perform equally well.

(11) If there is intelligent life on Mars, then provided it takes a form we can recognize, we should have discovered it by now. If we are unable to recognize intelligent Martians, that must be because they are very different from ourselves. Since we have not, in fact, discovered any intelligent form of life on Mars, it appears, therefore, that either there are no intelligent Martians, or there are some, but they are very unlike us.

(12) The one thing that would defeat the Mayor is a financial scandal involving members of the council. After all, his survival is conditional upon the support of the urban middle class, which in turn depends upon the council's reputation for financial integrity. (D, S, U)

5 The syntax of the formal language

Before leaving the topic of symbolizing English we should be more exact about such notions as the *scope* of a connective in a formula and of the *main connective* of a formula, concepts we have already used without being completely precise about their application. Precise definitions of these concepts are possible only when we are equally precise about the *grammar* of the language into which we are translating English. Of course, it is easy to recognize that such a sequence of symbols as '~& B C' is ungrammatical, since there is no grammatical English into which it could be translated, even given a dictionary for the sentence-letters. But that is not an *intrinsic* account of the problem with '~& B C'. In general, one does not classify the sentences of a language other than English as grammatical or ungrammatical according to whether or not they can be translated into grammatical English. Other languages possess their *own* grammars, and it is with respect to these that judgements of grammaticality and ungrammaticality should be made. We are going to regard the *formal* language of our symbolizations as an independent language with its own grammar, just like other natural languages (Sanskrit, Spanish etc.). We will call this formal language the *Language of Sentential Logic*, or LSL for short.

There is one respect in which LSL is unlike a natural language. In natural language, the basic elements, words, have their own meanings, but the sentence-letters of LSL do not have their own meanings. A natural language is said to be an *interpreted* language, while a formal language is said to be *uninterpreted* (apart from its connectives). We can regard the dictionaries of our translations as temporarily bestowing a meaning on particular sentence-letters, so that it makes sense, within the context of a given problem, to ask if a complex sentence of LSL and a sentence of English mean the same. But the purpose of translation from a natural language into LSL is not the usual purpose of translation between two natural languages, which is to find a way of expressing the same meaning in both. The main purpose of symbolization, rather, is to exhibit the logical form of the English by rendering it in a notation that abstracts away from aspects of the English that may obscure its logical properties.

The collection of basic elements of a language is called the language's *lexicon*. The lexicon of English is the set of all English words, but the lexicon of LSL will differ from a natural language's lexicon in two respects. First, the basic elements of formulae of LSL are sentence-letters, not words. Second, in any natural language the number of words is finite, but we will allow the lexicon of LSL to contain infinitely many sentence-letters.

The lexicon of LSL:

All sentence-letters 'A', 'B', 'C',..., 'A′', 'B′', 'C′',..., 'A″',...; connectives '~', '&', '∨', '→', '↔'; and punctuation marks '(' and ')'.

Next, we have to specify the grammatical rules, or syntax, of LSL. Formulae of LSL are constructed from sentence-letters and connectives with the aid of punctuation marks. A formula such as '(~A & ~B)' can be regarded as having been built up by stages. At the first stage, we choose the sentence-letters we are going to use, in this case 'A' and 'B'. At the next stage, we prefix each sentence-letter with the negation symbol. This gives us two grammatical formulae, '~A' and '~B', and we call the operation we apply at this stage *negation-formation*. Then we combine the two formulae into a single formula using '&' and parentheses to obtain '(~A & ~B)'; we call the operation used here *conjunction-formation* (the other syntactic formation operations we have at our disposal are disjunction-formation, conditional-formation and biconditional-formation). Each syntactic operation applies to grammatical formulae and produces a grammatical formula. Since the formulae which a construction process begins with are always grammatical formulae, sentence-letters being grammatical formulae by definition, and since at every subsequent step we apply some formation operation which always produces grammatical formulae if applied to grammatical formulae, it follows that at the end of a construction process we will always have a grammatical formula. So we can define the grammatical formulae as all and only the formulae produced by such a process.

Some terminology and abbreviations will be useful. Instead of 'grammatical formula' it is usual to employ the phrase 'well-formed formula', abbreviated 'wff', plural 'wffs'. The formation operations are commonly called *formation rules*, and we abbreviate the five as (*f*-~), (*f*-&), (*f*-∨), (*f*-→) and (*f*-↔). Each rule is a conditional: *if* it is given a wff or wffs as input, then such and such a wff results as output. As remarked, this means that since the construction process always begins with wffs, it always ends with a wff. A *symbol-string* is a sequence of symbols from the lexicon of LSL, including ungrammatical ones like '~&ABC'. We may now define the collection of wffs by the following stipulations:

The formation rules of LSL:

  (*f-sl*):   Every sentence-letter in the lexicon of LSL is a wff.
  (*f*-~):   For any symbol-string *p*, if *p* is a wff then the result of prefixing '~' to *p* is a wff.
  (*f*-&):   For any symbol-strings *p* and *q*, if *p* and *q* are wffs, the result of writing '(' followed by *p* followed by '&' followed by *q* followed by ')' is a wff.
  (*f*-∨):   For any symbol-strings *p* and *q*, if *p* and *q* are wffs, the result of writing '(' followed by *p* followed by '∨' followed by *q* followed by ')' is a wff.
  (*f*-→):   For any symbol-strings *p* and *q*, if *p* and *q* are wffs, the result of writing '(' followed by *p* followed by '→' followed by *q* followed by ')' is a wff.
  (*f*-↔):   For any symbol-strings *p* and *q*, if *p* and *q* are wffs, the result of writing '(' followed by *p* followed by '↔' followed by *q* followed by ')' is a wff.
  (*f!*):   Nothing is a wff unless it is certified as such by the previous rules.

In writing these rules we are again using the letters '*p*' and '*q*' as sentential variables so that we can state generalizations about LSL formulae, analogous to generalizations about numbers such as 'if x and y are even then $x + y$ is even'. Clause ($f!$) is called the *closure condition:* it ensures that *only* the symbol strings derivable by the rules count as wffs. The other rules themselves ensure that *all* symbol strings derivable by them are wffs. To say that a symbol string is 'certified' as a wff by the formation rules is just to say that the string can be constructed by a sequence of applications of the rules. Note that in applying the rules for the binary connectives, each new formula is formed with outer parentheses around it, while (f-~) does not require outer parentheses. This reflects the fact that we have the scope convention for '~' stated on page 15, so that outer parentheses are not needed to delimit a negation symbol's scope.

It is possible to draw a picture of the construction of a wff. Such a picture is given in the form of an (inverted) tree, called a 'parse tree' by linguists. In general, a tree is a collection of nodes joined by lines in a way that determines various paths through the tree from a unique *root* node to various end nodes, called *leaves*. In an *inverted* tree, the root node is at the top and the leaves are at the bottom, as in the tree displayed below on the left. The root node of this tree is ① and its leaves are nodes ④ and ⑥. A *path* is a route from one end of the tree to the other in which no node is traversed twice, so the paths in our example are ①-②-③-④ and ①-②-⑤-⑥. A *branch* in a tree is a segment of a path h beginning with a node on h where paths diverge and including all subsequent nodes on h to h's leaf. In the example on the left below, the branches are ②-③-④ and ②-⑤-⑥.

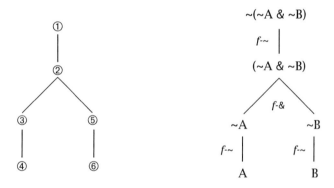

Parse trees are a special kind of tree in which the nodes are labeled with formulae in such a way that reading the tree from bottom to top recapitulates the construction of the formula. The sentence-letters label the leaves of the tree and the final formula labels the root. We also indicate which formation rule has been used in the passage up the tree to a particular node. On the right above we display a parse tree that results from labeling the root node of the tree on its left with '~(~A & ~B)', the leaves with 'A' and 'B' respectively, and the other nodes with the appropriate formulae.

According to our formation rules, any application of a rule for a *binary* connective produces a string of symbols with outer parentheses around it. This means that, strictly speaking, few of the formulae displayed on preceding pages which are not negations are well formed, since most of them lack outer parentheses. For example, though '(~A & ~B)' is a wff, '~A & ~B' is not. But outer parentheses do no work in determining the meaning of a string of symbols, and displaying them merely increases clutter. So we assume the *invisibility* convention for outer parentheses on formulae displayed on the page: the parentheses are really there, it is just that they are invisible. Hence, '~A & ~B' is a wff after all. Of course, if we embed '~A & ~B' as a subformula in a more complex formula, the invisible parentheses become visible again, as illustrated for example by 'C & ~(~A & ~B)'. It is only outer parentheses for a complete formula which may be invisible.

Here is the parse tree for formula (3.s) of §4, 'X → [~(~J ∨ (~M ∨ ~S)) → ~V]'. (We regard brackets as a permissible notational variant of parentheses.)

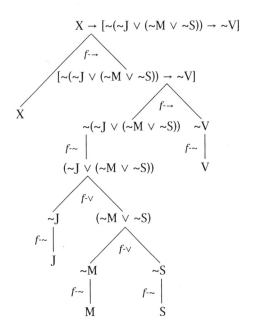

Choosing sentence-letters for the leaves of the tree, at the first stage in the construction of (4.3.s), is justified by the rule (*f-sl*). As in the previous example, we have used invisible outer parentheses at the root node, while at each intermediate node not generated by an application of (*f-~*), the formula formed at that node has visible outer parentheses. It is the use of these parentheses at intermediate nodes, together with the scope convention for '~', which guarantees that each string of symbols produced by the formation rules can be read in one and only one way. Note also that in arriving at (4.3.s), there was a point at which we made an arbitrary choice, concerning how to group the disjuncts of '~J ∨ ~M ∨ ~S' to make a two-disjunct disjunction. We chose the grouping '~J ∨ (~M ∨ ~S)', though '(~J ∨ ~M) ∨ ~S' would have served equally well. Had we made the latter choice, the difference in the parse tree would be that '~J' and '~M'

would be grouped by a first application of $(f\text{-}\vee)$, and then the result would be grouped with '~S' by a second application of $(f\text{-}\vee)$.

Parse trees can be read top down or bottom up. Reading them top down, we see how to break the formula into its syntactic components. Reading bottom up, we see how the formula can be built up from its atomic constituents. However, in actually constructing a parse tree, it is usually easier to work from the top down.

We can use parse trees or the formation rules to distinguish well-formed formulae from non-well-formed ones. Although some strings of symbols are obviously not well-formed, such as '~& B C', there are other non-well-formed strings with which the problem is more subtle. For example, in 'A → ((~B) & C)' the flaw is with the parentheses surrounding '~B'; the formation rule $(f\text{-}\sim)$ does not put parentheses around the negative formula formed by prefixing '~', so the parentheses in this example are incorrect. Non-well-formed strings like this can be diagnosed only by paying close attention to the exact requirements of the formation rules.

We are now in a position to give precise definitions of such concepts as 'main connective', 'scope' and their ilk. We begin with a preliminary definition, that of a *subformula* of a formula.

- If p is any wff, q is a *subformula* of p if and only if q labels some node of the parse tree for p.

By this definition, everything from the sentence-letters in p to p itself is a subformula of p. So we can list the subformulae of '~J ∨ (~M ∨ ~S)' as: 'J', 'M', 'S', '~J', '~M', '~S', '(~M ∨ ~S)', and '~J ∨ (~M ∨ ~S)'. The other definitions we promised can now be stated:

- If p is any wff and a binary connective c occurs in p, then the *scope* of that occurrence of c in p is the subformula of p which appears at the node in the parse tree of p at which that occurrence of c is introduced. The scope of an occurrence of '~' is the subformula at the node *immediately preceding* the one where that '~' is introduced.
- If p is any wff, the *main connective* of p is the connective introduced at the root node of p's parse tree. If the main connective of p is '~', the formula to which the '~' is prefixed is the *main subformula* of p. If the main connective of p is a binary connective, then the subformulae which it connects are the *main subformulae* of p.
- If c and c' are two occurrences of connectives in p, then c' is *within the scope of* c if and only if c' occurs in the subformula which is the scope of c.

These definitions give us precisely the right results. For example, the main connective of (3.s) on page 30 is the leftmost '→' which occurs in it, while the scope of the leftmost '~' is '(~J ∨ (~M ∨ ~S))'. But the rightmost '~' in (3.s) is not within the scope of the leftmost, since the rightmost '~' does not occur in the formula at the node preceding the one where the leftmost '~' is introduced. One slightly odd consequence of the definitions which is perhaps worth mentioning is that when a negation has within its scope a formula whose own main connective is binary, the subformula which is the scope of the negation and the subformula which is the scope of that binary connective are the same.

❏ Exercises

I   Write parse trees for the following LSL wffs:

(1) (~P & Q) → R
(2) ~(P & Q) → R
(3) ~[(P & Q) → R]
(4) A ↔ (~B → (C ∨ ~C))

((4) requires that 'C' be listed twice, at the bottom of two paths in the tree)

(a) List the subformulae of (4).
(b) What is the scope of the leftmost '~' in (4)?
(c) What is the scope of '→' in (4)?
(d) In (1), is '~' within the scope of '&' or vice versa?
*(e) Which is within the scope of which in (2) and (3)?
(f) In (4), which connective(s) is the leftmost '~' within the scope of?
(g) What about the other '~' in (4)?

II   Which of the following are not wffs of LSL? For each symbol-string you identify as a non-wff, briefly explain why it is impossible to assemble it with the formation rules (review the discussion of 'A → ((~B) & C)' on page 39).

(1) ((~P & Q) → R)
(2) (A ↔ ((B ↔ C)))
(3) (P → ~Q & R)
*(4) (~(~P & ~Q))
(5) (~S → (~T ↔ (U & (V ∨ W))))

6 Quotation, selective and otherwise

The remaining grammatical issue we have to deal with concerns the proper use of quotation marks, which to this point we have been using rather loosely. In English, a central use of quotation marks is to allow us to talk about words or symbols rather than the things for which they stand. For instance, we can contrast (1) and (2):

(1) Paris is the capital of France.

(2) 'Paris' contains five letters.

Without the quotation marks, (2) would make little sense, since it would say that Paris-the-city contains five letters. The distinction exhibited by (1) and (2) is called the *use/mention* distinction: in (1) we *use* the word 'Paris' to make a statement about the city it stands for, while in (2) we *mention* the word 'Paris' to say something about it (the word). Thus on previous pages, when we have wanted to say something about, that is, *mention*, particular formulae of LSL, we have put those formulae in quotes.

This works well enough so long as we only want to mention individual formulae, as in

(3) 'A & B' is a conjunction.

But ordinary quotes do not work if we want to make generalizations. Suppose we want to say that conjoining any two LSL wffs yields an LSL wff. Then we might try

(4) If p and q are LSL wffs, so is '$(p \& q)$'.

But this is incorrect: (4) is like

(5) If x and y are even numbers, so is '$(x + y)$'.

'$(x + y)$' is not a number at all, but simply a string of symbols. Similarly, '$(p \& q)$' is not an LSL wff, since 'p' and 'q' are not LSL sentence-letters (in the specification of the lexicon of LSL on page 35, sentence-letters were all defined to be uppercase roman). However, if the quotation marks are removed from (5), all is well; not so with (4), for in (4) we *are* trying to mention formulae, though not any *specific* formula. Some kind of quotation is still involved. What we are really trying to say with (4) is the following:

(6) If p and q are LSL wffs, so is the symbol-string which results from writing '(' followed by p followed by '&' followed by q followed by ')'.

which is exactly how we *did* say it in the syntax clause (*f*-&) on page 36. In these syntax clauses, and in (6), we use quotation marks *selectively*, putting them around the parentheses and the connectives but not around the sentential variables 'p' and 'q'.

The rationale for selective quotation is best understood in the light of the distinction between the *object language* and the *metalanguage*. In a given context, the object language is the language under discussion, and the metalanguage is the language in which the discussion is being carried out. Thus in the context of a French class, French is the object language and the metalanguage is usually the native language of the students (though in 'immersion' courses, French will be the metalanguage as well). At this point in this book, the object language is LSL and the metalanguage is the language in which the book is written, English augmented by some technical notation. Sometimes in discussing an object language we want to say things not about a particular object-language sentence but about a class of sentences characterized syntactically. The characterization may involve elements of the object language and of the metalanguage, as in (6), where we want to talk about the class of conjunctions in LSL. The letters '*p*' and '*q*' which we use to stand for arbitrary symbol-strings or wffs of LSL do not themselves belong to LSL. They are part of the metalanguage, augmented English, and for that reason are known as *metavariables* for sentences. But the symbol '&' is an LSL symbol, so it is something we mention, rather than use, in the metalanguage—throughout this book, whenever we want to use an expression for a connective in the metalanguage, we will use English word(s), thereby keeping object language and metalanguage apart. What should be used and what should be mentioned in the present context is made clear in (6), where ordinary quotes surround everything being mentioned. The trouble with (4) is that the ordinary quotes have the effect of creating a mention of the whole expression '(*p* & *q*)', where all we really want to mention are the conjunction symbol and the parentheses, that is, the items from the object language.

Rather than write out clauses like (6) every time we say something about a general class of formulae, we can abbreviate by using *corner quotes*, also called *selective* quotes. Corner quotes around a whole expression put ordinary quotes selectively around parts of that expression, those parts which are object language symbols. The corners simply abbreviate fully explicit formulations such as (6). Thus the shorter version of (6) is:

(7) If *p* and *q* are LSL wffs, so is ⌜(*p* & *q*)⌝.

The syntax clauses given on page 36 may themselves be abbreviated by clauses in the style of (7). Thus instead of writing

(*f*-↔) For any symbol-strings *p* and *q*, if *p* and *q* are wffs, then the result of writing '(' followed by *p* followed by '↔' followed by *q* followed by ')' is a wff

we could have written

(8) (*f*-↔): For any symbol-strings *p* and *q*, if *p* and *q* are wffs then ⌜(*p* ↔ *q*)⌝ is a wff.

The other syntax clauses may be abbreviated in the same way. We can contrast the effect of corners and ordinary quotes with an example. Suppose *p* is the LSL sentence '(A & B)', and *q* is the LSL sentence 'C'. Then the result of writing *p* followed by '↔' followed by *q* is '(A & B) ↔ C'; that is, ⌜*p* ↔ *q*⌝ = '(A & B) ↔ C'. But the result of writing '*p*' followed by '↔' followed by '*q*' is just '*p* ↔ *q*'; the fact that '*p*' and '*q*' are metalanguage symbols which stand for LSL wffs is irrelevant when we mention them by putting ordinary quotes around them. So if we want to use certain metalanguage symbols to stand for pieces of the object language, corners must be employed.

❑ Exercises

I   Rewrite the formation rules (*f*-~) and (*f*-→) in abbreviated form using corners (compare (7) of this section).

II   (1) Let *p* be the English phrase 'snow is' and let *q* be the word 'white'.

*(a)   Write out the result of writing *p* followed by *q*, that is, write out ⌜*p q*⌝.
*(b)   Write out the result of writing *p* followed by '*q*'.

(2) Let *p* be '(A & B)' and let *q* be 'R'.

(a)   Write out ⌜*p* ∨ *q*⌝.
(b)   Write out '*p*' followed by *q*.

III   On page 21 and page 23 rules for symbolizing sentences with 'only if' and 'unless' are stated, but strictly speaking incorrectly. Use corners to state them correctly.

IV   Which of the following statements contain misuses of English quotation marks or corner quotes, or omissions where use is required? For each such statement which you identify, explain briefly what is wrong with it and how it could be made correct.

*(1)   Rome is the name of a city.
(2)   'Rome' is the largest city in Italy.
(3)   Superman's other name is Clark Kent.
*(4)   If *p* is an indicative English sentence, ⌜it is not the case that *p*⌝ is also an indicative English sentence.
(5)   If *p* is an indicative sentence of French, so is ⌜*p*, n'est ce pas?⌝.
(6)   If *p* is a wff of LSL so is ⌜~*p*⌝.
(7)   If ⌜(*p* & *q*)⌝ is a true conjunction of LSL then *p* is true and *q* is true.
(8)   In French, Vive and la and France can go together to form ⌜Vive la France⌝.
(9)   In LSL, A and B can go together to form ⌜A & B⌝.
(10)   If '*p*' is a complete sentence of English then it contains a subject and a verb.
*(11)   If *p* is an indicative sentence of English then it is obvious to every English speaker that *p* is an indicative sentence of English.

(Note: Allowing for syntactic error, (11) is ambiguous—on one reading it is syntactically correct but for the other requires quotes. Explain.)

V   In which of the statements under IV is English both the object language and the metalanguage? (You may count corners and the use of metavariables for sentences as part of English.) In IV.5, what is the object language and what is the metalanguage?

7 Summary

- Many connectives such as 'only if' and 'unless' can be expressed in terms of our basic five.
- To symbolize an argument, we identify its premises and conclusion and the atomic sentences which occur in it. We then create a dictionary and substitute letters for atomic sentences as an intermediate step.
- The resulting formulae belong to a language, LSL, with its own syntax.
- The syntactic rules determine parse trees for well-formed formulae, that is, sentences of LSL, and syntactic concepts like 'main connective' can be defined using parse trees.
- We use corners around phrases which characterize classes of object language expressions and which contain metalanguage symbols being used and object language symbols (usually connectives) being mentioned.

3 Semantics for Sentential Logic

1 Truth-functions

Now that we know how to recover the sentential logical form of an English argument from the argument itself, the next step is to develop a technique for testing argument-forms for validity. The examples we have already considered indicate that whether or not a logical form is valid depends at bottom on the meanings of the sentential connectives which occur in it. For example, in Chapter 1 we considered the two cases

  A: P & Q
    ∴ P

and

  B: P ∨ Q
    ∴ P

the first of which is valid and the second invalid. Since the only difference between the two is that one has '&' where the other has '∨', it must be the difference in meaning between these two connectives that explains the difference in validity status between the two argument-forms. So a technique for testing argument-forms for validity must be based upon a precise specification of the meanings of the connectives.

First, some terminology. If a sentence is true, then it is said to *have the truth-value* TRUE, written 'T'. If a sentence is false, then it is said to *have the truth-value* FALSE, written '⊥'. We make the following assumption, often called the *Principle of Bivalence:*

> There are exactly two truth-values, T and ⊥. Every meaningful sentence, simple or compound, has one or other, but not both, of these truth-values.

We already remarked that classical sentential logic is so called in part because it is the logic of the *sentential* connectives. What makes it *classical* is the fact that the Principle of Bivalence is embodied in the procedure for giving meaning to sentences of LSL. (By implication, therefore, there are other kinds of sentential logic based on different assumptions.) Granted the Principle of Bivalence, we can precisely specify the meaning, or *semantics*, of a sentential connective in the following way. A connective attaches to one or more sentences to form a new sentence. By the principle, the sentence(s) to which it attaches already have a truth-value, either ⊤ or ⊥. The compound sentence which is formed must also have a truth-value, either ⊤ or ⊥, and which it is depends both on the truth-values of the simpler sentence(s) being connected and on the connective being used. A connective's semantics are precisely specified by saying what will be the truth-value of the compound sentence it forms, given all the truth-values of the constituent sentences.

Negation. The case of negation affords the easiest illustration of this procedure. Suppose *p* is some sentence of English whose truth-value is ⊤ ('2 + 2 = 4'). Then the truth-value of ⌜it is not the case that *p*⌝ is ⊥. In the same way, if *p* is some sentence of English whose truth-value is ⊥ ('2 + 2 = 5'), the truth-value of ⌜it is not the case that *p*⌝ is ⊤. Hence the effect of prefixing 'it is not the case that' to a sentence is to *reverse* that sentence's truth-value. This fact exactly captures the meaning of 'it is not the case that', at least as far as logic is concerned, and we want to define our symbol '~' so that it has this meaning. One way of doing so is by what is called a *truth-table*. The truth-table for negation is displayed below.

$$
\begin{array}{c|c}
p & \sim p \\
\hline
\top & \bot \\
\bot & \top
\end{array}
$$

The symbol '~' forms a compound wff by being prefixed to some wff *p*. *p* has either the truth-value ⊤ or the truth-value ⊥, and these are listed in the column headed by *p* on the left. On the right, we enter on each row the truth-value which the compound formula ⌜~*p*⌝ has, given the truth-value of *p* on that row.

Any sentential connective whose meaning can be captured in a truth-table is called a *truth-functional* connective and is said to *express a truth-function*. The general idea of a function is familiar from mathematics: a function is something which takes some object or objects as input and yields some object as output. Thus in the arithmetic of natural numbers, the function of squaring is the function which, given a single number as input, produces its square as output. The function of adding is the function which, given two numbers as input, produces their sum as output. A truth-function, then, is a function which takes a truth-value or truth-values as input and produces a truth-value as output. We can display a truth-function simply in terms of its effect on truth-values, abstracting from sentences. Thus the truth-function expressed by '~' could be written: ⊤⇒⊥, ⊥⇒⊤, which says that when the input is a truth the output is a falsehood (⊤⇒⊥), and when the input is a falsehood, the output is a truth

($\bot \Rightarrow \top$). Notice that the input/output arrow '$\Rightarrow$' we use here is different from the arrow '$\rightarrow$' we use for the conditional. '$\sim$' expresses a *one-place* or *unary* truth-function, because the function's input is always a single truth-value. In the same way, squaring is a one-place function of numbers, since it takes a single number as input.

Conjunction. The truth-table for conjunction is slightly more complicated than that for negation. '&' is a two-place connective, so we need to display two for-mulae, p and q, each of which can be $\top$ or $\bot$, making four possibilities, as in (a) below. The order in which the possibilities are listed is by convention the stan-dard one. (a) is empty, since we still have to decide what entries to make on each row of the table. To do this, we simply consider some sample English con-junctions. The assertion 'salt is a solid and water is a liquid' is one where both conjuncts are true, and this is enough to make the whole conjunction true. Since this would hold of any other conjunction as well, $\top$ should go in the top row of the table. But if one of the conjuncts of a conjunction is false, be it the first or the second, that is enough to make the whole conjunction false: consid-er 'salt is a gas and water is a liquid' and 'water is a liquid and salt is a gas'. Finally, if both conjuncts are false, the result is false as well. So we get the table in (b). We can write the resulting truth-function in the arrow notation, as in (c),

p	q	p & q
$\top$	$\top$	
$\top$	$\bot$	
$\bot$	$\top$	
$\bot$	$\bot$	

(a)

p	q	p & q
$\top$	$\top$	$\top$
$\top$	$\bot$	$\bot$
$\bot$	$\top$	$\bot$
$\bot$	$\bot$	$\bot$

(b)

&

$\top\top \Rightarrow \top$
$\top\bot \Rightarrow \bot$
$\bot\top \Rightarrow \bot$
$\bot\bot \Rightarrow \bot$

(c)

&	$\top$	$\bot$
$\top$	$\top$	$\bot$
$\bot$	$\bot$	$\bot$

(d)

though this time the function is *two-place*, since the appropriate input is a *pair* of truth-values (addition is an example of a two-place function on numbers, since it takes two numbers as input). There is also a third way of exhibiting the meaning of '&', which is by a matrix, as in (d). The values in the side column represent the first of the two inputs while the values in the top row represent the second of the two inputs. A choice of one value from the side column and one from the top row determines a position in the matrix, where we find the value which conjunction yields for the chosen inputs. (b), (c) and (d), therefore, all convey the same information.

Disjunction. To exhibit the semantics of a connective, we have to write out the truth-function which the connective expresses in one of the three formats just illustrated. What truth-function does disjunction express? Here matters are not as straightforward as with negation and conjunction, since there tends to be some disagreement about what to enter in the top row of the truth-table for '$\lor$', as we noted in connection with Example 2.2.10 on page 17. Recall also

(1) Either I will watch television this evening or read a good book

on the supposition that I do *both*, so that both disjuncts are true. But we have declared the policy of always treating disjunction as meaning inclusive disjunction, so (1) is true if I do both, and therefore the top row of the table for '∨' contains ⊤.

The remaining rows of the truth-table are unproblematic. If I only watch television (row 2 of the table below) or only read a good book (row 3), then (1) is clearly true, while if I do neither (row 4), it is false. So in the inclusive sense, a disjunction is true in every case except where both disjuncts are false. This leads us to the following representations of the meaning of '∨':

p	q	$p \lor q$
⊤	⊤	⊤
⊤	⊥	⊤
⊥	⊤	⊤
⊥	⊥	⊥

(a)

∨

⊤⊤ ⇒ ⊤
⊤⊥ ⇒ ⊤
⊥⊤ ⇒ ⊤
⊥⊥ ⇒ ⊥

(b)

∨	⊤	⊥
⊤	⊤	⊤
⊥	⊤	⊥

(c)

The difference between the actual table (a) for '∨' and the one we would have written had we chosen to use the symbol for exclusive disjunction (one or the other but not both) would simply be that the top row would contain ⊥ instead of ⊤.

The Conditional. This leaves us still to discuss the conditional and the biconditional. Since the biconditional was defined as a conjunction of conditionals (page 24), we will be able to *calculate* its truth-table once we have the table for '→', since we already know how to deal with conjunctions. However, the table for '→' turns out to be a little problematic. There is one row where the entry is clear. The statement

(2) If Smith bribes the instructor then Smith will get an A

is clearly false if Smith bribes the instructor but does not get an A. (2) says that bribing the instructor is sufficient for getting an A, or will lead to getting an A, so if a bribe is given and an A does not result, what (2) says is false. So we can enter a ⊥ in the second row of the table for '→', as in (a) below.

p	q	$p \to q$
⊤	⊤	
⊤	⊥	⊥
⊥	⊤	
⊥	⊥	

(a)

p	q	$p \to q$
⊤	⊤	⊤
⊤	⊥	⊥
⊥	⊤	⊤
⊥	⊥	⊤

(b)

⊤⊤ ⇒ ⊤
⊤⊥ ⇒ ⊥
⊥⊤ ⇒ ⊤
⊥⊥ ⇒ ⊤

(c)

→	⊤	⊥
⊤	⊤	⊥
⊥	⊤	⊤

(d)

But what of the other three rows? Here are three relevant conditionals:

(3) If Nixon was U.S. president then Nixon lived in the White House.

(4) If Agnew was British prime minister then Agnew was elected.

(5) If Agnew was Canadian prime minister then Agnew lived in Ottawa.

(3) has a true antecedent and a true consequent, (4) a false antecedent and a true consequent, and (5) a false antecedent and a false consequent, but all three of the conditionals are true (only elected members of Parliament can be British prime minister, but unelected officials can become U.S. president). Relying just on these examples, we would complete the table for '→' as in (b) of the previous figure, with equivalent representations (c) and (d).

The trouble with (b), (c) and (d) is that they commit us to saying that *every* conditional with a true antecedent and consequent is true and that *every* conditional with a false antecedent is true. But it is by no means clear that this is faithful to our intuitions about ordinary indicative conditionals. For example,

(6) If Moses wrote the Pentateuch then water is H_2O

has an antecedent which is either true or false—most biblical scholars would say it is false—and a consequent which is true, and so (6) is true according to our matrix for '→'. But many people would deny that (6) is true, on the grounds that there is no relationship between the antecedent and the consequent: there is no sense in which the nature of the chemical composition of water is a *consequence* of the identity of the author of the first five books of the Bible, and if (6) asserts that it *is* a consequence, then (6) is false, not true.

As in our discussion of 'or', there are two responses one might have to this objection to our table (b) for 'if...then...'. One response is to distinguish two senses of 'if...then...'. According to this response, there is a sense of 'if...then...' which the table correctly encapsulates, and a sense which it does not. The encapsulated sense is usually called the *material* sense, and '→' is said to express the *material conditional*. Indeed, even if it were held that in English, 'if...then...' never expresses the material conditional, we could regard the table (b) above as simply a *definitional introduction* of this conditional. There would then be no arguing with table (b); the question would be whether the definitionally introduced meaning for the symbol '→' which is to be used in translating English indicative conditionals is adequate for the purposes of sentential logic. And it turns out that the answer to this question is 'yes', since it is the second row of the table which is crucial, and the second row is unproblematic. An alternative response to the objection is to say that the objector is confusing the question of whether (6) is literally true with the question of whether it would be appropriate to assert (6) in various circumstances. Perhaps it would be inappropriate for one who knows the chemical composition of water to assert (6), but such inappropriateness is still consistent with (6)'s being literally true, according to this account.

The parallel with the discussion of 'or' is not exact, and these are issues we will return to in §8 of this chapter. But whatever position one takes about the meaning of 'if...then...' in English, the reader should be assured that it is adequate for the purposes of sentential logic to translate English indicative conditionals into LSL using the material conditional '→', even if one does regard this conditional as somewhat artificial. The artificiality will not lead to intuitively valid arguments being assessed as invalid, or conversely.

The Biconditional. For any LSL wffs p and q, the biconditional ⌜$p \leftrightarrow q$⌝ simply abbreviates the corresponding conjunction of conditionals ⌜$(p \to q) \& (q \to p)$⌝, according to our discussion in §3 of Chapter 2. It follows that if we can work out the truth-table for that conjunction, given the matrices for '&' and '→', we will arrive at the truth-function expressed by '↔'. What is involved in working out the table for a formula with more than one connective? In a formula of the form ⌜$(p \to q) \& (q \to p)$⌝, p and q may each be either true or false, leading to the usual four possibilities. The truth-table for ⌜$(p \to q) \& (q \to p)$⌝ should tell us what the truth-value of the formula is for each of these possibilities. So we begin by writing out a table with the formula along the top, as in (a) below:

p	q	$(p \to q) \& (q \to p)$	p	q	$(p \to q) \& (q \to p)$	p	q	$(p \to q) \& (q \to p)$
T	T		T	T	T	T	T	T	T	T
T	⊥		T	⊥	⊥	T	T	⊥	⊥	⊥	T
⊥	T		⊥	T	T	⊥	⊥	T	T	⊥	⊥
⊥	⊥		⊥	⊥	T	T	⊥	⊥	T	T	T
		(a)			(b)			(c)

The formula is a conjunction, so its truth-value in each of the four cases will depend upon the truth-value of its conjuncts in each case. The next step is therefore to work out the truth-values of the conjuncts on each row. The first conjunct is the (material) conditional ⌜$(p \to q)$⌝ whose truth-table we have already given, so we can just write those values in. The second conjunct is ⌜$q \to p$⌝. We know from our discussion of '→' above that the only case where a material conditional is false is when it has a true antecedent and false consequent, and this combination for ⌜$q \to p$⌝ occurs on row 3 (*not* row 2); so under ⌜$q \to p$⌝ we want ⊥ on row 3 and T elsewhere. This gives us table (b) above. We have now calculated the truth-value of each conjunct of ⌜$(p \to q) \& (q \to p)$⌝ on each row, so it remains only to calculate the truth-value of the whole conjunction on each row. Referring to the tables for conjunction, we see that a conjunction is true in just one case, that is, when both conjuncts are true. In our table for ⌜$(p \to q) \& (q \to p)$⌝, both conjuncts are true on rows 1 and 4, so we can complete the table as in (c) above. Notice how we highlight the column of entries under the main connective of the formula. The point of doing this is to distinguish the final answer from the other columns of entries written in as intermediate steps.

The example of ⌜$(p \to q) \& (q \to p)$⌝ illustrates the technique for arriving at the truth-table of a formula with more than one occurrence of a connective in it, and it also settles the question of what truth-function '↔' expresses. We complete our account of the meanings of the connectives with the tables for the

p	q	$p \leftrightarrow q$
T	T	T
T	⊥	⊥
⊥	T	⊥
⊥	⊥	T

$$\leftrightarrow$$

TT ⇒ T
T⊥ ⇒ ⊥
⊥T ⇒ ⊥
⊥⊥ ⇒ T

$\leftrightarrow$	T	⊥
T	T	⊥
⊥	⊥	T

(a)   (b)   (c)

biconditional displayed above. Note that, by contrast with $\ulcorner p \rightarrow q \urcorner$ and $\ulcorner q \rightarrow p \urcorner$, $\ulcorner p \leftrightarrow q \urcorner$ and $\ulcorner q \leftrightarrow p \urcorner$ have the same truth-table; this bears out our discussion of Examples 2.4.6 and 2.4.7 on page 32, where we argued that the order in which the two sides of a biconditional are written is irrelevant from the logical point of view.

In order to acquire some facility with the techniques which we are going to introduce next, it is necessary that the meanings of the connectives be memorized. Perhaps the most useful form in which to remember them is in the form of their function-tables, so here are the truth-functions expressed by all five connectives:

	&	∨	→	↔
~	TT ⇒ T	TT ⇒ T	TT ⇒ T	TT ⇒ T
	T⊥ ⇒ ⊥	T⊥ ⇒ T	T⊥ ⇒ ⊥	T⊥ ⇒ ⊥
T ⇒ ⊥	⊥T ⇒ ⊥	⊥T ⇒ T	⊥T ⇒ T	⊥T ⇒ ⊥
⊥ ⇒ T	⊥⊥ ⇒ ⊥	⊥⊥ ⇒ ⊥	⊥⊥ ⇒ T	⊥⊥ ⇒ T

The information represented here can be summarized as follows:

- Negation reverses truth-value.
- A conjunction is true when and only when both conjuncts are true.
- A disjunction is false when and only when both disjuncts are false.
- A conditional is false when and only when its antecedent is true and its consequent is false.
- A biconditional is true when and only when both its sides have the same truth-value.

These summaries should also be memorized.

There are some entertaining puzzles originated by Raymond Smullyan which involve manipulating the notions of truth and falsity in accordance with the tables for the connectives. In a typical Smullyan setup, you are on an island where there are three kinds of inhabitants, Knights, Knaves and Normals. Knights always tell the truth and Knaves always lie, while a Normal may sometimes lie and sometimes tell the truth. You encounter some people who make certain statements, and from the statements you have to categorize each of the people as a Knight, a Knave, or a Normal. Here is an example, from Smullyan:

You meet two people, A and B, each of whom is either a Knight or a Knave. Suppose A says: 'Either I am a Knave or B is a Knight.' What are A and B?

We reason to the solution as follows. There are two possibilities for A, either Knight or Knave. Suppose that A is a Knave. Then what he says is false. What he says is a disjunction, so by any of the tables for '∨', both disjuncts of his statement must be false. This would mean that A is a Knight and B is a Knave. But A cannot be a Knight if he is a Knave (our starting supposition). Thus it follows that he is not a Knave. So by the conditions of the problem, he is a Knight and what he says is true. Since he is a Knight, the first disjunct of his statement is false, so the second disjunct must be true. Hence B is a Knight as well.

❏ Exercises

The following problems are from Smullyan.[1] In each case explain the reasoning that leads you to your answer in the way just illustrated.

(1)   There are two people, A and B, each of whom is either a Knight or a Knave. A says: 'At least one of us is a Knave.' What are A and B?

(2)   With the same conditions as (1), suppose instead A says: 'If B is a Knight then I am a Knave.' What are A and B? [Refer to the truth-table for '→'.]

(3)   There are three people, A, B and C, each of whom is either a Knight or a Knave. A and B make the following statements:

A: 'All of us are Knaves.'
B: 'Exactly one of us is a Knight.'

What are A, B and C?

*(4)   Two people are said to be of the *same type* if and only if they are both Knights or both Knaves. A and B make the following statements:

A: 'B is a knave.'
B: 'A and C are of the same type.'

On the assumption that none of A, B and C is Normal, can it be determined what C is? If so, what is he? If not, why not?

---

[1]   © 1978 by Raymond Smullyan. Reprinted by permission of Simon and Schuster.

(5) Suppose A, B and C are being tried for a crime. It is known that the crime was committed by only one of them, that the perpetrator was a Knight, and the only Knight among them. The other two are either both Knaves, both Normals, or one of each. The three defendants make the statements below. Which one is guilty?

> A: 'I am innocent.'
> B: 'That is true.'
> C: 'B is not Normal.'

(6) A, who is either a Knight or a Knave, makes the following statement:

> A: 'There is buried treasure on this island if and only if I am a Knight.'

(i) Can it be determined whether A is a Knight or a Knave?

(ii) Can it be determined whether there is buried treasure on the island?

2 Classifying formulae

Any formula of LSL has a truth-table, for every formula is constructed from a certain number of sentence-letters and each sentence-letter can be either $\top$ or $\bot$. So there are various possible combinations of truth-values for the sentence-letters in the formula, and for each of those possible combinations, the formula has its own truth-value. Here are truth-tables for three very simple formulae, 'A → A', 'A → ~A' and '~(A → A)':

A	A → A
$\top$	$\top$
$\bot$	$\top$

(a)

A	A → ~A
$\top$	$\top$ $\boxed{\bot}$ $\bot$
$\bot$	$\bot$ $\boxed{\top}$ $\top$

(b)

A	~(A → A)
$\top$	$\boxed{\bot}$ $\top$
$\bot$	$\boxed{\bot}$ $\top$

(c)

In table (a), on the first row, 'A → A' is $\top \to \top$, which by the function-table or matrix for '→' is $\top$, and on the second row, $\bot \to \bot$, which is also $\top$. In table (b) we have $\top \to \bot$ on the top row, which gives $\bot$, and $\bot \to \top$ on the bottom, which gives $\top$ (the final answer is in the box). In table (c), we have the negation of the formula of table (a), so in its final answer column, table (c) should have $\bot$ where table (a) has $\top$, and $\top$ where (a) has $\bot$. We enter the column for the subformula 'A → A' first, directly under the main connective of this subformula, and then apply the truth-function for '~' to that column. *We always display the final answer under the main connective of the whole formula.*

These three formulae exhibit the three possibilities for any formula: that in its truth-table the final answer column contains nothing but ⊤s, or a mixture of ⊤s and ⊥s, or nothing but ⊥s. There is a technical term for each of these three kinds of formula:

- A formula whose truth-table's final answer column contains only ⊤s is called a *tautology.*
- A formula whose truth-table's final answer column contains only ⊥s is called a *contradiction.*
- A formula whose truth-table's final answer column contains both ⊤s and ⊥s is said to be *contingent.*

Thus 'A → A' is a tautology, '~(A → A)' is a contradiction, and 'A → ~A' is a contingent formula.

When a formula only has a single sentence-letter in it, as in the three formulae just exhibited, there are only two possibilities to consider: the sentence-letter is either ⊤ or ⊥. And as we have seen in giving the truth-tables for the binary connectives, when there are two sentence-letters there are four possibilities, since each sentence-letter can be either ⊤ or ⊥. The number of possibilities to consider is determined by the number of sentence-letters in the formula, not by the complexity of the formula. Thus a truth-table for '(A ↔ B) → ((A & B) ∨ (~A & ~B))' will have only four rows, just like a table for 'A → B', since it contains only two sentence-letters. But it will have many more *columns:*

A B	(A ↔ B)	→	((A & B)	∨	(~A & ~B))
⊤ ⊤	⊤	⊤	⊤	⊤	⊥ ⊥ ⊥
⊤ ⊥	⊥	⊤	⊥	⊥	⊥ ⊥ ⊥ ⊤
⊥ ⊤	⊥	⊤	⊥	⊥	⊤ ⊥ ⊥
⊥ ⊥	⊤	⊤	⊥	⊤	⊤ ⊤ ⊤ ⊤
	1		2	6 3	5 4

Here the numbers indicate the order in which the columns are computed. We begin by entering the values for 'A ↔ B' and 'A & B', simply taking these from the truth-tables for '↔' and '&'; this gives us the columns numbered 1 and 2. Then we compute the entries for '~A' and '~B' by applying the negation truth-function to the entries under the two sentence-letters on the far left; this gives us columns 3 and 4. Using columns 3 and 4, we next compute the values for the conjunction '~A & ~B', which gives us column 5, since it is only on the bottom row that columns 3 and 4 both contain ⊤. We then use columns 2 and 5 to compute the entries for the disjunction '(A & B) ∨ (~A & ~B)', yielding column 6 under the disjunction symbol. Lastly, we use columns 1 and 6 to compute the final answer for the whole formula. Inspecting the final column reveals that the formula is a tautology. There is some flexibility about the order in which we compute columns—the main constraint is that before computing the column for any subformula *q* of a given formula *p*, we must first compute the columns for all *q*'s subformulae.

What of formulae with more than two sentence-letters, for example, '(A → (B ∨ C)) → (A → (B & C))'? The first question is how many different combinations of truth-values have to be considered when there are three sentence-letters. It is not difficult to see that there are eight combinations, as the following argument shows: A can be either ⊤ or ⊥, and in each of these two cases, B can be either ⊤ or ⊥, giving us four cases, and in each of these four, C can be either ⊤ or ⊥, giving eight cases in all. More generally, if there are n sentence-letters in a formula, then there will be 2^n cases, since each extra sentence-letter doubles the number of cases. In particular, we need an eight-row truth-table for the formula '(A → (B ∨ C)) → (A → (B & C))'.

There is a conventional way of listing the possible combinations of truth-values for any n sentence-letters. Once the sentence-letters are listed at the top left of the table, we alternate ⊤ with ⊥ under the innermost (rightmost) letter until we have 2^n rows. Then under the next-to-innermost, we alternate ⊤s and ⊥s in *pairs* to fill 2^n rows. Continuing leftward we alternate ⊤s and ⊥s in fours, then eights, then sixteens, and so on, until every sentence-letter has a column of 2^n truth-values under it. This procedure guarantees that all combinations are listed and none are listed twice. For our example, '(A → (B ∨ C)) → (A → (B & C))', we will therefore have ⊤ followed by ⊥ iterated four times under 'C', two ⊤s followed by two ⊥s followed by two ⊤s followed by two ⊥s under 'B', and four ⊤s followed by four ⊥s under 'A'. We then proceed to compute the table for the formula in the usual way:

A B C	(A → (B ∨ C))	→	(A → (B & C))		
⊤ ⊤ ⊤	⊤		⊤	⊤	⊤
⊤ ⊤ ⊥	⊤		⊥	⊥	⊥
⊤ ⊥ ⊤	⊤		⊥	⊥	⊥
⊤ ⊥ ⊥	⊥	⊥	⊤	⊥	⊥
⊥ ⊤ ⊤	⊤		⊤	⊤	
⊥ ⊤ ⊥	⊤		⊤	⊤	
⊥ ⊥ ⊤	⊤		⊤	⊤	
⊥ ⊥ ⊥	⊤		⊤	⊤	
	4	3	2	1	

This formula is contingent, since it has a mixture of ⊤s and ⊥s. Notice also that the truth-table has not been completely filled in. When the number of rows in a truth-table is large, it is advisable to look for shortcuts in arriving at the final column. So in column 1, for example, we do not compute the bottom four rows, since we know that a material conditional with a false antecedent is true, and hence 'A → (B & C)' will be true on the bottom four rows since 'A' is false there (refer to A's column on the extreme left); the values of 'B & C' are therefore irrelevant on these rows. Similarly in column 3, we can ignore all but row 4, since the falsity of 'A → (B ∨ C)' requires the truth of 'A' and the falsity of 'B ∨ C', which in turn requires the falsity of both 'B' and 'C'. 'B' and 'C' are both ⊥ only on rows 4 and 8, and by inspection of these two rows, we see that 'A' is ⊤ only on row 4. Hence it is only on row 4 that the condition for 'A → (B ∨ C)' to be ⊥

is satisfied, and we can fill in ⊤ on all the other rows in column 4, as we have done in the displayed table.

Up to this point we have referred to combinations of truth-values listed on the left of a truth-table as 'cases' and have described a truth-table for a formula as giving the truth-value of the formula in each of the 'possible cases'. The more usual word for 'case' is *interpretation*. Thus a formula of LSL with n sentence-letters has 2^n possible interpretations. However, the term 'interpretation' is used in every kind of system of logic: an interpretation is a way of giving meaning to the sentences of the language appropriate for the kind of logic in question, and with more complex languages, this involves more than specifying truth-values for sentence-letters. So for each kind of logic, we have to say explicitly what kind of thing an interpretation of a formula of the language for that logic is. For sentential logic, we have:

> An *interpretation of a formula p of LSL* is an assignment of truth-values to the sentence-letters which occur in p.

So in a truth-table for p we find on the left a list of all the possible interpretations of p, that is, all the possible assignments of truth-values to the sentence-letters in p. A single interpretation of a formula is given by specifying the truth-values which its sentence-letters have on that interpretation. For example, the third interpretation in the table on the previous page assigns ⊤ to 'A' and 'C' and ⊥ to 'B', and this interpretation makes the formula $p = $ '(A → (B ∨ C)) → (A → (B & C))' false.

Because the term 'interpretation' has application in every kind of logic, technical concepts of sentential logic defined using it will also have wider application. For example, we have previously spoken of formulae being 'equivalent' or of their 'meaning the same' in a loose sense. Exactly what this amounts to is spelled out in the following:

> Two formulae p and q are said to be *logically equivalent* if and only if, on any interpretation assigning truth-values to the sentence-letters of both, the truth-value of the first formula is the same as the truth-value of the second.

Here we have not explicitly restricted the definition to formulae of LSL, since the same notion of logical equivalence will apply to formulae of any system of logic in which formulae have truth-values or something analogous. In the special case of LSL, for formulae with exactly the same sentence letters, logical equivalence amounts to having the same truth-table. So by inspecting the table on the following page, we see that 'A & B' and '~(~A ∨ ~B)' are logically equivalent, and that '~(A ∨ B)' and '~A & ~B' are logically equivalent. For each interpretation gives the same truth-value to 'A & B' as it does to '~(~A ∨ ~B)', and each gives the same truth-value to '~(A ∨ B)' as it does to '~A & ~B'.

A B	A & B	~(~A v ~B)		~(A v B)		~A & ~B
T T	T	T	⊥	⊥	T	⊥
T ⊥	⊥	⊥	T	⊥	T	⊥
⊥ T	⊥	⊥	T	⊥	T	⊥
⊥ ⊥	⊥	⊥	T	T	⊥	T

In our discussion of the connective 'unless' in §3 of Chapter Two, the conclusion we reached was that ⌜p unless q⌝ can be symbolized as ⌜~q → p⌝, for any formulae p and q. Suppose that p and q are sentence-letters, say 'A' and 'B'. Then what we find is that 'unless' is just 'or', for we have the following table,

A B	A v B	~B → A
T T	T	T
T ⊥	T	T
⊥ T	T	T
⊥ ⊥	⊥	⊥

which shows that '~B → A' is logically equivalent to 'A v B'. In testing English arguments for validity, it is an advantage to symbolize the English with formulae which are as simple as possible, so at this point we will change our policy as regards 'unless'. Henceforth, we symbolize 'unless' using 'v':

- ⌜p unless q⌝ is symbolized ⌜p v q⌝.

Reflection on the meaning of 'unless' should indicate that this policy is intuitively correct: if the company will go bankrupt unless it receives a loan, that means that either the company will go bankrupt or (if it does not) then it receives (i.e. must have received) a loan.

❑ Exercises

I   Construct complete truth-tables for the following formulae and classify each as tautologous, contradictory or contingent. Be sure to mark your final column clearly and place it directly under the main connective of the formula.

    (1) A → (B → (A & B))
    (2) ~R → (R → S)
    (3) R → (S → R)
  *(4) (A ↔ B) & (A & ~B)
    (5) ((F & G) → H) → ((F v G) → H)
    (6) (A ↔ (B v C)) → (~C → ~A)
    (7) (A ↔ B) & ((C → ~ A) & (B → C))

II   Use truth-tables to determine which formulae in the following list are logically equivalent to which. State your results.

(1) A ∨ B   (2) A → B
(3) ~(A & ~B)   (4) ~(~A & ~B)
(5) ~A ∨ B   (6) A ∨ ~A
(7) (A → (A & ~A)) → ~A

III If *p* is a sentence of LSL which is not a tautology, does it follow that ⌜~*p*⌝ *is* a tautology? Explain.

3 Testing for validity by exhaustive search

We are now in a position to present the first technique for testing an argument-form for validity. Recall our opening examples of a valid and an invalid English argument from §1 of Chapter 1:

A: (1) If our currency loses value then our trade deficit will narrow.
    (2) Our currency will lose value.
    (3) ∴ Our trade deficit will narrow.

B: (1) If our currency loses value then our trade deficit will narrow.
    (2) Our trade deficit will narrow.
    (3) ∴ Our currency will lose value.

Concerning argument A, we said that the truth of the conclusion (3) is 'guaranteed' by the truth of the two premises (1) and (2), but we did not explain exactly what the guarantee consists in. The (sentential) invalidity of argument B we explained in the following way: even if (1) in B is true, its truth is consistent with there being other conditions which are sufficient for a narrowing of our trade deficit, so even given the truth of (2) in B, we cannot conclude (3), since it may have been one of those other conditions which has brought about the truth of (2) without our currency having lost value at all. Hence it is incorrect to say that the truth of (1) and (2) in B *guarantees* the truth of (3) (even if in fact (1), (2) and (3) *are* all true).

Reflecting on this explanation of B's invalidity, we see that we demonstrate the lack of guarantee by describing how circumstances could arise in which both premises would be true while the conclusion is false. The point was not that the premises are in fact true and the conclusion in fact false, but merely that for all that the premises and conclusion *say*, it would be *possible* for the premises to be true and the conclusion false. And if we inspect argument A, we see that this is exactly what *cannot* happen in its case. Thus the key to the distinction between validity and invalidity in English arguments appears to have to do with whether or not there is a *possibility* of their having true premises and a false conclusion. Yet we also saw that validity or invalidity is fundamentally a property of argument-forms, not the arguments themselves. The sentential logical forms of A and B are respectively

$$C:\ F \to N$$
$$F$$
$$\therefore N$$

and

$$D:\ F \to N$$
$$N$$
$$\therefore F$$

What then would it mean to speak of the 'possibility' of C or D having true premises and a false conclusion?

We can transfer the notion of the possibility of having true premises and false conclusion from English arguments to the LSL argument-forms which exhibit the English arguments' sentential forms, by using the concept of interpretation explained on page 56. To say that it is possible for an LSL form to have true premises and a false conclusion is to say that *there is at least one interpretation* of the LSL form on which its premises are true and its conclusion false. In sentential logic, an interpretation is an assignment of truth-values, so whether or not an LSL argument-form is valid depends on whether or not some assignment of truth-values makes its premises true and its conclusion false. We render this completely precise as follows:

An *interpretation of an LSL argument-form* is an assignment of truth-values to the sentence-letters which occur in that form.

An argument-form in LSL is *valid* if there is no interpretation of it on which its premises are true and its conclusion false, and *invalid* if there is at least one interpretation of it on which its premises are true and its conclusion false.

An English argument (or argument in any other natural language) is sententially valid if its translation into LSL yields a valid LSL argument-form, and is sententially invalid if its translation into LSL yields an invalid form.

Since there is no question of discerning finer structure in an LSL argument-form (as opposed to an English argument) using a more powerful system of logic, judgements of validity and invalidity for LSL forms are absolute, not relative to sentential logic. We can test an LSL form for validity by exhaustively listing all its interpretations and checking each one to see if any makes the form's premises all true while also making its conclusion false. Interpretations are listed in truth-tables, so we can use the latter for this purpose, by writing the argument-form out along the top. For example, we can test the two arguments C and D on the previous page with the following four-row table:

F N	F → N	F	∴ N	F → N	N	∴ F
T T	T	T	T	T	T	T
T ⊥	⊥	T	⊥	⊥	⊥	T
⊥ T	T	⊥	T	T	T	⊥
⊥ ⊥	T	⊥	⊥	T	⊥	⊥

The table shows that C is valid according to our definition, since none of the four interpretations listed makes the two premises 'F → N' and 'F' true while at the same time making the conclusion 'N' false. The table also shows that argument-form D is invalid, since inspection of the entries for the third interpretation (highlighted), on which 'F' is false and 'N' is true, shows that D's premises 'F → N' and 'N' are true on this interpretation while its conclusion 'F' is false. In terms of the original English argument, the truth-values ⊥ for 'our currency will lose value' and T for 'our trade deficit will narrow' are exactly the ones which would obtain in a situation where our trade deficit narrows for some other reason while our currency stays the same or rises, which is the kind of situation whose possibility we mentioned in order to show the sentential invalidity of B. The third interpretation in the table, therefore, expresses what is common to all situations which show that a given English argument with the same form as B is sententially invalid.

To summarize, this technique of testing an LSL argument-form for validity consists in listing all its possible interpretations and exhaustively inspecting each one. If one is found which makes the premises of the argument true and its conclusion false, then the LSL argument-form is invalid; if no interpretation which does this is found, the LSL argument-form is valid.

In applying this test to more complex LSL argument-forms, with large numbers of sentence-letters, it is important to exploit as many short cuts as possible. For example, in §4 of Chapter 2, we considered the argument:

E: If God exists, there will be no evil in the world unless God is unjust, or not omnipotent, or not omniscient. But if God exists then He is none of these, and there is evil in the world. So we have to conclude that God does not exist.

To test this English argument for sentential validity, we translate it into LSL and examine each of the interpretations of the LSL argument-form to see if any makes all the premises true and the conclusion false. If none do, the LSL argument-form is valid. This means that the English argument E is sententially valid and therefore valid absolutely. But if some interpretation does make the premises of the LSL argument-form all true and the conclusion false, then the LSL argument-form is invalid and so the English argument is sententially invalid.

The symbolization at which we arrived was:

F: X → [~(~J ∨ (~M ∨ ~S)) → ~V]
    [X → (~~J & (~~M & ~~S))] & V
    ∴ ~X

F contains five sentence-letters and therefore has 2^5 interpretations. Consequently, we would appear to need a truth-table with thirty-two rows to conduct an exhaustive check of whether or not F is valid. However, we can use the following table to test it for validity:

X V J M S	X → [~(~J ∨ (~M ∨ ~S)) → ~V]	[X → (~~J & (~~M & ~~S))] & V	∴ ~X
T T T T T	⊥ T T	[T] T	⊥
T T T T ⊥		T	⊥
T T T ⊥ T		T	⊥
T T T ⊥ ⊥		T	⊥
T T ⊥ T T		T	⊥
T T ⊥ T ⊥		T	⊥
T T ⊥ ⊥ T		T	⊥
T T ⊥ ⊥ ⊥		T	⊥
T ⊥ T T T		⊥	⊥
T ⊥ T T ⊥		⊥	⊥
T ⊥ T ⊥ T		⊥	⊥
T ⊥ T ⊥ ⊥		⊥	⊥
T ⊥ ⊥ T T		⊥	⊥
T ⊥ ⊥ T ⊥		⊥	⊥
T ⊥ ⊥ ⊥ T		⊥	⊥
T ⊥ ⊥ ⊥ ⊥		⊥	⊥

Two features of this table are immediately striking. The first is that it only has sixteen rows, instead of the advertised thirty-two. Where are the missing sixteen? The answer is that we are able to discount sixteen rows because we are only trying to discover whether there is an interpretation (row) on which all the premises of the argument-form are true while its conclusion is false. If we see that the conclusion is *true* on a certain interpretation, it would be a waste of effort to compute the truth-values of the premises on that interpretation, for clearly, *that* interpretation will not be one where all the premises are true and the conclusion *false*. To apply this point, observe that the conclusion of our argument-form is '~X', which is true on every interpretation which makes 'X' false, and the missing sixteen rows are exactly those on which 'X' is false. We deliberately made 'X' the first sentence-letter in the listing at the top of the table, so that the column beneath it would contain sixteen Ts followed by sixteen ⊥s, since this allows us to ignore the bottom sixteen rows, these being the interpretations where the conclusion is true and where we are consequently uninterested in the values ascribed to the premises.

The other striking feature of the table is that only the top row has been completed. The justification for this is comparable to that for ignoring the bottom sixteen rows. Just as we are not interested in interpretations which make the conclusion true, so we are not interested in interpretations which make one of the premises false, since those interpretations will not be ones where the conclusion is false and *all* the premises true. And it is easy to see from our table that every interpretation except the first makes premise 2 false; interpretations 9–16 make the second conjunct of the premise, 'V', false, while interpretations 2–8 make the first conjunct, 'X → (~~J & (~~M & ~~S))', false (because they make 'X' true and '(~~J & (~~M & ~~S))' false, since they make at least one of 'J' or 'M' or 'S' false. So only interpretation 1 makes premise 2 of F true, and so it is only its value for premise 1 that we are interested in computing. Hence, whether or not F is valid comes down to whether or not interpretation 1 makes premise 1 true. A simple calculation shows that in fact it makes it false. It follows that F is valid: no interpretation makes all the premises true and the conclusion false.[2] Consequently, E is sententially valid, and therefore valid absolutely. (The reader should study the reasoning of this and the previous paragraph for as long as is necessary to grasp it fully.)

This example nicely illustrates how with a little ingenuity we can save ourselves a lot of labor in testing for validity using the method of exhaustive search. But the method is still unwieldy, and completely impractical for LSL arguments which contain more than five sentence-letters. Given our definition of validity, then, the next step is to try to develop a more efficient way of testing for it.

❏ Exercises

Use the method of exhaustive search to test the arguments symbolized in the exercises for 2.4 for sentential validity. Display your work and state the result you obtain.

4 Testing for validity by constructing interpretations

A faster way of determining the validity or invalidity of an LSL argument-form is to attempt an explicit construction of an interpretation which makes the premises true and the conclusion false: if the attempt succeeds, the LSL form is invalid, and if it breaks down, then (provided it has been properly executed) the

---

[2] To repeat a point from Chapter 2, this does not mean that we have proved that God does not exist. We have shown merely that the original English argument is sententially valid, not that it is sound (recall that a sound argument is one which has all its premises true as well). In traditional Christian theology, the first premise would be disputed: it would be argued that the existence of evil is consistent with the existence of a just, omnipotent and omniscient God, since evil would be said to be a consequence of the free actions of human and supernatural beings, and God, it is held, is obliged not to interfere with the outcome of freely chosen actions. In other religions, or other versions of Christianity, different premises would be disputed. For instance, in some Eastern religions, it would be denied that there is evil in the world, on the grounds that all suffering is 'illusion'.

LSL form is valid. To test an English argument for sentential validity in this way, we first exhibit its form by translating it into LSL, and then we make assignments of truth-values to the sentence-letters in the conclusion of the LSL form so that the conclusion is false. We then try to assign truth-values to the remaining sentence-letters in the premises so that all the premises come out true. For example, we can test the argument G from §4 of Chapter 2, repeated here as

A: We can be sure that Jackson will agree to the proposal. For otherwise the coalition will break down, and it is precisely in these circumstances that there would be an election; but the latter can certainly be ruled out.

We begin by translating A into LSL, which results in the LSL argument-form

B: $(\sim J \rightarrow C) \& (E \leftrightarrow C)$
    $\sim E$
    $\therefore J$

(reread the discussion in §4 of Chapter 2 if necessary) and then we try to find an interpretation of the three sentence-letters in B which makes the conclusion false and both premises true. To make the conclusion 'J' false, we simply stipulate that 'J' is false. Taking the simpler of the two premises first, we stipulate that 'E' is false, making '~E' true. The question now is whether there is an assignment to 'C' on which premise 1 comes out true. Since 'E' is false, we need 'C' to be false for the second conjunct of premise 1 to be true, but then '~J → C' is ⊤ → ⊥, which is ⊥, making premise 1 false. There are no other options, so we have to conclude that no interpretation makes all the premises of B true and the conclusion false. Thus B is a valid LSL argument-form, and therefore A is a sententially valid English argument, and so valid absolutely.

Here is another application of the same technique, to the LSL argument-form

C: $A \rightarrow (B \& E)$
    $D \rightarrow (A \vee F)$
    $\sim E$
    $\therefore D \rightarrow B$

To make the conclusion false we stipulate that 'D' is true and 'B' is false. The simplest premise is the third, so next we make it true by stipulating that 'E' is false. This determines the truth-value of 'A' in the first premise if that premise is to be true: if 'E' is false then 'B & E' is false, so we require 'A' to be false for the premise to be true. So far, then, we have shown that for the conclusion to be false while the first and third premises are true, we require the assignment of truth-values ⊤ to 'D', ⊥ to 'B', ⊥ to 'E' and ⊥ to 'A'. The question is whether this assignment can be extended to 'F' so that premise 2 comes out true. With 'D' being true and 'A' false, premise 2 is true when 'F' is true and false when 'F' is false. Since we are free to assign either truth-value to 'F', we obtain an interpretation which shows C to be invalid by stipulating that 'F' is true. In other words, the interpretation

D	B	E	A	F
⊤	⊥	⊥	⊥	⊤

makes all the premises of C true and its conclusion false, so C is invalid.

This technique is a significant improvement over drawing up the thirty-two row truth-table that would be required to test C for validity by the method of exhaustive search. However, the two examples we have just worked through contain a simplifying feature that need not be present in general, for in both B and C there is only one way of making the conclusion false. How do we proceed when there is more than one? An example in which this situation arises is the following:

> D: ~A ∨ (B → C)
> E → (B & A)
> C → E
> ∴ C ↔ A

We deal with this by distinguishing cases. There are two ways of making the conclusion false, and we investigate each case in turn to see if there is any way of extending the assignment to make all the premises true:

Case 1: 'C' is true, 'A' is false. Then for premise 3 to be true, we require 'E' to be true, and so for premise 2 to be true, we require 'B & A' to be true, but we already have 'A' false. Consequently, there is no way of extending the assignment 'C' true, 'A' false, to the other sentence-letters so that all the premises are true. *But this does not mean that D is valid.* For we still have to consider the other way of making the conclusion false.

Case 2: 'C' is false, 'A' is true. Then for premise 1 to be true, 'B → C' must be true, which requires 'B' to be false, since we already have 'C' false. (Why take premise 1 first this time? Because the assignment of ⊥ to 'C' and ⊤ to 'A' does not determine the truth-value of 'E' in the simplest premise, premise 3, while it does determine the truth-value of 'B' in premise 1. When nothing can be deduced about the truth-values of the sentence-letters in a premise, given the assignments already made, we look for another premise where something *can* be deduced.) We now have 'C' false, 'A' true and 'B' false. Hence for premise 2 to be true we must have 'E' false, and this also makes premise 3 true. So in Case 2 we arrive at an interpretation on which D's premises are true and its conclusion false.

Our overall conclusion, therefore, is that D is invalid, as established by the following interpretation:

E	B	C	A
⊥	⊥	⊥	⊤

Notice that we do not say that D is 'valid in Case 1' and 'invalid in Case 2'. Such locutions mean nothing. Either there is an interpretation which makes D's premises true and conclusion false or there is not, and so D is either invalid or valid *simpliciter*. The notions of validity and invalidity do not permit relativization to cases. What we find in Case 1 is not that D is 'valid in Case 1', but rather that Case 1's way of making the conclusion false does not lead to a demonstration of invalidity for D.

Example D suggests that judicious choice of order in which to consider cases can reduce the length of the discussion, for if we had taken Case 2 first there would have been no need to consider Case 1: as soon as we have found a way of making the premises true and the conclusion false we can stop, and pronounce the argument-form invalid. It is not always obvious which case is the one most likely to lead to an interpretation that shows an invalid argument-form to be invalid, but with some experience we can make intelligent guesses.

When an argument-form is valid we say that its premises *semantically entail* its conclusion, or that its conclusion is a *semantic consequence* of its premises. So in example B we have established that '(~J → C) & (E ↔ C)' and '~E' semantically entail 'J', or that 'J' is a semantic consequence of '(~J → C) & (E ↔ C)' and '~E'; and also, in example D, that '~A ∨ (B → C)', 'E → (A & B)' and 'C → E' do not semantically entail 'C ↔ A'. There are useful abbreviations for semantic entailment and non-entailment in a formal language like LSL: for entailment we use the symbol '⊨', known as the *double-turnstile*, and for non-entailment we use the symbol '⊭', known as the *cancelled double-turnstile*. The turnstiles also implicitly put quotes around formulae where they are required. So we can express the results of this section as follows:

(1) (~J → C) & (E ↔ C), ~E ⊨ J
(2) A → (B & E), D → (A ∨ F), ~E ⊭ D → B
(3) ~A ∨ (B → C), E → (B & A), C → E ⊭ C ↔ A.

Thus semantic entailment is essentially the same notion as validity, and semantic nonentailment essentially the same notion as invalidity. If we are asked to evaluate an expression such as (1), which says that '(~J → C) & (E ↔ C)' and '~E' semantically entail 'J', we simply use one or another technique for determining whether or not there is an interpretation which makes the premises of (1) true and the conclusion false. If there is, (1) is false, if there is not, (1) is true. Note the correct use of 'valid' versus 'true'. Arguments are valid or invalid, not true or false. But (1) is a statement *about* some premises and a conclusion: it says that the conclusion follows from the premises, and this statement itself is either true or false. Statements like (1), which contain the double-turnstile, are called *semantic sequents*.

Since entailment is essentially the same as validity, the formal definition of the symbol '⊨' for LSL is just like the definition of 'valid argument-form of LSL':

> For any formulae $p_1,...,p_n$ and q of LSL, $p_1,...,p_n \vDash q$ if and only if there is no interpretation of the sentence-letters in $p_1,...,p_n$ and q under which $p_1,...,p_n$ are all true and q is false.

In the special case where there are no $p_1,...,p_n$—or as it is sometimes put, where $n = 0$—we delete from the definition the phrases which concern $p_1,...,p_n$. This leaves us with 'For any formula q of LSL, $\vDash q$ if and only if there is no interpretation under which q is false'. If there is no interpretation on which q is false, this means q is true on every interpretation, in other words, that q is a tautology, and so we read '$\vDash q$' as 'q is a tautology'. '$\nvDash q$', then, means that q is either contingent or a contradiction.

❑ Exercises

I   Use the method of constructing interpretations to determine whether the following statements are correct. Explain your reasoning in the same way as in the worked examples, and if you claim a sequent is incorrect, exhibit an interpretation which establishes this.

(1)   A → B, B → (C ∨ D), ~D ⊨ A → C
(2)   (A & B) → C, B → D, C → ~D ⊨ ~A
(3)   A → (C ∨ E), B → D ⊨ (A ∨ B) → (C → (D ∨ E))
*(4) A → (B & C), D → (B ∨ A), C → D ⊨ A ↔ C
(5)   A ∨ (B & C), C ∨ (D & E), (A ∨ C) → (~B ∨ ~D) ⊨ B & D
(6)   A → (B → (C → D)), A & C, C → B ⊨ ~B ↔ (D & ~D)
(7)   (A ↔ B) & (B ↔ C) ⊨ (A ∨ ~A) & ((B ∨ ~B) & (C ∨ ~C))
(8)   (A ↔ B) ∨ (B ↔ C) ⊨ A ↔ (B ∨ C)
(9)   (~A & ~B) ∨ C, (A → D) & (B → F), F → (G ∨ H) ⊨ ~G → (H ∨ C)

II   Test the following English arguments for sentential validity by translating them into LSL and testing each of the resulting LSL arguments for validity, using the method of constructing interpretations. Give a complete dictionary for each argument and be sure not to use different sentence-letters of LSL for what is essentially the same simple sentence of English. Explain your reasoning in the same way as in the worked examples, and if you claim an argument is invalid, exhibit an interpretation which establishes this.

(1)   The next president will be a woman only if the party that wins the next election has a woman leader. Since no party has a woman leader at the moment, then unless some party changes its leader or a new party comes into being, there will be no female president for a while. Therefore, unless a new party comes into being, the next president will be a man.

(2)   B is a Knave, since if he is a Knight then what he says is false and in that case he is not a Knight. (In symbolizing this argument, assume everyone is either a Knight or a Knave.)

(3)   A theory which has been widely accepted in the past is always refuted eventually. This being so, unless we are much more intelligent than our predecessors, the Theory of Relativity is bound to be refuted. If we are more intelligent, the human brain must have increased in size, which requires that heads be bigger. You say the human head has not got bigger in historical times. If you are right about this, the Theory of Relativity will be refuted.

*(4)   If Yossarian flies his missions then he is putting himself in danger, and it is irrational to put oneself in danger. If Yossarian is rational he will ask to be grounded, and he will be grounded only if he asks. But only irrational people are grounded, and a request to be grounded is proof of rationality. Consequently, Yossarian will fly his missions whether he is rational or irrational. (F, D, R, A) (In symbolizing this argument, treat statements about people in general as if they concerned Yossarian specifically; e.g., symbolize 'only irrational people are grounded' as a statement about Yossarian.)

(5)   If the safe was opened, it must have been opened by Smith, with the assistance of Brown or Robinson. None of these three could have been involved unless he was absent from the meeting. But we know that either Smith or Brown was present at the meeting. So since the safe was opened, it must have been Robinson who helped open it.

(6)   If God is willing to prevent evil but is unable to do so, He is impotent. If God is able to prevent evil but unwilling to do so, He is malevolent. Evil exists if and only if God is unwilling or unable to prevent it. God exists only if He is neither impotent nor malevolent. Therefore, if God exists then evil does not exist.

(7)   We don't need a space station except if we need people in orbit, and we only need people in orbit if there are going to be manned expeditions to other planets, and then only if launch technology doesn't improve. A space station is a pointless extravagance, therefore, since interplanetary exploration will all be done by machines if we don't find better ways of getting off the ground.

(8)   The Mayor will win if the middle class votes for her. To prevent the latter, her rivals must credibly accuse her of corruption. But that charge won't stick if she isn't corrupt. So honesty assures the Mayor of victory.

(9)   A decrease in crime requires gun control. So the Mayor's only winning strategy involves banning guns, because he won't win without the middle-class vote and he'll lose that vote unless crime goes down. (4 sentence-letters)

(10) The Mayor's three problems are crime, corruption and the environment. He can't do anything about the first and he won't do anything about the third. So he won't be reelected, since winning would require that he solve at least two of them.

5 Testing for validity with semantic tableaux

The arguments which we gave in the previous section to establish semantic consequence and failure of semantic consequence, though rigorous, are rather unstructured: at various points it is left to the reasoner to decide what to do next, and it is not always obvious what that should be. In this section we briefly describe a format for testing argument-forms for validity (semantic sequents for correctness) which imposes rather more structure on the process, a format known as a *semantic tableau*. To determine whether or not $p_1,...,p_n \vDash q$ we build a *search tree* which grows downward—like parse trees, search trees are inverted—as we extend our search for an assignment of truth-values which makes $p_1,...,p_n$ all true and q false. The reader may wish to review the tree terminology on page 37 of Chapter 2.5.

A search tree begins at the top or root node with a list of *signed formulae*. A signed formula is a formula preceded by either the characters 'T:' or the characters 'F:'. 'T:' may be thought of as abbreviating 'it is true that' and 'F:' as abbreviating 'it is false that'. There is a T-rule and an F-rule for each connective, and the rule reflects that connective's truth-table. The tree is extended downward by applying T-rules and F-rules to the main connectives of formulae at nodes on it. For example, the T-rule for a conjunction ⌜p & q⌝ allows us to extend a tree with ⌜T:p & q⌝ at a node n by adding a new node to the bottom of every path on which node n lies; the new node is labeled with the signed formulae ⌜T:p⌝ and ⌜T:q⌝. We call this rule T-&. The corresponding F-rule for a conjunction ⌜p & q⌝ allows us to extend a tree which has ⌜F:p & q⌝ at a node n by splitting the tree at the bottom node of every path on which n lies, the new left node on each such path holding the signed formula ⌜F:p⌝ and the new right node holding the signed formula ⌜F:q⌝. We call this rule F-&. The collection of rules we get by providing a T-rule and an F-rule for each connective is known as the collection of *tableau rules*. These rules reflect the requirements for a formula of the relevant sort to have the stated truth-value. Thus the rule T-& reflects the fact that for ⌜p & q⌝ to be true, p must be true and q must be true, while the rule F-& reflects the fact that for ⌜p & q⌝ to be false, either p must be false or q must be false. The disjunction here corresponds to the splitting in the search tree (and to the distinguishing of cases in some of the informal arguments in the previous section).

To search for an interpretation on which $p_1,...,p_n$ are all true and q false, we construct a search tree at whose root node we list the signed formulae T:p_1...T:p_n and F:q. We then extend the tree downward by applying the tableau rules. We may apply the rules to any signed formulae in any order, but whichever order we choose, one or the other of two outcomes is inevitable. To describe these outcomes, it is useful to define the notion of a path in a semantic tableau or search tree being *closed*:

> A path Π in a semantic tableau is *closed* if and only if there is a formula s and nodes n_1 and n_2 of Π such that ⌜T: s⌝ occurs at n_1 and ⌜F: s⌝ occurs at n_2. Π is *open* if and only if it is not closed.

The first of the two possible outcomes is that at some stage of the search, every path in the tableau closes. In this case the tableau is itself said to be closed and the search for an interpretation making $p_1,...,p_n$ all true and q false has failed. So $p_1,...,p_n \vDash q$. The second possible outcome is that we have eventually applied all possible rules but at least one path in the tableau is still open, that is, the path has no nodes n_1 and n_2 such that for some formula s, $\ulcorner \mathbb{T}: s \urcorner$ occurs at n_1 and $\ulcorner \mathbb{F}: s \urcorner$ occurs at n_2. In this case the search for an interpretation on which $p_1,...,p_n$ are all true and q false has succeeded, for as we shall see, if we take an open path Π at this stage and assign $\top$ to every atomic sentence p such that $\mathbb{T}: p$ is on Π and $\bot$ to every atomic sentence p such that $\mathbb{F}: p$ is on Π, we obtain an interpretation which makes all of $p_1,...,p_n$ true and q false. So in this situation, we can conclude $p_1,...,p_n \nvDash q$.

The tableau rules for the connectives of LSL are as follows:

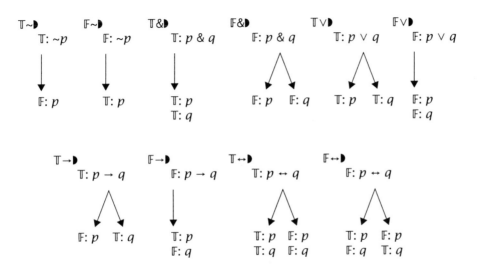

In interpreting these rule diagrams, remember that an occurrence of a formula at a node in a tree may lie on many paths, since branching may occur below that node and a path is a total route through a tree from top to bottom. With this in mind, think of a rule with a single arrow as an instruction: to apply it to an occurrence of a formula of the form at the tail of the arrow (i.e. the upper formula), we extend all currently open paths on which this formula-occurrence lies by adding a new node to each, labeling the new node with the signed formula(e) at the head of the arrow. To apply a rule with a branching arrow to an occurrence of a formula of the form at the tail of the arrow, we extend all currently open paths containing the formula occurrence by branching to two new nodes, labeled as indicated.

We illustrate the technique by repeating two examples from §4, this time using tableaux. First, we establish the validity of the argument-form B of §4 by showing that a tableau for it closes.

Example 1: Show (~J → C) & (E ↔ C), ~E ⊨ J.

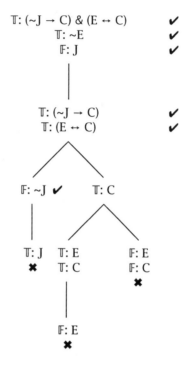

Each node in this tree is generated by applying one of the tableau rules to a signed formula at a previous node, though not necessarily the node immediately above. Once we have applied a tableau rule to a signed formula *s*, we check *s* with the dingbat '✔'. We mark a closed path by positioning the dingbat '✖' under its leaf. In this example there are three paths (because there are three bottom nodes), and all three are marked as closed. Path 1, the path down the tree which terminates in the node labeled 'T: J', is closed because it also has 'F: J' on it; path 2, the path down the tree which terminates in the node labeled 'F: E', is closed because it also has 'T: E' on it (the last rule used on this path is T~, which is applied to the middle formula at the root node, producing the occurrence of 'F: E' which closes the path); and path 3, the path down the tree which terminates in the node labeled 'F: C', is closed because it also has 'T: C' on it. It is worth noting that in general, the size of a tree can be kept to a minimum by applying rules like T~ which do not cause branching before applying rules like T↔ which do cause branching. However, in Example 1 only one path is still open at the point at which T~ is applied to the root node, and so only one copy of a node labeled with 'F: E' has to be added to the tree.

The next example is argument-form C in §4, which we already know to be invalid. So a tree search for an interpretation establishing invalidity should succeed, that is, the tree should have at least one open path. Here is the tree.

Example 2: Determine whether A → (B & E), D → (A ∨ F), ~E ⊨ D → B.

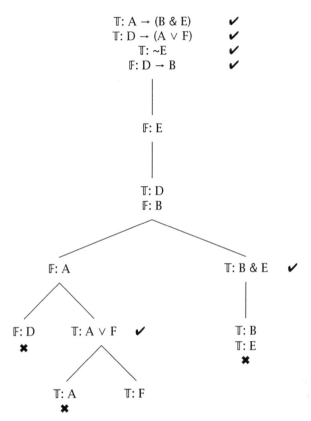

Paths 1, 2 and 4 are closed. However, path 3 does not satisfy the condition for being closed, and all nonatomic signed formulae above its leaf are checked, so there is nothing more we can do to attempt to close the path. Thus our search ends with the path still open, which shows that A → (B & E), D → (A ∨ F), ~E ⊭ D → B. As we already indicated, we can derive an interpretation which establishes this failure of semantic consequence by looking at the signed atomic formulae on the open path, in this case 'F: E', 'T: D', 'F: B', 'F: A' and 'T: F'. Reading truth-value assignments off the signatures, we obtain exactly the interpretation demonstrating invalidity that we arrived at through applying the method of constructing interpretations to argument-form C in §4, as exhibited on page 64.

❑ Exercises

Repeat the exercises of I of §4 using semantic tableaux rather than the method of constructing interpretations.

6 Properties of semantic consequence

One advantage of the double-turnstile notation is that it enables us to raise in a convenient form certain questions about relationships among argument-forms. For example, in symbolizing A of §4 as B (page 63), we treated all of the first sentence after 'for' as a single premise, a conjunction. Would it have made any difference if we had symbolized that piece of English as two separate premises? This question is a general one, and we can put it in the following way: for any formulae p, q and r of LSL,

Example 1: If $(p \mathbin{\&} q) \vDash r$, does it follow that $p, q \vDash r$?

Example 2: If $p, q \vDash r$, does it follow that $(p \mathbin{\&} q) \vDash r$?

The answer to both questions is 'yes'. In other words, as far as LSL validity is concerned, there is no significant difference between multiple premises versus a conjunction with multiple conjuncts.

It is not too difficult to see that the answer to both questions is 'yes'. But there is a general strategy for answering questions like these which it is useful to be able to illustrate with simple examples. Let us begin with Example 1. We reason as follows:

> *Example 1:* If $(p \mathbin{\&} q) \vDash r$ then by definition of '$\vDash$' this means that no interpretation makes $\ulcorner p \mathbin{\&} q \urcorner$ true and r false. So by the table for '&', no interpretation makes p true, q true and r false. But if $p, q \nvDash r$ then some interpretation makes p true, q true and r false, which is what we have just ruled out. Consequently, if $(p \mathbin{\&} q) \vDash r$ it cannot be that $p, q \nvDash r$, so it follows that $p, q \vDash r$.

> *Example 2:* If $p, q \vDash r$ then this means that no interpretation makes p true, q true and r false. On the other hand, if $(p \mathbin{\&} q) \nvDash r$ then some interpretation makes $\ulcorner p \mathbin{\&} q \urcorner$ true and r false, which means that it makes p true, q true and r false. But that is what we have just ruled out. Hence it cannot be that $(p \mathbin{\&} q) \nvDash r$, so it follows that $(p \mathbin{\&} q) \vDash r$.

In problems like these, we are using the metavariables 'p', 'q' and 'r' to abstract from sequents with LSL formulae as their premises and conclusions, and are considering instead sequents with metalanguage formulae that describe patterns which sequents with LSL formulae may instantiate. One such abstract sequent is *given*, and another is *queried*. The strategy for solving the problem is to begin by using the definition of '$\vDash$' to determine which kinds of interpretation are ruled out by the given sequent. To determine whether the queried sequent is correct, we then ask what kinds of interpretation would show that it is *not* correct. If all interpretations of these kinds have been ruled out by the given sequent, the queried sequent does follow from the given one. But if they have not all been ruled out, the queried sequent does not follow from the given one. For some specific formulae p, q and r, the resulting queried sequent may of course be semantically correct. The issue is whether the correctness of the given sequent guarantees this for all formulae p, q and r.

Here are two further examples, where p, q and r are any sentences of LSL:

Example 3: If $(p \leftrightarrow q) \vDash (p \& q)$, does it follow that $(p \vee q) \vDash (p \& q)$?

> *Answer:* If $(p \leftrightarrow q) \vDash (p \& q)$ then no interpretation makes $\ulcorner p \leftrightarrow q \urcorner$ true and $\ulcorner p \& q \urcorner$ false. Thus the interpretations which are ruled out are those which make p false and q false. (Remember that p and q are any LSL wffs, not necessarily atomic ones, so there might be more than one interpretation which results in p false, q false.) If $(p \vee q) \nvDash (p \& q)$ then some interpretation makes $\ulcorner p \vee q \urcorner$ true and $\ulcorner p \& q \urcorner$ false. Any such interpretation must either have p true and q false or vice-versa. In fact, neither of these combinations has been ruled out, so it does not follow from $(p \leftrightarrow q) \vDash (p \& q)$ that $(p \vee q) \vDash (p \& q)$.

Example 4: If $\vDash (p \rightarrow (q \rightarrow r))$, does it follow that $\sim r \vDash (\sim p \rightarrow \sim q)$?

> *Answer:* If $\vDash (p \rightarrow (q \rightarrow r))$ then no interpretation makes p true, q true and r false. If $\sim r \nvDash (\sim p \rightarrow \sim q)$ then some interpretation makes $\ulcorner \sim r \urcorner$ true and $\ulcorner \sim p \rightarrow \sim q \urcorner$ false, that is, it makes r false, p false and q true. Such an interpretation has not been ruled out, so from $\vDash (p \rightarrow (q \rightarrow r))$ it does not follow that $\sim r \vDash (\sim p \rightarrow \sim q)$.

The semantic consequence relation is the fundamental concept of modern logic. It is just as important to be able to reason about it, as in Examples 1–4 above, as it is to be able to execute a technique for detecting when it holds and when it fails, like the ones developed in §3–§5 of this chapter.

❑ Exercises

In the following, p, q, r and s are any sentences of LSL. In (1)–(6), explain your reasoning in the manner illustrated by the examples of this section.

(1) If $\vDash p \rightarrow (q \rightarrow r)$, does it follow that $p, q \vDash r$? Does it follow that $r \vDash (p \rightarrow q)$?
(2) If $(p \rightarrow q) \vDash r$, does it follow that $p \vDash (q \rightarrow r)$?
*(3) If $p \vDash (q \& r)$, does it follow that $(p \leftrightarrow q) \vDash (p \leftrightarrow r)$?
(4) If $(p \leftrightarrow q) \vDash (r \vee s)$, does it follow that $(p \& q) \vDash r$?
(5) If $(p \rightarrow q) \vDash (\sim r \vee s)$, does it follow that $(p \& r) \vDash \sim q$?
(6) If $(p \& q) \vee (r \& s) \vDash (p \rightarrow r) \leftrightarrow (q \rightarrow s)$, does it follow that $(p \& \sim r), \sim q \vDash \sim s$?
(7) Although there is no sentence-letter in common between premise and conclusion, $(A \& \sim A) \vDash B$. Briefly explain why.
*(8) Although there is no sentence-letter in common between premise and conclusion, $B \vDash (A \vee \sim A)$. Briefly explain why.

7 Expressive completeness

At the end of §1 in Chapter 2 we claimed that our five sentential connectives '~', 'v', '&', '→' and '↔' are all we need in sentential logic, since other sentential connectives are either *definable* in terms of these five or else *beyond the scope* of sentential logic. When we say that a connective is beyond the scope of classical sentential logic, what we mean is that it is *non-truth-functional;* in other words, there is no truth-function that it expresses (see §1 of this chapter for a discussion of expressing a truth-function). In the next section we will consider various connectives of this sort. Meanwhile, we will concern ourselves with the definability of other truth-functional connectives.

An example of a truth-functional connective which is definable in terms of our five is 'neither...nor...', since for any English sentences p and q, ⌜neither p nor q⌝ is correctly paraphrased as ⌜not p and not q⌝ (see (11) on page 18). But this is just one example. How can we be confident that *every* truth-functional connective can be defined in terms of '~', 'v', '&', '→' and '↔'? Our confidence is based in the fact that our collection of connectives has a property called *expressive completeness*, which we now explain.

At the end of §1 of this chapter, we listed the function-tables for the one-place, or *unary*, function expressed by '~', and the four two-place, or *binary*, functions expressed by the other connectives. However, there are many more unary and binary truth-functions than are expressed by the five connectives individually. For example, there are three other unary truth-functions:

T ⇒ T	T ⇒ T	T ⇒ ⊥
⊥ ⇒ T	⊥ ⇒ ⊥	⊥ ⇒ ⊥
(a)	(b)	(c)

To show that all unary truth-functional connectives are definable in terms of our five basic connectives, we establish the stronger result that all unary truth-functions are definable, whether or not they are expressed by some English connective. (While (b) is expressed by 'it is true that', neither (a) nor (c) has an uncontrived rendering.) Our question is therefore whether we can express all of (a), (b) and (c) in terms of our five chosen connectives. And in this case it is easy to see that (a) is captured by '...v ~...', (b) by '~~...', and (c) by '...& ~...', where in (a) and (c) the same formula fills both ellipses.

What about the other binary truth-functions? We have connectives for four, and we know how to define a fifth, the truth-function

$$TT \Rightarrow \bot$$
$$T\bot \Rightarrow \bot$$
$$\bot T \Rightarrow \bot$$
$$\bot\bot \Rightarrow T$$

which is expressed by 'neither...nor...' (to repeat, ⌜neither p nor q⌝ is true just in case both p and q are false, so we express it with ⌜~p & ~q⌝). But there are many more binary truth-functions, and again we are concerned to define all of them, not merely those which correspond to some idiomatic phrase like 'neither...nor...'. First, how many other binary functions are there? There are as many as there are different possible combinations of outputs for the four pairs of truth-values which are the inputs to binary truth-functions. Hence there are sixteen different binary truth-functions: the output for the first pair of input truth-values ⊤⊤ is either ⊤ or ⊥, giving us two cases, and in each of these cases, the output for the second pair of inputs ⊤⊥ is either ⊤ or ⊥, giving a total of four cases so far, and so on, doubling the number of cases at each step for a total of sixteen. So apart from examining each of the remaining eleven binary truth-functions one by one, is there a *general* reason to assert that they can all be defined by our five connectives?

Even supposing that we can give a general reason why all binary truth-functions should be definable in terms of our five, that would not be the end of the matter, since for every n, there are truth-functions of n places, though when $n > 2$ they rarely have a 'dedicated' English phrase which expresses them.[3] The claim that our five chosen connectives suffice for sentential logic is the claim that for any n, every truth-function of n places can be expressed by our five connectives. This is our explanation of the notion of expressive completeness of a collection of connectives, which we embody in a definition:

> A set of connectives S is *expressively complete* if and only if for every n, all n-place truth-functions can be expressed using only connectives in S.

The set of connectives which we wish to prove expressively complete is $\{\sim, \&, \lor, \rightarrow, \leftrightarrow\}$ (the curly parentheses, or *braces*, are used for sets, with all the members of the set exhibited between them). But what does it mean to say that a truth-function is *expressed* using connectives in this set? This means that there is a *formula* which expresses the truth-function and which is built up from sentence-letters and connectives in the set. And this in turn is explained using truth-tables. We observe that every truth-function corresponds to a truth-table (and conversely). For example, the three-place function

$$\top\top\top \Rightarrow \top$$
$$\top\top\bot \Rightarrow \bot$$
$$\top\bot\top \Rightarrow \bot$$
$$\top\bot\bot \Rightarrow \top$$
$$\bot\top\top \Rightarrow \top$$
$$\bot\top\bot \Rightarrow \top$$
$$\bot\bot\top \Rightarrow \top$$
$$\bot\bot\bot \Rightarrow \top$$

corresponds to the truth-table laid out on page 55. In general, given a function-

---

[3] The phrase 'if...then..., otherwise...' is an example of a locution expressing a three-place sentential connective. However, ⌜if p then q, otherwise r⌝ can be paraphrased as ⌜if p then q and if not-p then r⌝.

table, the corresponding truth-table is the table with the output of the function as its final column. We say that a function is *expressed* by a formula, or a formula *expresses* a function, if that formula's truth-table is the table corresponding to the function. So the three-place function just exhibited is expressed by the formula '(A → (B ∨ C)) → (A → (B & C))' from page 55.

Consequently, to show that every truth-function is expressible in terms of the five connectives of LSL, it suffices that we show how, given any truth-table, we can recover a formula which contains only LSL connectives and whose truth-table it is. In other words, we have to develop a technique that is the reverse of the one we have for constructing truth-tables, given formulae; the problem now is to construct formulae, given truth-tables.

There is a systematic way of doing this. A truth-table lists various possible interpretations, that is, assignments of truth-values to certain sentence-letters. Say that a formula *defines* an assignment I of truth-values to sentence-letters $\pi_1,...,\pi_n$ if and only if that formula is true on I and on no other assignment to $\pi_1,...,\pi_n$. Then given an assignment I to $\pi_1,...,\pi_n$, one can use $\pi_1,...,\pi_n$ to construct a formula in '&' and '~' which defines I as follows: take each sentence-letter which is assigned ⊤ and the negation of each which is assigned ⊥ and form the conjunction of these letters and negated letters. So, for example, the interpretation consisting in the assignment of ⊥ to 'C', ⊤ to 'D', ⊥ to 'B', ⊥ to 'A' and ⊤ to 'F' is defined by '~C & D & ~B & ~A & F', since this formula is true on that assignment, and only that one, to those sentence-letters. Now suppose we are given a randomly chosen truth-table with 2^n rows and a final column of entries, but no formula and no sentence-letters are specified. It is easy to find a formula for the table in the two special cases in which all interpretations lead to ⊤ or all to ⊥. Otherwise, to construct a formula for the table, we choose sentence-letters $\pi_1,...,\pi_n$ and use them to construct the formulae which define the interpretations where there is a ⊤ in the final column of the table, and disjoin these interpretation-defining formulae together. This produces a disjunction such that each disjunct is true on exactly one row of the table (the one it defines), making the whole disjunction true at that row. For each row where there is a ⊤ there is a disjunct in the constructed formula with this effect. And the formula has no other components. Therefore it is true on exactly the rows in the table where there is a ⊤. Consequently, this disjunction expresses the truth-function given by the table.

In sum, we have the following three-step procedure for constructing a formula for any truth-table:

- If there are no ⊤s in the final column, let the formula be 'A & ~A'; if there are no ⊥s, let it be 'A ∨ ~A'.
- Otherwise, using the appropriate number of sentence-letters 'A', 'B' and so on, for each interpretation which gives a ⊤ in the final column of the table construct a conjunction of sentence-letters and negated sentence-letters defining that interpretation.
- Form a disjunction of the formulae from the previous step.

Here are two applications of this technique.

Example 1:

A B	
⊤ ⊤	⊥
⊤ ⊥	⊤
⊥ ⊤	⊥
⊥ ⊥	⊤

⊤⊤ ⇒ ⊥
⊤⊥ ⇒ ⊤
⊥⊤ ⇒ ⊥
⊥⊥ ⇒ ⊤

Function   Corresponding table

The second and fourth interpretations (inputs) produce ⊤s in the final column of the table. The second interpretation is: ⊤ assigned to 'A', ⊥ to 'B', so its defining formula is 'A & ~B'. The fourth interpretation is: ⊥ assigned to 'A', ⊥ to 'B', so its defining formula is '~A & ~B'. Consequently, the formula we arrive at is '(A & ~B) ∨ (~A & ~B)', and a simple calculation confirms that this formula does indeed have the displayed truth-table. Of course, there are many other (in fact, infinitely many other) formulae which have this table. For example, the reader may have quickly noticed that the formula '(A ∨ ~A) & ~B' also has the table in Example 1. But to show that the truth-function is expressible, all we have to find is *at least one* formula whose table is the table corresponding to the function, and our step-by-step procedure will always produce one. Moreover, when we consider functions of three places or more it is no longer so easy to come up with formulae for their corresponding tables simply by inspecting the entries and experimenting a little. So it is best to follow the step-by-step procedure consistently, as in our next example.

Example 2:

A B C	
⊤ ⊤ ⊤	⊤
⊤ ⊤ ⊥	⊥
⊤ ⊥ ⊤	⊥
⊤ ⊥ ⊥	⊤
⊥ ⊤ ⊤	⊥
⊥ ⊤ ⊥	⊤
⊥ ⊥ ⊤	⊥
⊥ ⊥ ⊥	⊤

⊤⊤⊤ ⇒ ⊤
⊤⊤⊥ ⇒ ⊥
⊤⊥⊤ ⇒ ⊥
⊤⊥⊥ ⇒ ⊤
⊥⊤⊤ ⇒ ⊥
⊥⊤⊥ ⇒ ⊤
⊥⊥⊤ ⇒ ⊥
⊥⊥⊥ ⇒ ⊤

Function   Corresponding table

Here it is interpretations 1, 4, 6 and 8 which produce a ⊤. The four formulae defining these interpretations are respectively: 'A & B & C', 'A & ~B & ~C', '~A & B & ~C' and '~A & ~B & ~C'. Therefore a formula for the table, grouping disjuncts conveniently, is:

$$[(A \,\&\, (B \,\&\, C)) \vee (A \,\&\, (\sim B \,\&\, \sim C))] \vee [(\sim A \,\&\, (B \,\&\, \sim C)) \vee (\sim A \,\&\, (\sim B \,\&\, \sim C))].$$

It should be clear from this method that we are proving something stronger than that the set of LSL connectives $\{\sim,\&,\vee,\rightarrow,\leftrightarrow\}$ is expressively complete, since our procedure for finding a formula for an arbitrary table involves only the connectives in the subset $\{\sim,\&,\vee\}$. What we are showing, therefore, is that $\{\sim,\&,\vee\}$ is expressively complete, from which the expressive completeness of $\{\sim,\&,\vee,\rightarrow,\leftrightarrow\}$ follows trivially: if we can express any truth-function by some formula in $\{\sim,\&,\vee\}$, then the same formula is a formula in $\{\sim,\&,\vee,\rightarrow,\leftrightarrow\}$ which expresses the truth-function in question (the point is that p's being a formula in $\{\sim,\&,\vee,\rightarrow,\leftrightarrow\}$ requires that p contain no *other* connectives, but not that it contain occurrences of *all* members of $\{\sim,\&,\vee,\rightarrow,\leftrightarrow\}$). But can we do better than this? That is, is there an even smaller subset of $\{\sim,\&,\vee,\rightarrow,\leftrightarrow\}$ which is expressively complete? Given the expressive completeness of $\{\sim,\&,\vee\}$, a simple way to show that some other set of connectives is expressively complete is to show that *the connectives of the other set can define those of* $\{\sim,\&,\vee\}$.

What is it for one or more connectives to define another connective? By this we mean that there is a rule which allows us to replace every occurrence of the connective c to be defined by some expression involving the defining connectives. More precisely, if p is a formula in which there are occurrences of c, then we want to select every subformula q of p of which c is the main connective, and replace each such q with a formula q' logically equivalent to q but containing only the defining connectives. For instance, we already know that we can define '$\leftrightarrow$' using the set of connectives $\{\&,\rightarrow\}$, since every occurrence of '$\leftrightarrow$' in a formula p is as the main connective of a subformula $\ulcorner(r \leftrightarrow s)\urcorner$, and we have the following substitution rule:

- Replace each subformula q of p of the form $\ulcorner(r \leftrightarrow s)\urcorner$ with $\ulcorner((r \rightarrow s) \,\&\, (s \rightarrow r))\urcorner$.

Applying this substitution rule throughout p yields a logically equivalent formula p' which contains no occurrence of '$\leftrightarrow$'. For example, we eliminate every occurrence of '$\leftrightarrow$' from 'A $\leftrightarrow$ (B $\leftrightarrow$ C)' in two steps (it does not matter which '$\leftrightarrow$' we take first):

Step 1: $[A \rightarrow (B \leftrightarrow C)] \,\&\, [(B \leftrightarrow C) \rightarrow A]$
Step 2: $[A \rightarrow ((B \rightarrow C) \,\&\, (C \rightarrow B))] \,\&\, [((B \rightarrow C) \,\&\, (C \rightarrow B)) \rightarrow A]$

Using a similar approach, we can show that the set of connectives $\{\sim,\&\}$ is expressively complete. Given that $\{\sim,\&,\vee\}$ is expressively complete, the problem can be reduced to defining '$\vee$' in terms of '$\sim$' and '$\&$'. For a disjunction to be true, at least one disjunct must be true, which means that it is not the case that the disjuncts are both false. So the substitution rule should be:

- Replace every subformula of the form $\ulcorner(r \vee s)\urcorner$ with $\ulcorner\sim(\sim r \,\&\, \sim s)\urcorner$.

Since each substitution produces a logically equivalent formula, or as we say, since *substitution preserves logical equivalence*, then if we begin with a formula in {~,&,∨} for a given truth-table, and replace every occurrence of '∨' using the substitution rule, we end up with a formula in {~,&} for that same truth-table. Since every table has a formula in {~,&,∨}, it follows that every table has a formula in {~,&}, hence {~,&} is expressively complete. For example, we have already seen that '(A & ~B) ∨ (~A & ~B)' is a formula for the table of Example 1 above. Applying the substitution rule to the one occurrence of '∨' in this formula yields the logically equivalent formula

$$\sim[\sim(A \ \& \ \sim B) \ \& \ \sim(\sim A \ \& \ \sim B)]$$

which is therefore also a formula for the table in Example 1.

By the same technique we can show that {~,∨} and {~,→} are expressively complete (these are exercises). However, not every pair of LSL connectives is expressively complete; for example, {&,∨} and {~,↔} are not. This can be proved rigorously by a technique known as mathematical induction, but for our purposes it is enough to understand one example, {&,∨}, intuitively. The point is that negation cannot be expressed in terms of {&,∨}, since no formula built out of 'A' and '&' and '∨' can be false when 'A' is true. Since '~A' is false when 'A' is true, this means no formula built out of 'A' and '&' and '∨' has the same truth-table as '~A'. Thus {&,∨} is expressively *incomplete*.

There are two connectives which are expressively complete *individually*, though neither belongs to LSL. One is a symbol for 'neither...nor...', '↓', and the other is called Sheffer's Stroke, after its discoverer, and written '|'. Their function-tables are:

↓			
T T	⇒	⊥	
T ⊥	⇒	⊥	
⊥ T	⇒	⊥	
⊥ ⊥	⇒	T	

|			
T T	⇒	⊥	
T ⊥	⇒	T	
⊥ T	⇒	T	
⊥ ⊥	⇒	T	

To see that '↓' is expressively complete, we use the already established expressive completeness of {~,&}. Since every table has a formula whose only connectives are '~' and '&', we can derive a formula for any table by finding the formula in '~' and '&' for it and then using substitution rules to eliminate all occurrences of '~' and '&', replacing them with formulae containing only '↓'. What should the substitution rules be in this case? One can play trial and error with truth-tables, but it is not hard to see that ⌜not-p⌝ can be paraphrased, if awkwardly, as ⌜neither p nor p⌝. We also note that ⌜neither not-p nor not-q⌝ is true exactly when p and q are both true, which makes it equivalent to that conjunction; and we already know how to eliminate 'not' in ⌜not-p⌝ and ⌜not-q⌝. So we get the following substitution rules:

- Replace every subformula of the form ⌜~r⌝ with ⌜(r↓r)⌝.
- Replace every subformula of the form ⌜(r & s)⌝ with ⌜(r↓r)↓(s↓s)⌝.

It is easy to check the correctness of these rules using truth-tables. Hence '↓' is expressively complete by itself, and a similar argument shows Sheffer's Stroke to be expressively complete by itself (this is an exercise).

A final comment. We saw earlier that there are four unary truth-functions and sixteen binary ones. But for an arbitrary n, how many n-place truth-functions are there? To answer this we generalize the reasoning which gave us the answer 'sixteen' in the binary case. If a truth-function takes a sequence of n truth-values as input, there are 2^n different possible such input sequences for it: the first element of an input sequence may be ⊤ or ⊥, giving two cases, and in each of these cases the second element may be ⊤ or ⊥, giving a total of four cases, and so on, doubling the number of cases at each step, giving a total of 2^n cases at the nth element of the sequence. And if there are 2^n different possible input sequences to a truth-function, there are 2^{2^n} different possible combinations of truth-values which can be the function's output: the output for the first input may be either ⊤ or ⊥, giving two cases, and in each of these, the output for the second input may be ⊤ or ⊥, and so on, doubling the number of cases with each of the 2^n inputs, for a total of 2^{2^n} different possible outputs. Hence there are 2^{2^n} different n-place truth-functions.

❑ Exercises

I   Find formulae in $\{\sim,\&,\vee\}$ which express the truth-functions (1), (2) and (3) below. Then give formulae in $\{\sim,\&\}$ for (1) and *(2) (use the rule on page 78).

(1)		(2)		(3)	
⊤⊤⊤ ⇒ ⊤		⊤⊤⊤ ⇒ ⊥		⊤⊤⊤ ⇒ ⊤	
⊤⊤⊥ ⇒ ⊥		⊤⊤⊥ ⇒ ⊥		⊤⊤⊥ ⇒ ⊤	
⊤⊥⊤ ⇒ ⊥		⊤⊥⊤ ⇒ ⊤		⊤⊥⊤ ⇒ ⊤	
⊤⊥⊥ ⇒ ⊥		⊤⊥⊥ ⇒ ⊤		⊤⊥⊥ ⇒ ⊤	
⊥⊤⊤ ⇒ ⊤		⊥⊤⊤ ⇒ ⊥		⊥⊤⊤ ⇒ ⊥	
⊥⊤⊥ ⇒ ⊥		⊥⊤⊥ ⇒ ⊥		⊥⊤⊥ ⇒ ⊥	
⊥⊥⊤ ⇒ ⊥		⊥⊥⊤ ⇒ ⊤		⊥⊥⊤ ⇒ ⊥	
⊥⊥⊥ ⇒ ⊥		⊥⊥⊥ ⇒ ⊤		⊥⊥⊥ ⇒ ⊤	

II   Granted that $\{\sim,\&,\vee\}$ is expressively complete, explain carefully why each of the following sets of connectives is expressively complete (compare the explanation on page 78 for $\{\sim,\&\}$). State your substitution rules. In (3), '←' means 'if', so $\ulcorner p \leftarrow q \urcorner$ is read $\ulcorner p$ if $q \urcorner$.

(1) $\{\sim,\vee\}$ (2) $\{\sim,\rightarrow\}$ *(3) $\{\sim, \leftarrow\}$ (4) $\{\,|\,\}$

*III $\{\sim,\leftrightarrow\}$ is expressively incomplete. Can you think of a general pattern of distribution of ⊤s and ⊥s in the final column of a truth-table which would guarantee that there is no formula in $\{\sim,\leftrightarrow\}$ which has that table? Try to explain your answer (this is much harder than the expressive incompleteness of $\{\&,\vee\}$).

8 Non-truth-functional connectives

The five connectives of LSL have been shown to be adequate for all of truth-functional logic. What is being deliberately excluded at this point, therefore, is any treatment of non-truth-functional connectives, that is, connectives which do not express truth-functions. There are extensions of classical logic which accommodate non-truth-functional connectives, but at this point all we need to know is how to determine whether a given connective is truth-functional or non-truth-functional.

A truth-functional connective expresses a truth-function, which in turn can be written as a function-table, so a proof that a certain n-place connective is not truth-functional would consist in showing that its meaning cannot be expressed in a function-table. A function-table associates each of the possible 2^n inputs with a single output, either ⊤ or ⊥, so what we need to prove about a connective to show that it is non-truth-functional is that for at least one input, there is no single output that could be correctly associated with the connective. For if this is so, the truth-value of a sentence formed using the connective is not a function of the truth-values of the component sentences. Here are two examples.

Example 1: Philosophers distinguish two kinds of fact, or truth, those which are *contingent* and those which are noncontingent or *necessary*. Something is contingently the case if it might not have been the case, that is, if there are ways things could have gone in which it would not have been the case. Something is necessarily the case if there is no way things could have gone in which it would not have been the case. Note that this use of 'contingent' is much broader than its use to mean 'not a tautology and not a contradiction', which was the way we employed the term in §2. Understood in this new, broad sense, we can show that the connective 'it is a matter of contingent fact that...' is non-truth-functional. 'It is a matter of contingent fact that...' is a one-place connective which can be prefixed to any complete sentence. If it were truth-functional, then it would have a function-table ⊤ ⇒ ?, ⊥ ⇒ ? like negation, in which each query is replaced by either ⊤ or ⊥. It is easy to see that when the input is ⊥, the output is also ⊥. In other words, if the ellipsis in 'it is a matter of contingent fact that...' is filled by a false sentence, the result is a false sentence; for if p is false, then ⌜it is a matter of contingent fact that p⌝ is false as well, since it is not a fact at all that p. But there is a problem when the input to the connective is ⊤. If the connective is truth-functional the output must always be ⊤, or always be ⊥, yet we can show that neither of these alternatives correctly represents the meaning of 'it is a matter of contingent fact that'.

   (a) Let 'A' mean 'Gorbachev was president of the Soviet Union'. Then 'A' is true, and 'it is a matter of contingent fact that A' is true as well, since it is a contingent fact that Gorbachev was president of the Soviet Union, there being many other ways things could have gone in which he would not have achieved that office—for example, he might have been killed in World War II.

(b) Let 'A' mean 'all triangles have three angles'. Then 'A' is true, but 'it is a matter of contingent fact that A' is false, for it is not a *contingent* fact that all triangles have three angles. Having three angles is part of the meaning of '*triangle*', so it is necessary rather than contingent that all triangles have three angles: there are no alternative ways things could have gone in which there are triangles with fewer than or more than three angles.[4]

(a) and (b) together show that there is no correct way of completing the first entry ⊤ ⇒ ? in a function-table for 'it is contingent that...' (a) shows that it would be wrong to put ⊥, and (b) that it would be wrong to put ⊤. This illustrates the general technique for establishing that a connective is non-truth-functional: we find two examples which show that for a particular input there is no single correct output.

Example 2: The connective 'after' is not truth-functional. 'After' is a two-place connective, and can occur either between two complete sentences, as in ⌜p after q⌝, or at the beginning, as in the syntactic variant ⌜After q, p⌝. To show that 'after' is not truth-functional, we show that there is no correct output for the input ⊤,⊤.

(a) Let 'A' mean 'Thatcher was elected prime minister' and 'B' mean 'Nixon was elected president'. Then 'A after B' is true, since Thatcher was first elected in 1979 and Nixon last elected in 1972.

(b) Let 'A' mean 'Nixon was elected president' and 'B' mean 'Thatcher was elected prime minister'. Then 'A after B' is false.

Thus in any purported function-table for 'after' it would be impossible to complete the first entry ⊤,⊤ ⇒ ?, because (a) shows ⊥ is incorrect and (b) shows ⊤ is incorrect.

These two examples explain the terminology 'non-truth-functional', for their morals are that the truth-value of ⌜it is a matter of contingent fact that p⌝ is not a function merely of the *truth-value* of p, and the truth-value of ⌜p after q⌝ is not a function merely of the *truth-values* of p and q. Rather, the truth-value of ⌜it is contingent that p⌝ depends, for true p, on the nature of the *reason why p* is true, and the truth-value of ⌜p after q⌝, for true p and q, depends on the *temporal order* of the events reported in p and q. The connective 'it is a matter of contingent fact that' belongs to an extension of classical logic called *modal* logic (see Chapter 9), and the connective 'after' to an extension of classical logic called *tense* logic (see Burgess).

---

[4] Do not confuse the question (i) 'Could triangles have had more than or fewer than three angles?' with the question (ii) 'Could the word "triangle" have been defined differently?' Obviously, the word 'triangle' could have been defined to mean what we actually mean by 'square' ('tri' could have been used for 'four'), but that is irrelevant to question (i), where 'triangles' is used with its normal meaning. Given what (i) means, the correct answer to it is 'no'.

The alert reader will have noticed some parallels between our discussion of 'after' and our earlier discussion of 'if...then...' in §1 of this chapter. We have demonstrated that 'after' is not truth-functional by providing two examples in both of which 'after' is flanked by true sentences and in which the resulting 'after' sentences have *different* truth-values. But a comparable pair of examples for 'if...then...' as it is used in English can apparently be constructed. We used the sentence

(1)   If Nixon was president then Nixon lived in the White House

to motivate the entry $\top\top \Rightarrow \top$ in the function-table of the truth-function expressed by 'if...then...', but we also noted that many would judge that the conditional

(2)   If Moses wrote the Pentateuch then water is H_2O

is not true, on the grounds that the consequent is unrelated to the antecedent. Yet on the assumption that 'Moses wrote the Pentateuch' is true, (2) has a true antecedent and true consequent. Here one example with true antecedent and true consequent gives us a true conditional while another, also with true antecedent and true consequent, does not. Why do we not just conclude that the 'if...then...' of the indicative conditional is not truth-functional?

We remarked in our discussion in §1 that on one view, there are at least two senses of 'if...then...' and our truth-table for '→' captures only one of its senses, the material sense. The senses which are not captured are the non-truth-functional ones, where the truth-value of the conditional depends not just on the truth-values of its antecedent and consequent, but also on whether there is a certain kind of connection between antecedent and consequent. Different senses of the conditional would correspond to different kinds of connection, for instance, there would be a *causal* sense in which the truth of the consequent has to be *caused* by the truth of the antecedent. Unfortunately, it seems as though *all* the natural senses of 'if...then...' are non-truth-functional, so one could reasonably worry about the reliability of translations of English conditionals into a symbolism which cannot express these natural senses. This is why the case of the conditional is unlike the case of disjunction: even though we took inclusive disjunction as basic, the exclusive sense is also truth-functional and easily defined. But we cannot define any of the non-truth-functional senses of the conditional.

However, in the case of both disjunction and the conditional, there is an alternative to postulating two or more senses of the English connective. According to the alternative view, a conditional such as 'if Moses wrote the Pentateuch then water is H_2O' really is true, but sounds strange for the following reason. It is correct to assert conditionals only when we believe it is not the case that the antecedent is true and the consequent false, and the standard grounds for such a belief are either (i) we believe that the antecedent is false, or (ii) we believe that the consequent is true, or (iii) we believe that if the antecedent is true, that will in some way *make* the consequent true. But if our grounds for the belief that it is not the case that the antecedent is true and the consequent false are (i) or (ii), then asserting the conditional violates one of the maxims we have to observe if conversation is to be an efficient way of communicating information, the maxim *to be as informative as possible*. If our grounds are (i) or (ii), we should just deny the antecedent or assert the consequent, that is, we should make the more informative statement. However, this does not mean that if we assert the conditional instead, we say something false: the conditional is still true, but it is conversationally inappropriate. The problem with 'if Moses wrote the Pentateuch then water is H_2O', therefore, is that in the absence of any mechanism tying the authorship of the Pentateuch to the chemical composition of water, the most likely ground for asserting the conditional is that we believe that water is H_2O. But then that is what we should say, not something less informative. This means that we are left with (iii) as the only grounds on which it is normally appropriate to assert a conditional. So the suggestion is that when people deny that 'if Moses wrote the Pentateuch then water is H_2O' is true, they are failing to distinguish the question of truth from the question of appropriateness. All we can really object to about (2) is that in ordinary contexts it is likely to be conversationally inappropriate; but this is consistent with its being true.

The maxim to be as informative as possible needs to be qualified, for there are circumstances in which maximum informativeness would not be appropriate in the context. For example, in giving clues to children engaged in a treasure hunt, one may say 'if it's not in the garden then it's in the bedroom' so as to leave it open which should be searched, even though one knows the treasure is in the bedroom. This is a case where the conditional seems true despite the absence of a mechanism that brings about the treasure's being in the bedroom from its not being in the garden (the treasure is in the bedroom because that is where it was put). If we distinguish the appropriateness of a conditional from its truth, therefore, we can maintain that the English 'if...then...' has only one meaning, the truth-functional one. (This approach to the conditional and related matters was developed by Paul Grice; see Grice.)

❏ Exercises

Show that the following connectives are not truth-functional:

(1) 'It is necessary that...'
(2) '...before...'
*(3) 'It is surprising that...'
(4) '...because...'
*(5) '..., which means that...'
(6) 'At noon...'
(7) 'Someone knows that...' (Note: (7) is tricky.)

9 Summary

- Negation reverses truth-value; a conjunction is ⊤ when and only when both conjuncts are ⊤; a disjunction is ⊥ when and only when both disjuncts are ⊥; a conditional is ⊥ when and only when its antecedent is ⊤ and its consequent is ⊥; a biconditional is ⊤ when and only when both its sides have the same truth-value.
- Every formula of LSL is either a tautology (true on every interpretation), a contradiction, or contingent. Equivalent formulae have the same truth-value on each interpretation.
- An English argument has a valid argument-form if its translation in LSL is a valid argument-form.
- An argument-form in LSL is valid if and only if no interpretation makes its premises true and its conclusion false. LSL validity may be determined either by exhaustive search or by constructing an interpretation.
- Classical sentential logic is the logic of truth-functional connectives, all of which can be defined by '~', '&' and '∨'. Non-truth-functional sentential connectives require extensions of classical logic, such as modal logic and tense logic, to handle them.

4   Natural Deduction in Sentential Logic

1   The concept of proof

We have at least partly achieved the goal we set ourselves in Chapter 1, which was to develop a technique for evaluating English arguments for validity. However, there is a respect in which our approach to arguments differs from that of the typical person involved in a debate. The typical debater does not proceed by stating her premises, then her conclusion, and then defying the opposition to describe how it would be possible for the premises to be true and the conclusion to be false. Rather, one tries to *reason from* one's premises *to* one's conclusion. In other words, to show that the conclusion follows from the premises, one tries to *deduce* it *from* the premises. And there is nothing in the techniques of Chapter 3 which captures the idea of arguing for a conclusion by deducing it from premises.

Deducing the conclusion of an argument from the premises is ordinarily called *giving* an argument. By 'giving an argument' we mean to connote an activity which stands in contrast to dogmatic assertion. However, we have already reserved the term 'argument' for the listing of the premises and the conclusion, so we will call the new component of deducing the conclusion from the premises *giving a proof* of the argument. Here is an informal illustration of what is involved in giving a proof. In §3 of Chapter 1 we presented the following argument:

> If the safe was opened, it must have been opened by Smith, with the assistance of Brown or Robinson. None of these three could have been involved unless he was absent from the meeting. But we know that either Smith or Brown was present at the meeting. So since the safe was opened, it must have been Robinson who helped open it.

As remarked earlier, this is a valid argument, as we could now show by translating it into LSL and demonstrating the validity of its form. But it is also possible to show that the English argument is valid using a less formal method, by exhibiting how its conclusion can be deduced from its premises. To do this, we begin by listing the premises individually:

(1) If the safe was opened, it must have been opened by Smith, with the assistance of Brown or Robinson.
(2) None of these three could have been involved unless he was absent from the meeting.
(3) Either Smith or Brown was present at the meeting.
(4) The safe was opened.

From this we have to deduce that Robinson helped open the safe. Here is how we might proceed:

(5) Smith opened the safe and either Brown or Robinson helped (from (4) and (1)).
(6) Smith was absent from the meeting (from (5) and (2)).
(7) Brown was present at the meeting (from (6) and (3)).
(8) Brown did not help to open the safe (from (7) and (2)).
(9) Therefore Robinson did (from (8) and (5)).

This is set out more carefully than in ordinary conversation, but the mechanics are the same. We arrive at the conclusion by making a succession of small steps from what we have already established, or from what we are given in the premises. Each step is easily seen to be correctly inferred from previously established lines, and the earlier lines appealed to at each step are noted. Since the individual steps are unobjectionable, we have demonstrated that the conclusion follows from the premises, but we have done it by giving a proof, not by constructing interpretations.

We are now going to articulate the process of giving a proof. To do this we have to be completely explicit about what principles can be appealed to in making individual steps like (5)-(9) above. These principles are called *rules of inference* and the process of using them to deduce a conclusion from premises is called *natural deduction*. A *system of natural deduction* is simply a collection of rules of inference. Systems of natural deduction were first described and investigated by Gerhard Gentzen, and the system we shall present here is Gentzen's system NK (German: *Natürliche Kalkül*).

We formulate the rules of inference not for English but for LSL. The rationale for abstracting from English is the same as before: whether or not a conclusion can be deduced from some premises depends not on the subject-matter of the premises and conclusion but on their logical form. For example, one could replace the argument about who helped open the safe with a parallel argument about, say, which country will have the biggest reduction in inflation next year, and which the next biggest (see the exercise following). So long as the premises have the same logical forms as the premises of the argument about who Smith's accomplice was, a five-step proof of the conclusion of the new argument could be given in parallel with our five-step proof of 'Robinson helped open it', each step justified by the same rules applied to the same previous line numbers. That is, whether or not a conclusion is deducible from certain premises depends just on their forms. Since we exhibit the forms by translating into LSL, we may as well construct the proofs in LSL from the outset.

One final introductory remark. The argument displayed above is the sort of thing one might find in a Sherlock Holmes novel, but of course precise proof is most commonly found in mathematics. A mathematician once related being asked why we insist on proving things in mathematics—"Why can't we just trust each other?" the student wondered. Proof is important in mathematics primarily because it is the only way to extend mathematical knowledge, but this aspect of its importance affects only those few individuals capable of making mathematical discoveries. However, for the rest of us, *following* a proof has an effect that is not obtained if we simply trust professors of mathematics not to make mistakes: following a proof helps us to understand *why* a particular proposition is true. For example, one might just accept that there are infinitely many prime numbers, but to see why this is so, we need to grasp one of the (many) proofs of this result. This makes the nature of proof in mathematics a subject of universal interest in its own right, and the historical motivation for developing the systems of natural deduction to be presented in this book was simply to study the kinds of reasoning accepted in standard mathematics. Indeed, classical logic is sometimes defined as the logic used in classical mathematics. Still, mathematical reasoning is really only a more rigorous version of reasoning of the type of the detective's about who opened the safe, and so the applicability of the study of deduction is not restricted to mathematics.

❑ Exercise

The following argument has the same form as the example of this section:

> If some countries reduce inflation next year, Brazil will reduce it the most, with either Argentina or Britain next. None of these three will reduce inflation unless they increase interest rates. But either Brazil or Argentina will hold interest rates down. So since some countries will reduce inflation next year, Britain will be the second most successful.

Deduce the conclusion of this argument from the premises in a way that exactly parallels the deduction of 'Robinson helped open the safe' from the premises of the example in this section.

2 Rules for conjunction and the conditional

In order to begin a proof, one needs to have some lines to which rules of inference can be applied. These lines are usually the premises of the argument to be proved, and we think of the process of listing the premises, as in (1)-(4) of the example of the previous section, as itself involving the application of a rule of inference. For the moment, we will take this rule to say that we may begin a proof by writing down one or more lines and labeling them as premises. This rule is sometimes called the Rule of Assumptions.

In the system NK, each of our five connectives is governed by two rules of inference. For each connective *c*, there is a rule which says when a statement can be inferred from some already established statements, one of which has *c* as its main connective. This rule is called the *elimination rule* for *c*, since the formula which is inferred will have fewer occurrences of *c* than the formula of which *c* is the main connective. For example, the elimination rule for '&' allows us to infer 'A' from 'A & B', while the elimination rule for '→' allows us to infer 'B' from 'A' and 'A → B' (inspection of line 5 in the example of §1 indicates that it is justified by the elimination rule for '→', since line 1 is a conditional and line 4 is its antecedent). We can express the content of these two rules in words as: from a conjunction we can infer either conjunct, and from a conditional and its antecedent we can infer its consequent. Each connective *c* also has an *introduction rule*, which says when a formula with *c* as its main connective may be inferred from already established formulae. For example, the introduction rule for '&' allows us to infer 'A & B' from 'A' and 'B'; more generally, from any two formulae we may infer their conjunction. Since these three rules are the simplest, we begin our exposition of NK with them.

The following is a valid LSL argument-form:

Example 1:   A & B
             C & D
             (A & D) → H
             ∴ H

Here is what an NK-proof of the conclusion of Example 1 from its premises looks like. We abbreviate '&-Elimination' and '&-Introduction' by '&E' and '&I', and '→-Elimination' by '→E'.

1	(1)	A & B	Premise
2	(2)	C & D	Premise
3	(3)	(A & D) → H	Premise
1	(4)	A	1, &E
2	(5)	D	2, &E
1,2	(6)	A & D	4,5 &I
1,2,3	(7)	H	3,6 →E ◆

The intuitive idea behind the proof is straightforward. In order to obtain 'H', the consequent of (3), we should apply →E to (3), but to use →E we need a conditional *and* its antecedent, so we must first obtain the antecedent 'A & D' of (3). This formula is a conjunction, so we should try to obtain it by &I, which requires that we first derive its two conjuncts. The conjuncts occur separately in (1) and (2), so we obtain each on its own by using &E. On the right of the proof we have written the justification of each line: (1)-(3) are premises and (4)-(7) are justified in the way explained.

Our explanation of the proof of Example 1 also illustrates an important technique in constructing proofs. It is rarely efficient to begin by blindly applying rules to the premises to see what happens: it is much better to formulate a *strategy*. A strategy can be worked out by looking at the conclusion and asking what rule might deliver it at the last line of the proof. *If the conclusion contains connectives, the rule is likely to be the introduction rule for the main connective;* if it does not, as in our case, the relevant rule can often be gleaned by inspecting the premises. In Example 1 it seems plausible that at the last line, we will be applying →E using (3). This tells us that at some previous line we have to establish the antecedent of (3), 'A & D'. This latter formula has '&' as its main connective and so will likely be inferred by &-Introduction. This means in turn that we have to obtain 'A' by itself and 'D' by itself, and looking at the premises, it is clear how this should be done. Our procedure, therefore, is to start by setting ourselves the overall 'goal' of proving the conclusion of the argument-form, and then break down this task into a sequence of subgoals until we arrive at subgoals whose execution is obvious. We then put all the steps together in the appropriate order, marking the finish of the proof with '◆' or some similar symbol.

We have left an important feature of our sample proof unexplained, the line numbers on the left. *We use these numbers to keep track of which premises any given line 'depends' upon.* The premises themselves are not inferred from any other lines, so each premise has its own line number written on the left. At line 4 we apply a rule of inference to line 1 to obtain 'A', so (4) depends on (1), and similarly (5) depends on (2). At line 6 we conjoin a line that depends on (1) and a line that depends on (2), so the result depends on both (1) and (2); finally, at line 7 we use line 3 and line 6; (6) depends on (1) and (2), and (3) depends on (3), so (7) depends on (1), (2) and (3). In general, then, *the premise numbers which should be listed on the left are the premise numbers that the lines cited on the right depend upon.* Intuitively, the numbers on the left at a line j are the premises which are needed, or at least which have been used, to derive the formula at j. There may be other premises listed at the start of the proof, but if their numbers are not on the left at j, they were not used in deriving the formula at j.

One advantage of this 'bookkeeping' device is that it makes it easy to check at the end of a proof that we have indeed proved what was asked. In order to show that the conclusion of Example 1 follows from its three premises, we require not merely a proof which begins with those premises and whose last line is the desired conclusion, but a proof in which *the last line depends upon no lines other than premises of Example 1.* Had there been other numbers on the left of line 7, this would show that we had not derived 'H' from the right premises. In a correct proof of an argument-form, therefore, the formula at the last line is the conclusion of the argument-form, and the lines on which it depends are all premises which are listed at the start. Note that we do not say that the lines on which the conclusion depends are *all of* the premises which are listed: it is sufficient that the conclusion depend on a subset of the listed premises. For example, if we had somehow been able to derive 'H' without using, say, premise 2, then only the numbers 1 and 3 would be on the left at line 7. But we would still have a proof of Example 1; after all, if 'H' follows from (1) and (3), it follows from (1), (3) and any other statements whatsoever—the extra statements would simply be redundant.

Having seen the way in which proofs are written down in NK, we can give more exact statements of the four rules of inference we have introduced so far.

Rule of Assumptions (preliminary version): The premises of an argument-form are listed at the start of a proof in the order in which they are given, each labeled 'Premise' on the right and numbered with its own line number on the left. Schematically

$$j \quad (j) \qquad p \qquad\qquad \text{Premise}$$

Rule of &-Elimination: If a conjunction $\ulcorner p \mathbin{\&} q \urcorner$ occurs at a line j then at any later line k one may infer either conjunct, labeling the line 'j &E' and writing on the left all the numbers which appear on the left of line j. Schematically:

$$
\begin{array}{lll}
a_1,\ldots,a_n & (j) & p \mathbin{\&} q \\
& \vdots & \\
a_1,\ldots,a_n & (k) & p \qquad j \,\&E
\end{array}
\qquad \text{OR} \qquad
\begin{array}{lll}
a_1,\ldots,a_n & (j) & p \mathbin{\&} q \\
& \vdots & \\
a_1,\ldots,a_n & (k) & q \qquad j \,\&E
\end{array}
$$

Thus the rule of &E is really two rules, one which allows us to infer the first conjunct of a conjunction, the other of which allows us to infer the second. We choose which to apply depending on what we want to infer.

Rule of &-Introduction: For any formulae p and q, if p occurs at line j and q occurs at line k then the formula $\ulcorner p \mathbin{\&} q \urcorner$ may be inferred at line m, labeling the line 'j,k &I' and writing on the left all numbers which appear on the left of line j and all which appear on the left of line k. Here we may have $j < k$, $j = k$ or $j > k$. Schematically:

$$
\begin{array}{lll}
a_1,\ldots,a_n & (j) & p \\
& \vdots & \\
b_1,\ldots,b_u & (k) & q \\
& \vdots & \\
a_1,\ldots,a_n,b_1,\ldots,b_u & (m) & p \mathbin{\&} q \qquad j,k \,\&I
\end{array}
$$

It is important to master how schemata are used to explain the application of a rule. Compare the schema for &I with line 6 in the proof of Example 1. The formula p is 'A', the formula q is 'D', line j is line 4 and line k is line 5. Because the proof is very simple, lines 4 and 5 each depend on just one premise. So the application of the schema to lines 4 and 5 is for the case where $n = u = 1$; a_1 is the number 1 and b_1 is the number 2. Line m is line 6, and as the schema requires, the numbers 1 and 2 are mentioned on its left.

However, there are respects in which the application of a rule may be more flexible than its schema seems to indicate. In the schema for &I, the line with the first conjunct p of $\ulcorner p \mathbin{\&} q \urcorner$, line j, is a *separate* line occurring *before* the line with the second conjunct q, but this number and order of occurrence of lines are not necessary: the order of the conjuncts at line m need not reflect the order in which they occur earlier in the proof, and &I may be applied to a single line to yield a conjunction with the same formula as its two conjuncts; for example, if 'A ∨ B' is the formula at line r, &I applied to r and r again produces '(A ∨ B) & (A ∨ B)', a step we would label 'r,r &I'. This is what is meant by saying that we may have j < k, j = k or j > k.

> **Rule of →-Elimination:** For any formulae p and q, if $\ulcorner p \rightarrow q \urcorner$ occurs at a line j and p occurs at a line k then q may be inferred at line m, labeling the line 'j,k →E' and writing on the left all numbers which appear on the left of line j and all which appear on the left of line k. We may have j < k or j > k. Schematically:

$$
\begin{array}{llll}
a_1,\ldots,a_n & (j) & p \rightarrow q & \\
 & & \vdots & \\
b_1,\ldots,b_u & (k) & p & \\
 & & \vdots & \\
a_1,\ldots,a_n,b_1,\ldots,b_u & (m) & q & j,k \rightarrow E
\end{array}
$$

Here are two further examples of proofs in NK using these rules. First we prove the valid LSL argument

Example 2:   A & (B & C)
                ∴ C & (B & A)

The conclusion is a conjunction, so we will likely obtain it at the last line of our proof by an application of &I to two previous lines where the conjuncts appear. Our subgoals, therefore, are (i) to derive 'C' on a line by itself, and (ii) to derive 'B & A' on a line by itself. 'B & A' is another conjunction, so the subgoal for (ii) is to derive its two conjuncts 'B' and 'A' separately. In sum, then, we want to derive the individual sentence-letters 'A', 'B' and 'C' by themselves and then put them together in the right order using &I. To obtain the letters by themselves, we can apply &E to the premise.

1	(1)	A & (B & C)	Premise
1	(2)	A	1, &E
1	(3)	B & C	1, &E
1	(4)	B	3, &E
1	(5)	C	3, &E
1	(6)	B & A	4,2 &I
1	(7)	C & (B & A)	5,6 &I ◆

This proof illustrates the important point that an elimination rule for a connective *c can only be applied to a line if that line has an occurrence of c as its main connective, and the rule must be applied to that occurrence of the con-*

nective. So at line 2, for example, we cannot apply &E to the second occurrence of '&' in line 1. In this respect the rules of inference follow the syntactic structure of the formula. If you are applying an elimination rule for a connective *c* to a formula on a line and it is not the main connective of the formula which you are eliminating, then you have made a mistake (a very common one for beginners!).

Another example:

Example 3:   A → (B → (C → D))
               C & (A & B)
               ∴ D

Since the conclusion has no connectives, it will not be inferred by an introduction rule. Inspecting the premises, we see that the conclusion is the consequent of a conditional which is a subformula of the first premise, so we can obtain the conclusion if we can obtain that conditional, 'C → D', on a line by itself, and also its antecedent on a line by itself. The antecedent of 'C → D' comes from premise 2, so the main subgoal is to extract 'C → D' from premise 1. We can do this with two further applications of →E.

1	(1)	A → (B → (C → D))	Premise
2	(2)	C & (A & B)	Premise
2	(3)	A & B	2, &E
2	(4)	A	3, &E
1,2	(5)	B → (C → D)	1,4 →E
2	(6)	B	3, &E
1,2	(7)	C → D	5,6 →E
1	(8)	C	2, &E
1,2	(9)	D	7,8 →E ◆

Again, note that it is not possible to apply →E directly to (8) and (1) to obtain (9). In other words, the following four-line proof is *wrong:*

1	(1)	A → (B → (C → D))	Premise
2	(2)	C & (A & B)	Premise
1	(3)	C	2, &E
1,2	(4)	D	1,3 →E (NO!)

The main connective of (1) is the *leftmost* arrow in it, with antecedent 'A' and consequent 'B → (C → D)'. *It is only to the main connective of a line that →E can be applied (similarly for every other elimination rule)*, so we cannot apply →E to line 1 until we have 'A' on a line by itself, which we manage at line 4 in the correct proof above.

We said earlier that every connective has both an introduction rule and an elimination rule, but so far we have only stated an elimination rule for the conditional. What should the introduction rule be? In other words, what kind of reasoning do we use to *establish* a conditional? Suppose we want to prove that *if* our currency loses value, *then* our trade deficit will narrow. A natural way of arguing for this conditional would be the following:

> Suppose our currency loses value. Then foreign imports will cost more in this country and our exports will cost less abroad. So our demand for imports will fall while demand in other parts of the world for our goods will increase. Therefore our trade deficit will narrow.

This little argument appeals to many suppressed premises, for example, that if the price of a good increases, demand for it slackens ('perfect elasticity' in economists' jargon) and that our manufacturers can increase production to meet increased demand. These might be regarded as other premises from which the conditional 'if our currency loses value, then our trade deficit will narrow' is to be deduced. As can be seen, the argument works in the following way: to derive the conditional, we begin by supposing that its antecedent is true ('Suppose our currency loses value'). Then using the other implicit premises and various rules of inference, we derive the consequent ('Therefore our trade deficit will narrow'). This shows that the consequent follows from the antecedent and the other premises: the consequent is true *if* the antecedent and the other premises are true. Hence the conditional 'if our currency loses value, then our trade deficit will narrow' is true if those other premises by themselves are true. This last point is important: the truth of the whole conditional depends not on the truth of the other premises *and* the antecedent, but only on the other premises; one can certainly convince oneself of the conditional 'if our currency loses value then our trade deficit will narrow' by an argument such as the one displayed without holding that its antecedent 'our currency will lose value' is in fact true.

The rule of →-Introduction formalizes this procedure. In a formal proof, all premises are listed explicitly and we can appeal to them in the normal way; the strategy for deriving a conditional, therefore, is just to add an extra assumption after the premises, this assumption being the antecedent of the conditional which we want to derive. Once we have deduced the consequent of the conditional from the premises together with the antecedent of the conditional, we are entitled to assert that the conditional itself follows from the premises without the extra assumption.

Here is a simple illustration of the strategy, in a proof of the following LSL argument-form:

Example 4: $R \rightarrow (S \rightarrow T)$
                 S
               $\therefore R \rightarrow T$

We have to derive the conditional 'R → T', so we should add its antecedent 'R' to the premises as an extra assumption (like supposing our currency loses value). The goal is then to derive its consequent 'T' from the two premises and the extra assumption, after which we may assert that the conditional 'R → T' follows from the two premises alone. The NK-proof of Example 4 is:

1	(1)	R → (S → T)	Premise
2	(2)	S	Premise
3	(3)	R	Assumption
1,3	(4)	S → T	1,3 →E
1,2,3	(5)	T	4,2 →E
1,2	(6)	R → T	3,5 →I ◆

There are a number of features of this proof to which close attention should be paid. First, assumptions, like premises, depend on themselves, so at line 3 we put '3' on the left. Second, line 3 is justified by essentially the same rule as lines 1 and 2. What allows us to write the formulae at (1) and (2) is not that they are premises, though that is why we write these particular formulae. Lines like (1) and (2), and also (3), are unobjectionable because, in general, a line in a proof may be read as claiming that the formula on the line holds 'according to' the formulae whose numbers are on the left of the line. And, of course, any formula holds according to *itself*. The Rule of Assumptions, which relies on this trivial fact, is simply the rule that any formula may be written at any line in a proof if it is numbered to depend on itself. So it is not strictly obligatory to distinguish between 'assumption' and 'premise'. However, we shall continue to use 'premise' for a formula which is a premise of the LSL argument-form which is being proved and 'assumption' for an extra hypothesis made in the course of giving a proof.

Third, the numbering on the left of line 6 should be noted: the assumption number '3', which appears on the left of (5), has been dropped, and the assumption itself has become the antecedent of the conditional at line 6. We said earlier that once we have derived the consequent of a conditional from its antecedent together with other premises, we are 'entitled to assert that' the conditional follows from the premises by themselves. This is implemented in NK by dropping the assumption number and absorbing the assumption by making it the antecedent of the conditional. That is, at line 5 in the proof we show that 'T' follows from lines 1, 2 and 3, where 'R' is the formula at line 3, so at line 6 we say that 'R → T' follows just from 1 and 2, which is what was to be proved. Dropping the assumption number and making the assumption the antecedent of a conditional is called *canceling* or *discharging* the assumption. That is why the correct version of the Rule of Assumptions is not as liberal as it appears. To complete a proof we may make as many extra assumptions as we like, but if we use these assumptions they will eventually have to be discharged. For the moment, then, *the only situation in which we make an assumption is when we want to derive a conditional, and the assumption formula is always the antecedent of the conditional we want to derive*. Note that if we do not drop 3 from the left at 6, then the last line would be represented as still depending on 3, which is an assumption, not a premise of Example 4. Therefore the proof would not be a proof of Example 4, since not all numbers on the left of the last line would be numbers of premises of Example 4. It would instead be a proof of the easier problem that has the extra premise 'R'.

We can iterate applications of →I, as we must to prove the following argument-form:

Example 5:   ((A & B) & C) → D
                ∴ C → (B → (A → D))

To derive 'C → (B → (A → D))' we should assume 'C' with the goal of deducing
'B → (A → D)'; in turn, to derive 'B → (A → D)' we should assume 'B' with the goal
of deducing 'A → D'; in turn again, to derive 'A → D' we should assume 'A' with
the goal of deducing 'D'. To obtain 'D', we use our assumptions to construct the
antecedent 'A & (B & C)' of the premise, and then apply →E. The proof is:

1	(1)	((A & B) & C) → D	Premise
2	(2)	C	Assumption
3	(3)	B	Assumption
4	(4)	A	Assumption
3,4	(5)	A & B	4,3 &I
2,3,4	(6)	(A & B) & C	5,2 &I
1,2,3,4	(7)	D	1,6 →E
1,2,3	(8)	A → D	4,7 →I
1,2	(9)	B → (A → D)	3,8 →I
1	(10)	C → (B → (A → D))	2,9 →I ◆

When a proof has many assumptions, it is important to check that the last line
depends only on premises: the numbers on the left at the end should not
include any of the assumptions. We made three assumptions in this proof, and
all are used, but they are successively discharged at lines 8, 9 and 10, so that
the last line of the proof depends only on line 1, which is how we want it.

We now state the new rule of →I and the final version of Assumptions.

Rule of Assumptions: At any line j in a proof, any formula *p* may be
entered and labeled as an assumption (or premise, where appropriate).
The number j should then be written on the left. Schematically:

           j (j)   *p*   Assumption (or: Premise)

Rule of →-Introduction: For any formulae *p* and *q*, if *q* has been
inferred at a line k in a proof and *p* is an assumption or premise occurring at a line j, then at line m we may infer ⌜*p* → *q*⌝, labeling the line
'j,k →I', and writing on the left the same assumption numbers as occur
on the left of line k, except that we delete j if it is one of these numbers.
Here we may have j < k, j = k or j > k. Schematically:

                    j (j)   *p*   Assumption (or: Premise)
                       ⋮
         $a_1,...,a_n$ (k)   *q*
                       ⋮
    $\{a_1,...,a_n\}/j$ (m)   *p* → *q*   j,k →I

In this statement of the rule, the notation '$\{a_1,...,a_n\}/j$' indicates the set of numbers $\{a_1,...,a_n\}$ with j removed if it is one of them. Applications of →I when j = k yield conditionals with the same formula as antecedent and consequent. Notice that it is not a requirement of applying →I that j be one of the premise or assumption numbers on which line k depends, nor is it a requirement that line j occur before line k. Here is a correct proof in which k < j (the consequent appears in the proof before the antecedent!) and j does not appear on the left of line k, a proof of a valid LSL argument-form:

Example 6:   A
          ∴ B → A

The validity of this argument-form reflects the fact that a conditional with a true consequent is true. The proof is:

1	(1)	A	Premise
2	(2)	B	Assumption
1	(3)	B → A	2,1 →I ◆

In terms of our schema for →I, j = 2 and k = 1. So the consequent of (3) is at (1) and the antecedent at (2), and the antecedent is not used to derive the consequent. In §1 and §7 of Chapter 3 we noted that the function-table for '→' has some odd-looking consequences for the truth-values of English conditionals. However, it can be formally proved that the function-table for '→' is the correct account of the conditional *if* the rules of →E and →I are correct: the rules and the table go hand in hand. Someone who rejects the table as an account of the meaning of 'if...then...' therefore owes us a new account of 'if...then...'-introduction, presumably involving a requirement that the antecedent be *non-redundantly used* in deriving the consequent. There is a branch of logic known as *relevance logic* which develops this approach (see Read), but it turns out that to implement the idea of non-redundant use in a reasonably satisfactory way leads to systems that are significantly more complicated than classical logic.
    Our last example is a proof of the valid argument-form

Example 7:   A → (B → C)
          ∴ (A → B) → (A → C)

which is straightforward provided we pay attention to the form of the conclusion. The conclusion is a conditional with 'A → B' as antecedent, so in our proof we should assume 'A → B' and try to derive 'A → C'. 'A → C' is another conditional, so to derive it we use the same method over again of assuming the antecedent and deriving the consequent. That is, we assume 'A' and try to derive 'C'. We then discharge our assumptions on a 'last in, first out' basis. This strategy yields the proof set out over the page. And more generally, in a logically organized proof where there are a number of assumptions made with a view to applying →I, any given application of →I will discharge the undischarged assumption most recently made.

1	(1)	A → (B → C)	Premise
2	(2)	A → B	Assumption
3	(3)	A	Assumption
2,3	(4)	B	2,3 →E
1,3	(5)	B → C	1,3 →E
1,2,3	(6)	C	4,5 →E
1,2	(7)	A → C	3,6 →I
1	(8)	(A → B) → (A → C)	2,7 →I ◆

Here are important points to remember about →I. In doing the exercises which follow, *the reader should make frequent reference to this list.*

- Use →I only when you wish to *derive* a conditional $\ulcorner p \to q \urcorner$ (if one of your premises or assumptions is a conditional, this is *not* a reason to assume its antecedent, since premises or assumptions are not things which you are trying to *derive*).
- To derive $\ulcorner p \to q \urcorner$ using →I, assume the antecedent p and try to derive the consequent q. Do *not* assume anything other than p, and assume the *whole* of p. For example, if p is a conjunction, do not assume just one of the conjuncts.
- When a conditional $\ulcorner p \to q \urcorner$ is derived by →I, the antecedent p must always be a formula which you have assumed at a previous line; it cannot be a formula which you have derived from other premises and assumptions.
- When you apply →I, remember to discharge the assumption by dropping the assumption number on the left.
- Check that the last line of your proof does not depend on any extra assumptions you have made over and above your premises.

❑ Exercises

Give proofs of the following LSL argument-forms:

(1) A → B
 B → C
 ∴ A → C

*(2) A → B
 A → C
 ∴ A → (B & C)

(3) A
 ∴ B → (A & B)

(4) A → B
 ∴ (A & C) → (B & C)

*(5) (A & B) → C
      ∴ A → (B → C)

(6) A → (B → C)
     ∴ (A & B) → C

(7) (A & B) → C
     A
     ∴ B → C

(8) A → B
     ∴ (C → A) → (C → B)

(9) A & (A → (A & B))
     ∴ B

(10) A → (A → B)
      ∴ A → B

(11) (A & B) → (C & D)
      ∴ [(A & B) → C] & [(A & B) → D]

(12) C → A
      C → (A → B)
      ∴ C → (A & B)

(13) A → (B → C)
      (A & D) → E
      C → D
      ∴ (A & B) → E

(14) A → (B → C)
      ∴ B → (A → C)

(15) A → (B → C)
      D → B
      ∴ A → (D → C)

*(16) A → B
       ∴ A → (C → B)

(17) A & (B & C)
      ∴ A → (B → C)

(18) B & C
      ∴ (A → B) & (A → C)

3 Sequents and theorems

When the conclusion of an LSL argument-form can be deduced from its premises in the system NK, the argument-form is said to be *NK-provable*, its conclusion is said to be a *deductive consequence in NK* of its premises, and its premises are said to *deductively entail* its conclusion *in NK* (we often omit the qualifier 'in NK'). At the end of §4 of Chapter 3 we introduced the notion of *semantic* consequence and the double-turnstile symbol '⊨' which expresses it (page 66). The notion of *deductive* consequence is a companion concept, in the sense that it provides another way of explaining the idea of a conclusion 'following from' premises, and we introduce the *single turnstile* '⊢$_{NK}$' to express it. When a conclusion q has been deduced from premises $p_1,...,p_n$ in the system NK, we write '$p_1,...,p_n$ ⊢$_{NK}$ q', which is read '$p_1,...,p_n$ deductively entail q in NK' or 'q is a deductive consequence of $p_1,...,p_n$ in NK'. An expression '$p_1,...,p_n$ ⊢$_{NK}$ q' is often called a *syntactic sequent*, so in the previous section we established the following syntactic sequents:

(1) A & B, C & D, (A & D) → H ⊢$_{NK}$ H
(2) A & (B & C) ⊢$_{NK}$ C & (B & A)
(3) A → (B → (C → D)), C & (A & B) ⊢$_{NK}$ D
(4) R → (S → T), S ⊢$_{NK}$ R → T
(5) ((A & B) & C) → D ⊢$_{NK}$ C → (B → (A → D))
(6) A ⊢$_{NK}$ B → A
(7) A → (B → C) ⊢$_{NK}$ (A → B) → (A → C)

Note that any particular sequent itself asserts that its conclusion can be derived in NK from its premises: that is the effect of '⊢$_{NK}$'. So, strictly speaking, syntactic sequents are true or false rather than provable or unprovable (*arguments* are provable or unprovable). But since showing a syntactic sequent to be true is done just by giving a proof, we will allow ourselves to speak of syntactic sequents as provable or unprovable. In general we omit the qualifiers 'syntactic' and 'semantic' where it is clear what kind of sequent is being discussed.

We make the meaning of '⊢$_{NK}$' absolutely precise by the following definition, which should be compared with the definition of '⊨' on page 66:

> $p_1,...,p_n$ ⊢$_{NK}$ q if and only if there is a numbered sequence of lines whose last line is q, each line in the sequence is either one of the premises $p_1,...,p_n$, or an assumption (in both cases with appropriate numbering on the left), or is correctly inferred from previous lines by some rule of NK, and every number on the left of the last line is the number of a line where one of $p_1,...,p_n$ appears as a premise.

Notice that since we do not require that the last line of a proof have on its left numbers of lines where *all* of $p_1,...,p_n$ appear as premises, every proof is a proof of infinitely many sequents, since there are infinitely many different ways of augmenting the premises actually used in a proof with redundant premises.

In defining semantic consequence we singled out the special case in which $n = 0$, that is, the case where there are no $p_1,...,p_n$. This left us with '$\models q$', which means that q is a tautology. With deductive consequence we may also consider the case where $n = 0$. We read '$\vdash_{NK} q$' as: q is a *theorem* of NK. Inspecting the definition of '$\vdash_{NK}$' and deleting material having to do with $p_1,...,p_n$, we see that '$\vdash_{NK} q$' means that q can be derived from no premises. In other words, q can be derived just from assumptions, all of which are discharged so that the last line of the proof depends on nothing at all. Here is the simplest possible example— we show $\vdash_{NK} A \rightarrow A$:

Example 1: Show $\vdash_{NK} A \rightarrow A$.

1	(1)	A	Assumption
	(2)	A → A	1,1 →I ◆

Referring to our statement of the rule of →I on page 96, Example 1 illustrates the case where j = k. When the assumption 1 is discharged at line 2, nothing remains on the left.

It should be clear from the way in which →I works that for every provable sequent there is a corresponding theorem with embedded conditionals that have the various premises of the sequent as antecedents; in general, if we have constructed a proof of $p_1,...,p_n \vdash_{NK} q$ then it is straightforward to give a proof of $\vdash_{NK} p_1 \rightarrow (p_2 \rightarrow...(p_n \rightarrow q)...)$, since we can append successive applications of →I to our proof of $p_1,...,p_n \vdash_{NK} q$. For example, a solution to Exercise 2.1, A → B, B → C $\vdash_{NK}$ A → C, can be easily converted into a proof which establishes that $\vdash_{NK} (A \rightarrow B) \rightarrow ((B \rightarrow C) \rightarrow (A \rightarrow C))$ by making the premises assumptions and appending enough applications of →I. Here $n = 2$ and $p_1 = $ 'A → B', $p_2 = $ 'B → C' and $q = $ 'A → C' so

$$\vdash_{NK} p_1 \rightarrow (p_2 \rightarrow q)) \quad = \quad \vdash_{NK} (A \rightarrow B) \rightarrow ((B \rightarrow C) \rightarrow (A \rightarrow C)).$$

Example 2: Show $\vdash_{NK} (A \rightarrow B) \rightarrow ((B \rightarrow C) \rightarrow (A \rightarrow C))$.

1	(1)	A → B	Assumption
2	(2)	B → C	Assumption
3	(3)	A	Assumption
1,3	(4)	B	1,3 →E
1,2,3	(5)	C	2,4 →E
1,2	(6)	A → C	3,5 →I
1	(7)	(B → C) → (A → C)	2,6 →I
	(8)	(A → B) → ((B → C) → (A → C))	1,7 →I ◆

The premises of Exercise 2.1 on page 98 are listed here as assumptions, since in the current problem they are antecedents of conditionals which we are trying to deduce. The assumptions we make are discharged in reverse order of listing by repeated uses of →I. Note again that discharging an assumption *always* creates a conditional with that assumption as antecedent.

❏ Exercises

Show the following:

(1) ⊢$_{NK}$ (A & B) → (B & A)
(2) ⊢$_{NK}$ (A → B) → (A → B)
(3) ⊢$_{NK}$ A → (B → A)
*(4) ⊢$_{NK}$ [A → (B & C)] → [(A → B) & (A → C)]
(5) ⊢$_{NK}$ [(A → A) → B] → B

4 Rules for negation

A very common strategy in ordinary reasoning and especially in mathematics is known as *reductio ad absurdum.* Say that two formulae are *explicitly contradictory* if and only if one is of the form q and the other of the form $\ulcorner{\sim}q\urcorner$, that is, if one is the negation of the other. Then the principle upon which *reductio* depends is the following: if explicitly contradictory formulae follow from a collection of premises and assumptions, then one of the collection may be rejected. There are many famous illustrations of the strategy in mathematics, but here is a very simple one:

Theorem: For any number $x > 1$, the least factor of x other than 1 is a prime number.
  Proof:

(a) Let j be the least factor of x other than 1.
(b) Suppose for *reductio* that j is composite.
(c) Then $j = yz$, $1 < y \leqslant z < j$.
(d) So $x = jk = (yz)k = y(zk)$.
(e) Thus y is a factor of x less than j and other than 1.
(f) (e) contradicts (a).
(g) The contradiction is generated by (b), so (b) should be rejected.
(h) Hence j is not composite (which means it is prime).

The explicitly contradictory formulae at which we arrive in this argument are (a) and (e). The assumption we made that led to this contradiction is 'j is composite'. Of course, the proof also makes unacknowledged use of other principles of arithmetic as premises, for example, in deriving (d). But if these other principles are things we know, the 'blame' for the contradiction about j rests on the assumption (b) that j is composite. So we can reject this assumption and conclude that j is not composite.

How might this strategy be implemented in NK? Our proof above has two stages. In the first stage we make an assumption, (b), and then reason so that we end up with explicitly contradictory statements, (a) and (e), as lines in the proof. In the second stage we infer from the fact that we have obtained explicitly contradictory statements that the assumption must be wrong and so we

reject it. These two stages will correspond in NK to ~-Elimination and ~-Intro-
duction respectively. The second stage involves ~-Introduction because *rejecting* an assumption is denying or negating it, so a 'not' is introduced. The first
stage is labeled in NK as ~-Elimination for the following reason. When we infer
the explicitly contradictory formulae q and $\ulcorner{\sim}q\urcorner$, we know that the collection
of premises and assumptions which we have used cannot all be correct, since
they lead to something absurd, such as (e) and (a) in the example. Using the
symbol '$\curlywedge$' to stand for the idea of something absurd, we make this explicit in
the proof by writing down '$\curlywedge$' on a line by itself, putting on its left all the numbers of the premises and assumptions on which q and $\ulcorner{\sim}q\urcorner$ depend; so the line
'says' that these premises and assumptions lead to an absurdity (this roughly
corresponds to line f in the example). We label the line 'j,k ~E', where j and k
are the lines where the mutually contradictory statements appear. The step is
called ~-Elimination since line j contains a '~' as its main connective, while '$\curlywedge$'
contains no occurrence of '~'. After we have written down '$\curlywedge$', we can go ahead
and apply ~I, inferring the negation of one of the premises or assumptions in
the proof. '$\curlywedge$' is called the *absurdity symbol.*
      Here is an illustration of these two rules:

Example 1: Show A → B, ~B ⊢_{NK} ~A.

1	(1)	A → B	Premise
2	(2)	~B	Premise
3	(3)	A	Assumption
1,3	(4)	B	1,3 →E
1,2,3	(5)	$\curlywedge$	2,4 ~E
1,2	(6)	~A	3,5 ~I ◆

This proof should be studied carefully. The conclusion is '~A', with main connective '~', hence we try to obtain it by ~I. This requires us to assume 'A' with
the subgoal of deriving '$\curlywedge$'. Deriving '$\curlywedge$' requires deriving the two halves of an
explicit contradiction, and looking at our premises and assumption, we see
how to obtain two such formulae. Once we have obtained them at lines 2 and
4, we can 'eliminate' the negation symbol in line 2 by writing '$\curlywedge$' at line 5. The
line where we write '$\curlywedge$' depends on whatever the explicitly contradictory lines
depend on, so (5) depends on lines 1, 2 and 3. And now that '$\curlywedge$' has been inferred, we can reject one of the premises or assumptions on which it depends.
To reject a formula means writing it down *prefixed by a negation symbol.* Thus
we arrive at (6); on the right we label our inference '~I', attaching the line numbers of the assumption being rejected and of the relevant occurrence of '$\curlywedge$',
while on the left we drop the line number of the assumption we are rejecting,
in this case 3. The justification for dropping the line number is that the assumption is, after all, being *rejected:* in deriving absurdity, we assume (3)'s
truth, but we do not assume its truth when rejecting it. Note that at (6) we can
write down the negation of *any* of the premises and assumptions on which (5)
depends; the one we choose is dictated by the formula we are trying to infer.
      The negation rules are therefore as follows:

Rule of ~-Elimination: For any formula q, if $\ulcorner\sim q\urcorner$ has been inferred at a line j in a proof and q at a line k, j < k or k < j, then we may infer '$\curlywedge$' at line m, labeling the line 'j,k ~E' and writing on its left the numbers on the left at j and on the left at k. Schematically (with j < k):

$$
\begin{array}{llll}
a_1,\ldots,a_n & \text{(j)} & \sim q \\
& \vdots \\
b_1,\ldots,b_u & \text{(k)} & q \\
& \vdots \\
a_1,\ldots,a_n,b_1,\ldots,b_u & \text{(m)} & \curlywedge & \text{j,k ~E}
\end{array}
$$

Rule of ~-Introduction: If '$\curlywedge$' has been inferred at line k in a proof and $\{a_1,\ldots,a_n\}$ are the premise and assumption numbers '$\curlywedge$' depends upon, then if p is an assumption (or premise) at line j, $\ulcorner\sim p\urcorner$ may be inferred at line m, labeling the line 'j,k ~I' and writing on its left the numbers in $\{a_1,\ldots,a_n\}/j$. We may have j < k or k < j or j = k. Schematically (with j < k):

$$
\begin{array}{llll}
\text{j} & \text{(j)} & p & \text{Assumption} \\
& \vdots & & \text{(or: Premise)} \\
a_1,\ldots,a_n & \text{(k)} & \curlywedge \\
& \vdots \\
\{a_1,\ldots,a_n\}/j & \text{(m)} & \sim p & \text{j,k ~I}
\end{array}
$$

Recall that $\{a_1,\ldots,a_n\}/j$ is $\{a_1,\ldots,a_n\}$ with j removed if it is one of them (as with →I, j does not *have* to be in $\{a_1,\ldots,a_n\}$).

It must be admitted that the sense in which ~E *eliminates* a negation is rather notational. On some other approaches, rather than a ~-E rule, we would use &I to infer $\ulcorner q \,\&\sim q\urcorner$ when we have q and $\ulcorner\sim q\urcorner$ on separate lines, and then reject a premise or assumption using a rule of *reductio*. Here the explicit contradiction $\ulcorner q \,\&\sim q\urcorner$ plays the role of '$\curlywedge$', and we do not need a new rule to obtain it from its conjuncts, since we already have &I. But there are certain technical and philosophical reasons to formulate the negation rules using '$\curlywedge$'. A technical consideration is that some sequents require roundabout proofs if explicit contradictions are used in place of '$\curlywedge$'. For instance, in the next section of this chapter, we give a ten-line proof of the sequent $\sim A \,\&\sim B \vdash_{NK} \sim(A \vee B)$. The reader who, after mastering the $\vee$-rules, attempts to prove the sequent using explicit contradictions in place of '$\curlywedge$', will find that the inability to derive the *same* explicit contradiction from 'A' and 'B' without extra steps complicates the proof. A philosophical consideration arises from the perspective that rules of inference, not truth-tables, constitute the primary way of explaining the meanings of connectives. The version of ~I which uses explicit contradictions is unsatisfactory from this point of view, since negation occurs in the formula $\ulcorner q \,\&\sim q\urcorner$ to which the introduction rule for negation is applied, and so we would be explaining negation in terms of itself. Of course, whether the formulation of ~I in terms of '$\curlywedge$' is an improvement depends on whether absurdity can be understood independently of understanding negation.

Since the combination of ~E and ~I always produces a formula whose main connective is '~', these rules are used to derive negative formulae, as in this example:

Example 2: Show ~B ⊢_{NK} ~(A & B).

1	(1)	~B	Premise
2	(2)	A & B	Assumption
2	(3)	B	2 &E
1,2	(4)	⋏	1,3 ~E
1	(5)	~(A & B)	2,4 ~I ◆

Since we wish to derive '~(A & B)' we expect to conclude the proof with an application of ~I, which yields the subgoal of proving '⋏'. To derive '⋏' we make an assumption which will lead to explicitly contradictory formulae, an assumption from whose rejection (negation) we can obtain '~(A & B)'. The simplest such assumption is of course 'A & B', whose rejection *is* '~(A & B)'.

~I is like →I in that it allows us to discharge assumptions. Here is an example in which both rules are used.

Example 3: Show ~(A & B) ⊢_{NK} A → ~B.

In order to derive 'A → ~B' we will use →I, which means we assume 'A' with the subgoal of proving '~B'. '~B' has '~' as main connective, so we can likely obtain it by ~I, which means we should assume 'B', as at line 3 below, with the subgoal of proving '⋏'. And with our two assumptions in place, it is easy to see how to generate explicitly contradictory formulae.

1	(1)	~(A & B)	Premise
2	(2)	A	Assumption
3	(3)	B	Assumption
2,3	(4)	A & B	2,3 &I
1,2,3	(5)	⋏	1,4 ~E
1,2	(6)	~B	3,5 ~I
1	(7)	A → ~B	2,6 →I ◆

But suppose that we had been asked instead to prove the sequent ~(A & ~B) ⊢_{NK} A → B.

Example 4: Show ~(A & ~B) ⊢_{NK} A → B.

The first two lines are clear:

1	(1)	~(A & ~B)	Premise
2	(2)	A	Assumption

The problem is now to deduce 'B' so that →I gives us the conclusion formula

'A → B'. But here we face a difficulty, for 'B' has no main connective, so our usual guide as to what I-rule should be used to infer the goal formula is not applicable. In such a case we turn to the premises, to see if there is any obvious way of applying E-rules to them. But this time there is not. So it is unclear how 'B' is to be obtained. However, there is a formula which is intuitively equivalent to 'B' and which does have a main connective, so that an I-rule is suggested by which to infer it. This formula is '~~B', the *double negation* of 'B'. If we could derive '~~B' and if we also have a rule which allows us to *cancel* a double negative '~~', we would thereby arrive at 'B' and could then finish with →I to get 'A → B'. As far as deriving '~~B' is concerned, its main connective is '~', so we would expect to get it by ~I. Hence we should assume '~B' and aim to deduce '⅄' by ~E. We get immediate confirmation that this strategy is on the right lines by looking at the premise and assumption already listed. The premise has '~' as main connective and so its role will presumably be to figure in an application of ~E. Such an application can be made if we can derive 'A & ~B', which is simple if we are going to assume '~B'. Let us use 'Double Negation', 'DN' for short, for the rule which allows canceling of '~~'. Then the proof is:

1	(1)	~(A & ~B)	Premise
2	(2)	A	Assumption
3	(3)	~B	Assumption
2,3	(4)	A & ~B	2,3 &I
1,2,3	(5)	⅄	1,4 ~E
1,2	(6)	~~B	3,5 ~I
1,2	(7)	B	6 DN
1	(8)	A → B	2,7 →I ◆

We state the new rule as follows:

> **Rule of Double Negation:** For any formula p, if $\ulcorner\sim\sim p\urcorner$ has been inferred at a line j in a proof, then at line k we may infer p, labeling the line 'j DN' and writing on its left the same numbers as on the left of line j. Schematically:

$$a_1,...,a_n \quad (j) \qquad \sim\sim p$$
$$\vdots$$
$$a_1,...,a_n \quad (k) \qquad p \qquad\qquad \text{j DN}$$

If there is a rule that allows canceling '~~', then, since this looks rather like an elimination rule, it may seem that there ought to be a companion rule allowing addition of two '~~'s. But DN has a technical character that is different from elimination rules, and we will see later that, anyway, a rule for adding '~~' would be redundant. However, the rule of DN is certainly not redundant. Without it, there is no way of proving the previous sequent just using the other rules we have currently introduced. In fact, the rule of DN, or some rule or group of rules which has the same effect, is in a sense *characteristic* of classical logic, in that certain nonclassical alternatives to classical logic differ from classical logic precisely in rejecting this rule and its equivalents. For example, the main non-classical alternative is *intuitionistic logic*, in which there is no way in general of canceling a double negation. Intuitionistic logic is described in some detail in Chapter 10; see also Van Dalen.

One aspect of the rule of ~I is worth elaborating upon. It may seem strange that we can appeal to '⅄' to reject an assumption which is not used in deriving '⅄' (referring to the schema for ~I on page 104, this is the case where j is not in $\{a_1,...,a_n\}$). However, the following proof is technically correct:

Example 5: Show B, ~B ⊢$_{NK}$ A.

1	(1)	B	Premise
2	(2)	~B	Premise
3	(3)	~A	Assumption
1,2	(4)	⅄	1,2 ~E
1,2	(5)	~~A	3,4 ~I
1,2	(6)	A	5 DN ◆

What this means is that from contradictory premises, any formula at all may be deduced; for though we derive 'A' in this example, the same sequence of steps would allow us to derive any other wff *p* of LSL, beginning with the assumption ⌜~*p*⌝ at line 3. Intuitively, this corresponds to the thought that, if 'B' and '~B' were *both* true, then *anything at all* would be true.[1]

A final comment concerns the new symbol '⅄' which has appeared in this section. What kind of an item is it? Syntactically, it behaves like an object lan-guage sentence-letter, since other connectives attach to it to form compound formulae like '~~⅄' and 'A → ⅄'. Semantically, however, it is more like a truth-functional connective. It is governed by the stipulation that every interpreta-tion assigns ⊥ to '⅄', which means that we can think of it as expressing a *con-stant zero-place* truth-function, the one which takes an empty input and outputs ⊥. So in the future we will include it among sentence-letters when describing the syntax of a language, but give it its own evaluation clause, like a connective, when describing a language's semantics.

Here are some useful points to bear in mind when using the negation rules:

- If you are trying to derive a formula with '~' as its main connective, use ~I to obtain it. This means that you should assume the formula within the scope of the main '~' and try to derive '⅄' by ~E.
- When you apply ~I, the formula which you infer must be the nega-tion of a premise or assumption. It cannot be the negation of a for-mula which has been inferred from other lines by an I or E rule for a connective.

---

[1] The sequent B, ~B ⊢$_{NK}$ A is sometimes embodied in classical logic as a rule of inference, from con-tradictory lines (equivalently, from '⅄') to infer any formula, pooling premises and assumptions. In systems with DN, this rule, known as *ex falso quodlibet*, is redundant, but there are other ways of for-mulating classical logic in which it is required. See further §10 of this chapter.

- If one of your premises or assumptions has '~' as its main connective, it is likely that its role in the proof will be to be one of a pair of contradictory formulae in an application of ~E. You should therefore consider trying to derive the formula within the scope of the '~' to get the other member of the contradictory pair.
- If you are trying to deduce a sentence-letter and there is no obvious way to do it, consider trying to derive its double negation and then use DN. Any double negation has '~' as its main connective. Now refer to the first point in this list.
- At this point in the development of NK, *you should only assume a formula p if you are trying to deduce its negation or trying to deduce a conditional with p as antecedent.* A common beginner's error is to make assumptions just for the sake of being able to apply an E-rule (particularly →E). At this stage, *only* make an assumption when you have reasoned how you will use ~I or →I to discharge it.

❏ Exercises

Show the following ('$p \dashv\vdash_{NK} q$' abbreviates '$p \vdash_{NK} q$ and $q \vdash_{NK} p$'):

*(1) $A \rightarrow \sim B \vdash_{NK} B \rightarrow \sim A$
(2) $\vdash_{NK} \sim(A \& \sim A)$
(3) $A \vdash_{NK} \sim\sim A$
(4) $\sim\sim A \rightarrow B, \sim B \vdash_{NK} \sim\sim\sim A$
(5) $A \rightarrow B \vdash_{NK} \sim(A \& \sim B)$
(6) $\sim(A \& B), A \vdash_{NK} \sim B$
(7) $A \vdash_{NK} \sim(B \& \sim(A \& B))$
*(8) $\vdash_{NK} (A \& \sim A) \rightarrow B$
(9) $B \rightarrow (A \& \sim A) \vdash_{NK} \sim B$
(10) $A \rightarrow B, A \rightarrow \sim B \vdash_{NK} \sim A$
(11) $(A \& B) \rightarrow \sim A \vdash_{NK} A \rightarrow \sim B$
(12) $(A \& \sim B) \rightarrow B \vdash_{NK} A \rightarrow B$
*(13) $A \rightarrow B, B \rightarrow \sim A \vdash_{NK} \sim A$
(14) $A \rightarrow (B \rightarrow C) \vdash_{NK} \sim C \rightarrow \sim(A \& B)$
(15) $A \rightarrow \sim(B \& C), B \rightarrow C \vdash_{NK} A \rightarrow \sim B$
(16) $A, A \rightarrow C, \sim B \rightarrow \sim C \vdash_{NK} \sim(A \rightarrow \sim B)$
(17) $\vdash_{NK} (\sim A \rightarrow A) \rightarrow A$
*(18) $(A \& \sim B) \rightarrow \sim A \vdash_{NK} A \rightarrow B$
(19) $\vdash_{NK} \sim\lambda$
(20) $\sim A \dashv\vdash_{NK} A \rightarrow \lambda$ (do *both* directions)
(21) $\sim A \vdash_{NK} A \rightarrow B$
(22) $\sim(A \rightarrow B) \dashv\vdash_{NK} A \& \sim B$ (do *both* directions)
(23) $\sim(A \& B), \sim(A \& \sim B) \vdash_{NK} \sim A$
(24) $\sim(A \rightarrow \sim A), \sim(B \rightarrow \sim B), \sim(A \& B) \vdash_{NK} \lambda$
(25) $A \rightarrow \sim(B \rightarrow C) \vdash_{NK} (A \rightarrow B) \& (A \rightarrow \sim C)$
(26) $\vdash_{NK} \sim\{[\sim(A \& B) \& \sim(A \& \sim B)] \& [\sim(\sim A \& B) \& \sim(\sim A \& \sim B)]\}$

5 Rules for disjunction

The remaining connectives for which we require rules are disjunction and the biconditional. The introduction rule for '∨' is extremely simple: if a formula p occurs on a line then we may infer the disjunction $\ulcorner p \lor q \urcorner$ or the disjunction $\ulcorner q \lor p \urcorner$ at a new line, depending on the same premises and assumptions. For example, if we have established that water is H_2O, then we can infer the weaker statement that either water is H_2O or water is H_4O_2 (this would be useful if we knew of something else that followed given that water is one or the other). Indeed, the new disjunct need have no relation to the one which has been estab- lished: if we have established that water is H_2O, then we can infer that either water is H_2O or Moses was the author of the Pentateuch—though it is less clear how *this* could be useful!

Here is a simple example of ∨I:

Example 1: Show $(A \lor (D \,\&\, E)) \to B \vdash_{NK} A \to B$.

Our strategy is to use →I to derive 'A → B', so we assume 'A' and pursue the subgoal of deducing 'B'. 'B' is the consequent of (1), so we can obtain it by →E if we can obtain the antecedent of (1), which we accomplish at (3)

1	(1)	$(A \lor (D \,\&\, E)) \to B$	Premise
2	(2)	A	Assumption
2	(3)	$A \lor (D \,\&\, E)$	2 ∨I
1,2	(4)	B	1,3 →E
1	(5)	A → B	2,4 →I ◆

The intuitive justification for line 3 is that we already have 'A' at (2), and if 'A' holds according to the premises and assumptions on which line 2 depends, then by the truth-table for '∨', *any* disjunction with 'A' as a disjunct will hold according to those premises and assumptions. We state the rule as follows:

Rule of ∨-Introduction: For any formula p, if p has been inferred at line j, then for any formula q, either $\ulcorner p \lor q \urcorner$ or $\ulcorner q \lor p \urcorner$ may be inferred at line k, labeling the line 'j, ∨I' and writing on its left the same premise and assumption numbers as on the left of j. Schematically:

$a_1,...,a_n$ (j) p $a_1,...,a_n$ (j) p

                        ⋮   OR   ⋮

$a_1,...,a_n$ (k) $p \lor q$ j ∨I $a_1,...,a_n$ (k) $q \lor p$ j ∨I

The rule of ∨-Elimination is somewhat more complex. What can we infer from a disjunction $\ulcorner p \lor q \urcorner$? Certainly not p by itself, nor q by itself. To see what is possible, let $\ulcorner p \lor q \urcorner$ be the sentence 'either water is H_2O or water is H_4O_2', and suppose that we are in the early days of chemistry and do not know which disjunct is true, though (somehow) we know the disjunction. One thing which follows from the disjunction is that water contains hydrogen, and the reason it follows is that it follows from *both* disjuncts, so that it does not matter which one of the two is true: assuming that it is the first disjunct which is true, then water contains hydrogen, and assuming that it is the second which is true, then water still contains hydrogen. We could set this out semi-formally as follows:

1	(1)	Water is H_2O or water is H_4O_2	Premise
2	(2)	Water is H_2O	Assumption
2	(3)	Water contains hydrogen	From (2)
4	(4)	Water is H_4O_2	Assumption
4	(5)	Water contains hydrogen	From (4)
1	(6)	Water contains hydrogen	From (1)

Lines (2)–(5) are numbered on the left as would be expected, but at line 6 we revert to whatever (1) depends on. The reason is that once we have shown that 'water contains hydrogen' follows from each disjunct separately, we have shown that it follows from the disjunction itself, so we repeat the statement, as at (6), asserting that it depends only on whatever the disjunction depends upon. Therefore we discharge the assumptions of the individual disjuncts: the statement 'water contains hydrogen' at line 6 no longer depends on assuming the truth of (2) or (4).

On the other hand, if all we know is 'either water is H_2O or water is some other substance with oxygen in its chemical composition', then we cannot infer 'water contains hydrogen' since we cannot infer it from the second disjunct, 'water is some other substance with oxygen in its chemical composition'. This example is paradigmatic of when something cannot be deduced from a disjunction, so along with the previous example, we can make a first approximation to a rule of ∨-Elimination, that a statement *r* follows from a disjunction ⌜*p* ∨ *q*⌝ if and only if *r* follows from *p* and *r* follows from *q*. Consequently, if we are trying to derive a statement *r* from a disjunction ⌜*p* ∨ *q*⌝ (*r* is called the *target* formula) we have to show that *r* follows from *p* and that it follows from *q*. We do this by assuming *p* and deriving *r* and then by assuming *q* and deriving *r*; this done, we can say that *r* follows from ⌜*p* ∨ *q*⌝. At the end of the process, *r* depends on whatever ⌜*p* ∨ *q*⌝ depends on. In particular, though we assumed *p* in order to show that *r* follows from it, and likewise assumed *q*, at the end of the process where we claim to have shown that *r* follows from the disjunction in virtue of following from each disjunct separately, *r* does not depend on either of the assumptions *p* and *q*, but only on the disjunction (or whatever it depends upon) itself.

However, our example of a successful inference from a disjunction is rather special, in that no other information is needed to infer *r* ('water contains hydrogen') from each disjunct ('water is H_2O' and 'water is H_4O_2'), above and beyond the information in that disjunct. A different example might involve an *r* which follows from each disjunct only with the aid of further premises from chemistry, for example, *r* might be 'water contains the simplest of all the elements'. In that case, deducing *r* would require the extra premise 'hydrogen is the simplest of all elements' so r would depend both on the disjunction and on this premise. Thus the correct account of what r depends on is that when r is shown to follow from p and to follow from q, this means that it follows from $\ulcorner p \vee q \urcorner$ and depends on whatever premises and assumptions $\ulcorner p \vee q \urcorner$ depends upon, but also on whatever other premises and assumptions are used in obtaining r from p, except p itself, and in obtaining r from q, except q itself.

Here is a simple proof which illustrates reasoning from a disjunction:

Example 2: Show (A & B) ∨ (A & C) ⊢$_{NK}$ A.

To derive the target formula 'A' from the disjunction '(A & B) ∨ (A & C)', we show that 'A' follows from each disjunct separately. This involves assuming the first disjunct, deriving 'A' from it, and then assuming the second disjunct and deriving 'A' from it. Once we have done that we repeat 'A' and adjust premise and assumption numbers to indicate that 'A' has been shown to follow from the disjunction itself; it is this step that is labeled ∨-Elimination, '∨E' for short.

1	(1)	(A & B) ∨ (A & C)	Premise
2	(2)	A & B	Assumption
2	(3)	A	2 &E
4	(4)	A & C	Assumption
4	(5)	A	4 &E
1	(6)	A	1,2,3,4,5 ∨E ◆

This proof exhibits the five-number format for labeling lines inferred by ∨E. The first number is the number of the disjunction itself, 1; the second is the number of the line, 2, where the first disjunct is assumed; the third is the number of the line, 3, where the target formula is inferred from the first disjunct (perhaps with the aid of other premises and assumptions, though not in this example); the fourth number is the number of the line, 4, where the second disjunct is assumed; and the fifth number is the number of the line, 5, where the target formula is derived from the second disjunct (perhaps with the aid of other premises and assumptions, though not in this example). The line labeled by '∨E' is the line where we assert the target formula a third time, altering premise and assumption numbers to indicate that we are now claiming that the target formula follows from the disjunction itself. The rule for writing numbers on the left of a line labeled '∨E' is this: first, write in the numbers from the left of the line where the disjunction itself occurs (in our example, this means we write the number 1 on the left of (6)); second, write in all numbers from the left of the line where the target formula is derived from the first disjunct of the disjunction, but *excluding* the number of the line where that disjunct is assumed (in our example, this step adds no numbers on the left of (6)); third, write in all numbers from the left of the line where the target formula is derived from the second disjunct of the disjunction, but *excluding* the number of the line where that disjunct is assumed (in our example, this step also adds no numbers on the left of (6)). Following this procedure in our example means that only the number 1 appears on the left of (6).

Generalizing this example, we may state the rule of ∨E as follows:

Rule of ∨-Elimination: If a disjunction $\ulcorner p \vee q \urcorner$ occurs at a line g, p is assumed at line h, r is derived at line i, q is assumed at line j and r is derived at line k, then at line m we may infer r, labeling the line 'g,h,i,j,k ∨E' and writing on its left every number on the left at line g, and at line i (except h), and at line k (except j). Schematically:

$$
\begin{array}{lll}
a_1,\ldots,a_n & (g) & p \vee q \\
 & \vdots & \\
h & (h) & p \qquad\qquad\qquad \text{Assumption} \\
 & \vdots & \\
b_1,\ldots,b_u & (i) & r \\
 & \vdots & \\
j & (j) & q \qquad\qquad\qquad \text{Assumption} \\
 & \vdots & \\
c_1,\ldots,c_w & (k) & r \\
 & \vdots & \\
X & (m) & r \qquad\qquad\qquad \text{g,h,i,j,k } \vee\text{E}
\end{array}
$$

where X is the set $\{a_1,\ldots,a_n\} \cup \{b_1,\ldots,b_u\}/h \cup \{c_1,\ldots,c_w\}/j$.

In this statement of the rule, the symbol '∪' stands for set-theoretic *union*. Y ∪ Z is the set consisting in all the members of Y and all the members of Z, so $\{a_1,\ldots,a_n\} \cup \{b_1,\ldots,b_u\}/h \cup \{c_1,\ldots,c_w\}/j$ is the set of all numbers on the left at line g, together with all numbers on the left at line i except h if it is one of them, together with all numbers on the left at line k except j if it is one of them. h need not be one of $b_1,\ldots,b_u$ nor j one of $c_1,\ldots,c_w$; this occurs when the assumed disjunct is not used in deriving the target formula.

The rule allows h = i or j = k, which occurs when the target formula is itself one of the disjuncts. A case where h = i is illustrated by the shortest proof of the sequent A ∨ B, ~B ⊢$_{NK}$ A, which embodies a familiar way in which disjunction and negation interact in inference—if a disjunction is true and one of its disjuncts is false, it must be the other disjunct which is true (recall Example K, page 8).

Example 3: Show A ∨ B, ~B ⊢$_{NK}$ A.

1	(1)	A ∨ B	Premise
2	(2)	~B	Premise
3	(3)	A	Assumption
4	(4)	B	Assumption
5	(5)	~A	Assumption
2,4	(6)	⅄	2,4 ~E
2,4	(7)	~~A	5,6 ~I
2,4	(8)	A	7 DN
1,2	(9)	A	1,3,3,4,8 ∨E ◆

The proof works by using ∨E on line 1, taking 'A' as target formula. That is, the idea is to infer 'A' from the first disjunct of (1) and then again from the second. Since 'A' *is* the first disjunct of (1), that part of the process goes quickly. Inferring 'A' from the second disjunct 'B' requires that we use the other premise of the sequent, '~B'. The reasoning in this part of the ∨E is the same as that in Example 4.5 (page 107).

Here is a slightly more complex example, which additionally illustrates a common exception to our rule of thumb that the last line of a proof will be inferred by the introduction rule for the main connective of the conclusion.

Example 4: Show A & (B ∨ C) ⊢_{NK} (A & B) ∨ (A & C).

A first thought about strategy might be that we should try to derive one or other of the conclusion-disjuncts 'A & B' or 'A & C' from the premise, and then use ∨I to obtain the target formula '(A & B) ∨ (A& C)'. If this is to work, then the first line of the proof is

      1   (1) A & (B ∨ C)   Premise

and the second-last line of the proof has to be either (*) or (**) below:

     1   (n–1) A & B   (*)

     1   (n–1) A & C   (**)

if ∨I is to produce '(A & B) ∨ (A& C)' on the last line, depending just on the premise of the sequent. But this will not work. The reason is that our rules of inference are deliberately chosen to ensure that *we cannot use them to prove a sequent unless the conclusion of the sequent is a semantic consequence of the premises of the sequent*. In our example, therefore, the suggested ∨I strategy will work only if either 'A & B' or 'A & C' is a semantic consequence of 'A & (B ∨ C)', that is, only if (†) or (††) holds:

     A & (B ∨ C) ⊨ (A & B)   (†)

     A & (B ∨ C) ⊨ (A & C)   (††)

But by considering the assignment of ⊤ to 'A' and 'C', ⊥ to 'B', we see that A & (B ∨ C) ⊭ (A & B), and by considering the assignment ⊤ to 'A' and 'B', ⊥ to 'C', we see that A & (B ∨ C) ⊭ (A & C). Consequently, in a correct proof of Example 4, the last line will not be obtained by ∨I.

However, one of the conjuncts of the premise is a disjunction, 'B ∨ C', on which we can use ∨E, and ∨I will often give us the target formula *within* an application of ∨E. That is, we should try to derive the target formula from 'B', and then from 'C', using the first conjunct 'A' of the premise where necessary, and this will show that the target formula follows from 'B ∨ C', together with 'A'. Here is the proof:

1	(1)	A & (B ∨ C)	Premise
1	(2)	A	1 &E
1	(3)	B ∨ C	1 &E
4	(4)	B	Assumption
1,4	(5)	A & B	2,4 &I
1,4	(6)	(A & B) ∨ (A & C)	5 ∨I
7	(7)	C	Assumption
1,7	(8)	A & C	2,7 &I
1,7	(9)	(A & B) ∨ (A & C)	8 ∨I
1	(10)	(A & B) ∨ (A & C)	3,4,6,7,9 ∨E ◆

To repeat the idea: we are trying to derive '(A & B) ∨ (A & C)' and we have lines 2 and 3 to work with. Since (3) is a disjunction, ∨E is called for. This means that we need to show that '(A & B) ∨ (A & C)' can be inferred from the first disjunct of (3), and also from the second, in each case using any other lines which are helpful. So at line 4 we assume the first disjunct of (3) and at line 6 we show that '(A & B) ∨ (A & C)' follows from it, with the help of (2). At line 7 we assume the second disjunct and show at line 9 that '(A & B) ∨ (A & C)' follows from it, also with the help of (2). At line 10, therefore, we may assert that '(A & B) ∨ (A & C)' follows from the disjunction at (3), with the help of (2), labeling (10) with the five relevant numbers, as given by the schema for ∨E (page 112). We then adjust assumption numbers, using the rule stated above: on the left of (10) we write the numbers the disjunction depends upon, the numbers on the left of (6) except for 4, and the numbers on the left of (9) except for 7.

With this rule we have assembled all the rules for the first four of our connectives (we have still to discuss '↔'). We are now in a position to construct some rather more complex proofs:

Example 5: Show ∼A & ∼B ⊢_NK ∼(A ∨ B).

Since the conclusion is negative, we will try to obtain it by using ∼I, which means that we should assume 'A ∨ B' with a view to proving 'λ' by ∼E, since we can then derive '∼(A ∨ B)' by ∼I. However, since 'A ∨ B' is a disjunction, we need to use ∨E to obtain 'λ' from it, which means we need to obtain 'λ' from each disjunct separately, with the assistance of the premise, '∼A & ∼B', in whatever way it may be helpful. Here is the proof:

1	(1)	∼A & ∼B	Premise
2	(2)	A ∨ B	Assumption
3	(3)	A	Assumption
1	(4)	∼A	1 &E
1,3	(5)	λ	3,4 ∼E
6	(6)	B	Assumption
1	(7)	∼B	1 &E
1,6	(8)	λ	6,7 ∼E
1,2	(9)	λ	2,3,5,6,8 ∨E
1	(10)	∼(A ∨ B)	2,9 ∼I ◆

Our next example looks trivial but actually involves something quite tricky, an application of ∨E within another application of ∨E.

Example 6: Show A ∨ (B ∨ C) ⊢_{NK} (A ∨ B) ∨ C.

Like Example 4, neither disjunct of the conclusion is by itself a semantic consequence of the premise (why not?), so the last line of the proof cannot be obtained straightforwardly by ∨I from one or the other disjunct. Instead, we have to use ∨I within an application of ∨E. So our subgoals are to obtain '(A ∨ B) ∨ C' from each disjunct of the premise by itself. The extra complication is that the second disjunct of the premise *is another disjunction,* so ∨E will have to be used on it as well. Here is the proof of Example 6:

1	(1)	A ∨ (B ∨ C)	Premise
2	(2)	A	Assumption
2	(3)	A ∨ B	2 ∨I
2	(4)	(A ∨ B) ∨ C	3 ∨I
5	(5)	B ∨ C	Assumption
6	(6)	B	Assumption
6	(7)	A ∨ B	6 ∨I
6	(8)	(A ∨ B) ∨ C	7 ∨I
9	(9)	C	Assumption
9	(10)	(A ∨ B) ∨ C	9 ∨I
5	(11)	(A ∨ B) ∨ C	5,6,8,9,10 ∨E
1	(12)	(A ∨ B) ∨ C	1,2,4,5,11 ∨E ◆

At line 4 we obtain the target formula from the first disjunct of (1) (which is assumed at line 2). At line 5 we assume the second disjunct of (1) and obtain the target formula from it at line 11, but since (5) is a disjunction, this second part of the main application of ∨E involves a subsidiary ∨E using the disjuncts of (5): the first disjunct is assumed at (6), the target formula derived at (8), the second is assumed at (9) and the target formula derived at (10). So line (11) is simultaneously the conclusion of the subsidiary ∨E and the completion of the second part of the main ∨E.

Finally, we prove another theorem:

Example 7: Show ⊢_{NK} A ∨ ~A.

This theorem is sometimes called the Law of Excluded Middle, and its theoremhood is a reflection of the truth-table for negation and the Principle of Bivalence ('every sentence is either true or false'). Since it is a theorem, we have to begin its proof with an assumption, and it seems that our only option is to assume '~(A ∨ ~A)' with a view to deriving 'ʌ' and then using ~I and DN. But '~(A ∨ ~A)' is a negative formula, so heeding the third point of the list at the end of §4, it is likely to be one of a pair of contradictory formulae for an application of ~E. We should therefore attempt to infer 'A ∨ ~A'. To do this we will have to make another assumption, one which we can also discharge in some way, and one which enables us to derive 'A ∨ ~A'. The assumption 'A' fits the description. Here is the proof:

1	(1)	~(A ∨ ~A)	Assumption
2	(2)	A	Assumption
2	(3)	A ∨ ~A	2 ∨I
1,2	(4)	⋏	1,3 ~E
1	(5)	~A	2,4 ~I
1	(6)	A ∨ ~A	5 ∨I
1	(7)	⋏	1,6 ~E
	(8)	~~(A ∨ ~A)	1,7 ~I
	(9)	A ∨ ~A	8 DN ◆

Notice that it would be wrong to stop at line 6, since (6) depends on (1), but our goal is to derive 'A ∨ ~A' depending on *no* premises or assumptions. Also, the use of DN at (9) cannot be avoided. Just as the rule of DN is characteristic of classical logic, so is the theoremhood of 'A ∨ ~A': in the alternatives to classical logic which reject DN, such as intuitionistic logic, the Law of Excluded Middle is not a theorem.

❑ Exercises

Show the following ('$p \dashv\vdash_{NK} q$' abbreviates '$p \vdash_{NK} q$ and $q \vdash_{NK} p$'):

   (1) (A ∨ B) → C ⊢$_{NK}$ A → C
   (2) A → B, (B ∨ C) → D, D → ~A ⊢$_{NK}$ ~A
   (3) A ∨ B ⊢$_{NK}$ B ∨ A
*(4) A ∨ ~~B ⊢$_{NK}$ A ∨ B
   (5) A ∨ A ⊢$_{NK}$ A
   (6) A → (B ∨ C), ~B & ~C ⊢$_{NK}$ ~A
   (7) A ∨ ⋏ ⊢$_{NK}$ A
   (8) A ∨ B, A → B, B → A ⊢$_{NK}$ A & B
   (9) (A & B) ∨ (A & C) ⊢$_{NK}$ A & (B ∨ C)
*(10) A ∨ B ⊢$_{NK}$ (A → B) → B
  (11) (A → ⋏) ∨ (B → ⋏), B ⊢$_{NK}$ ~A
  (12) ~(A → ~A), ~(A ∨ B) ⊢$_{NK}$ ⋏
*(13) A ∨ B ⊢$_{NK}$ ~(~A & ~B)
  (14) A & B ⊣⊢$_{NK}$ ~(~A ∨ ~B)
  (15) ~(~A & ~B) ⊢$_{NK}$ A ∨ B
  (16) ~A ∨ ~B ⊣⊢$_{NK}$ ~(A & B)
  (17) A → B ⊣⊢$_{NK}$ ~A ∨ B
  (18) ~(A → B) ⊣⊢$_{NK}$ ~(~A ∨ B)
  (19) (A ∨ B) ∨ C ⊢$_{NK}$ A ∨ (B ∨ C)
  (20) A ∨ (B & C) ⊣⊢$_{NK}$ (A ∨ B) & (A ∨ C)
  (21) ~(A ∨ (B & C)) ⊢$_{NK}$ ~(A ∨ B) ∨ ~(A ∨ C)
  (22) A → (~B ∨ ~C), ~C → ~D, B ⊢$_{NK}$ ~A ∨ ~D

6 The biconditional

The remaining connective for which we have to give rules is '↔'. We could provide the biconditional with analogues of →I and →E. ↔E would be: given a biconditional and one of its sides, infer the other, pooling assumptions; and ↔I would be: if q can be inferred from auxiliary premises and assumptions X and p as assumption, and if p can be inferred from auxiliary premises and assumptions Y and q as assumption, then $\ulcorner p \leftrightarrow q \urcorner$ can be inferred from auxiliary premises and assumptions X ∪ Y. However, when we provided '↔' with a truthtable in §1 of Chapter 3, we derived the table via the table for a conjunction of conditionals. Although in the syntax of LSL we took the shortcut of treating '↔' as if it were a *bona fide* binary connective in its own right, we shall otherwise work out its properties by relating it to the conditional form it abbreviates. So in NK, instead of the I and E rules described above, we shall employ a *rule of definition* which allows us to replace any LSL conjunction of conditionals with its notational abbreviation using '↔' (which we think of as an abbreviation in LSL) and to expand any abbreviation into the formula it abbreviates, in both cases carrying over the same premises and assumptions. This amounts to the following rule:

> **Rule of Definition for '↔':** If $\ulcorner (p \rightarrow q) \,\&\, (q \rightarrow p) \urcorner$ occurs as the entire formula at a line j, then at line k we may write $\ulcorner p \leftrightarrow q \urcorner$, labeling the line 'j, Df' and writing on its left the same numbers as are on the left of j. Conversely, if $\ulcorner p \leftrightarrow q \urcorner$ occurs as the entire formula at a line j, then at line k we may write $\ulcorner (p \rightarrow q) \,\&\, (q \rightarrow p) \urcorner$, labeling the line 'j, Df' and writing on its left the same numbers as are on the left of j.

Consequently, to infer a biconditional we would aim to deduce the corresponding conjunction of conditionals, normally using &I. This requires each conditional to be obtained by itself, so we would expect in the typical case to use →I twice: assume the antecedent of the first conditional, derive its consequent, apply →I, then repeat the process for the second conditional. Notice that the rule Df requires that the occurrence of '↔' being expanded be the main connective of the formula on line j, and that the '&' of a conjunction being abbreviated be the main connective. So expanding 'A & (B ↔ C)' into 'A & ((B → C) & (C → B))' would *not* be a legitimate application of Df—one must use &E first. Here is a simple example of a proof employing Df:

Example 1: Show ⊢$_{NK}$ A ↔ A.

1	(1)	A	Assumption
	(2)	A → A	1,1 →I
	(3)	(A → A) & (A → A)	2,2 &I
	(4)	A ↔ A	3 Df ◆

Since the problem is to derive the biconditional (4), we aim at proving the appropriate conjunction of conditionals (3).

Proofs involving biconditionals can often be long without requiring much ingenuity. The main problem is to keep track of where one is in the overall plan of execution of the proof. Here is an example:

Example 2: Show A ↔ ~B ⊢$_{NK}$ ~(A ↔ B).

Since the conclusion is a negative formula we are likely to obtain it by ~I, which requires that we assume 'A ↔ B' and infer '⅄', then use ~I to discharge the assumption (line 20 below). The premise and the assumption together give us four conditionals to work with (lines 3 and 4) but there is not much one can do with conditionals by themselves. So we must make a further assumption to get going, one which can also be discharged by ~I (line 11).

1	(1)	A ↔ ~B	Premise
2	(2)	A ↔ B	Assumption
2	(3)	(A → B) & (B → A)	2 Df
1	(4)	(A → ~B) & (~B → A)	1 Df
5	(5)	A	Assumption
2	(6)	A → B	3 &E
1	(7)	A → ~B	4 &E
2,5	(8)	B	5,6 →E
1,5	(9)	~B	5,7 →E
1,2,5	(10)	⅄	8,9 ~E
1,2	(11)	~A	5,10 ~I
2	(12)	B → A	3 &E
13	(13)	B	Assumption
2,13	(14)	A	12,13 →E
1,2,13	(15)	⅄	11,14 ~E
1,2	(16)	~B	13,15 ~I
1	(17)	~B → A	4 &E
1,2	(18)	A	16,17 →E
1,2	(19)	⅄	11,18 ~E
1	(20)	~(A ↔ B)	2,19 ~I ◆

Though lengthy, this proof is not devious. The overall strategy is to infer '⅄' (line 19) from 'A ↔ B' as assumption. For this to give '~(A ↔ B)' *depending only on line 1*, '⅄' must depend *only on lines 1 and 2*, as it does at (19), but not at (15) or (10). To obtain '⅄' depending just on lines 1 and 2, we use Df to expand (1) and (2). This gives us four conditionals, which inspection confirms cannot all be true. For the first conjuncts of (3) and (4) imply that 'A' is false, while the second conjuncts imply that 'A' is true. So if we assume 'A' we should be able to reach an explicit contradiction and thus '⅄' by ~E. The idea is to use the first conjuncts of (3) and (4) to reach '⅄' and so to reject 'A', yielding '~A' depending just on (1) and (2) (line 11) and then to use the second conjuncts to reach '⅄' again, still depending just on (1) and (2), so that we can reject (2) (lines 19 and 20). In order to get '⅄' the second time we have to make an inner application of ~I, as in lines 13-16.

This completes our core presentation of NK. We now turn to discussion of some simplifications and elaborations, after giving a summary of advice for doing proofs.

❏ Exercises

Show the following ('$p \dashv\vdash_{NK} q$' abbreviates '$p \vdash_{NK} q$ and $q \vdash_{NK} p$'):

(1) $A, A \leftrightarrow B \vdash_{NK} B$
(2) $A \leftrightarrow B \vdash_{NK} B \leftrightarrow A$
(3) $(A \& B) \leftrightarrow A \vdash_{NK} A \rightarrow B$
*(4) $(A \lor B) \leftrightarrow A \vdash_{NK} B \rightarrow A$
(5) $\sim A, A \leftrightarrow B \vdash_{NK} \sim B$
(6) $\sim A \leftrightarrow \sim B \dashv\vdash_{NK} A \leftrightarrow B$
(7) $\sim(A \leftrightarrow B) \dashv\vdash_{NK} \sim A \leftrightarrow B$
(8) $A \& B \vdash_{NK} A \leftrightarrow B$
(9) $A \leftrightarrow \sim A \dashv\vdash_{NK} A \& \sim A$
*(10) $(A \lor B) \lor C, B \leftrightarrow C \vdash_{NK} C \lor A$
(11) $A \rightarrow B, B \rightarrow C, C \rightarrow A \vdash_{NK} (A \leftrightarrow B) \& ((B \leftrightarrow C) \& (C \leftrightarrow A))$
(12) $A \leftrightarrow (B \lor C), B \rightarrow D, C \rightarrow E \vdash_{NK} D \lor (E \lor \sim A)$
(13) $(A \& B) \leftrightarrow (A \& C) \dashv\vdash_{NK} A \rightarrow (B \leftrightarrow C)$
(14) $\sim A \lor C, \sim B \lor \sim C \vdash_{NK} A \rightarrow \sim(A \leftrightarrow B)$
(15) $A \leftrightarrow (B \leftrightarrow C) \dashv\vdash_{NK} (A \leftrightarrow B) \leftrightarrow C$
(16) $A \leftrightarrow B \dashv\vdash_{NK} (A \& B) \lor (\sim A \& \sim B)$

7 Heuristics

Here are some rules of thumb ('heuristics') to remember when constructing proofs:

- If the conclusion of the sequent contains connectives, it is likely that the last line of the proof will be obtained by the introduction rule for the conclusion's main connective (the common exception is '∨'). This should indicate what assumptions, if any, need to be made, and what other formulae need to be derived. If these other formulae also contain connectives, iterate this rule of thumb on them. In this way, construct the proof backward from the conclusion as far as possible.
- When the above rule can no longer be applied, inspect the premises to see if they have any *obvious* consequences (e.g. if &E or →E can be applied to them).
- If you have applied the previous heuristics and still cannot obtain the formula p you want, try assuming $\ulcorner\sim p\urcorner$ and aim for $\ulcorner\sim\sim p\urcorner$ by ∼E, ∼I; then use DN.

- In constructing a proof, any assumptions you make must eventually be discharged, so you should only make assumptions in connection with the three rules which discharge assumptions. In other words, if you make an assumption p in a proof, you *must* be able to give one of the following reasons: (a) p is the antecedent of a conditional you are trying to derive; (b) you are trying to derive $\ulcorner{\sim}p\urcorner$, so you assume p with a view to using ~I; (c) you are using ∨E and p is one of the disjuncts of the disjunction to which you will be applying ∨E. *If you make an assumption and cannot justify it by one of (a), (b) or (c), you are almost certainly pursuing the wrong strategy.*

8 Sequent and Theorem Introduction

Logic, like other branches of knowledge, should be cumulative—we should be able to use what we already know in making new discoveries. At the moment, there is no way to do this. For instance, recall Example 4.1:

Example 1: Show A → B, ~B ⊢$_{NK}$ ~A.

1	(1)	A → B	Premise
2	(2)	~B	Premise
3	(3)	A	Assumption
1,3	(4)	B	1,3 →E
1,2,3	(5)	⋏	2,4 ~E
1,2	(6)	~A	3,5 ~I ◆

Now suppose we are asked to prove the following:

Example 2: Show C → D, ~D ⊢$_{NK}$ ~C.

Though Example 2 is not *literally* the same as Example 1, there is a clear sense in which it is 'essentially' the same. A proof of Example 2 can be obtained simply by going through Example 1's proof, substituting 'C' for 'A' and 'D' for 'B'. We will forego writing this proof out. Next, consider the following problem:

Example 3: Show R → (V → (S ∨ T)), ~(S ∨ T) ⊢$_{NK}$ R → ~V.

1	(1)	R → (V → (S ∨ T))	Premise
2	(2)	~(S ∨ T)	Premise
3	(3)	R	Assumption
1,3	(4)	V → (S ∨ T)	1,3 →E
5	(5)	V	Assumption
1,3,5	(6)	S ∨ T	4,5 →E
1,2,3,5	(7)	⋏	2,6 ~E
1,2,3	(8)	~V	5,7 ~I
1,2	(9)	R → ~V	3,8 →I ◆

This proof involves much reinventing of the wheel, since lines 2 and 4–8 recapitulate the proof of Example 1. These lines can be obtained from Example 1 by substituting 'V' for 'A' and 'S ∨ T' for 'B'. So it is inefficient to write them out: it ought to be possible to simplify the proof of a new sequent when part or all of its proof can be obtained by making substitutions in a proof already done.

The technique we need here is known as *Sequent Introduction*. First we define the idea of one sequent's being 'essentially the same' as another, or as we now put it, of its being a *substitution-instance* of another. The idea is that the sequents

(a) $C \rightarrow D, \sim D \vdash_{NK} \sim C$

and

(b) $(V \rightarrow (S \vee T)), \sim(S \vee T) \vdash_{NK} \sim V$

are essentially the same as the sequent

(c) $A \rightarrow B, \sim B \vdash_{NK} \sim A$

because each of (a) and (b) can be obtained from (c) by substituting specific formulae for the sentence-letters in (c). (a) is obtained by substituting 'C' for 'A' and 'D' for 'B' in (c) while (b) is obtained by substituting 'V' for 'A' and 'S ∨ T' for 'B' in (c). For this reason, (a) and (b) are called *substitution-instances* of (c). This general criterion is given in the following rather wordy definition:

> Let $\pi_1,...,\pi_t$ be the sentence-letters other than '$\wedge$' in the formulae $p_1,...,p_n$ and q. Then the sequent $r_1,...,r_n \vdash_{NK} s$ is said to be a *substitution-instance* of the sequent $p_1,...,p_n \vdash_{NK} q$ if and only if there are (not necessarily distinct) formulae $u_1,...,u_t$ such that if u_j replaces π_j throughout $p_1,...,p_n$ and q, $1 \leqslant j \leqslant t$, then the sequent $r_1,...,r_n \vdash_{NK} s$ results.

When $r_1,...,r_n \vdash_{NK} s$ is a substitution-instance of $p_1,...,p_n \vdash_{NK} q$, we also say that each r_i is a substitution-instance of the corresponding p_i and s is a substitution-instance of q. Thus (a) is a substitution-instance of (c), since we have $n = t = 2$, $\pi_1 = $ 'A', $\pi_2 = $ 'B', $p_1 = $ 'A → B', $p_2 = $ '∼B', $q = $ '∼A', $u_1 = $ 'C' and $u_2 = $ 'D'. Putting u_1 and u_2 for π_1 and π_2 respectively in sequent (c), A → B, ∼B $\vdash_{NK}$ ∼A, yields sequent (a), C → D, ∼D $\vdash_{NK}$ ∼C, so the latter is a substitution-instance of the former, as desired. Similarly, (b) is a substitution-instance of (c), since we have $n = t = 2$, $\pi_1 = $ 'A', $\pi_2 = $ 'B', $p_1 = $ 'A → B', $p_2 = $ '∼B', $q = $ '∼A', $u_1 = $ 'V' and $u_2 = $ 'S ∨ T'. Putting u_1 and u_2 for π_1 and π_2 respectively in (c) yields (b), which is what it means for (b) to be a substitution-instance of (c). Another example: the sequent

(d) $\sim(\sim(R \& S) \& \sim(U \leftrightarrow \sim V)) \vdash_{NK} (R \& S) \vee (U \leftrightarrow \sim V)$

is a substitution-instance of the sequent

(e) $\sim(\sim A \ \& \ \sim B) \vdash_{NK} A \lor B,$

since if we replace 'A' in (e) by '(R & S)' and 'B' by '(U ↔ ~V)', we obtain (d).

Notice that the definition of 'substitution-instance' allows some or all of the u_j to be the same; in other words, we can substitute the same formulae for different sentence-letters, though we cannot substitute different formulae for the same sentence-letter. Note also that we cannot replace 'ʎ'. For example, if we counted A ∨ B ⊢_{NK} A a substitution-instance of A ∨ ʎ ⊢_{NK} A, then putting 'B' for 'ʎ' throughout a proof of A ∨ ʎ ⊢_{NK} A (Exercise 5.7) should yield a proof of A ∨ B ⊢_{NK} A. But it does not. Indeed, there are no proofs of A ∨ B ⊢_{NK} A, since A ∨ B ⊬ A.

The method of sequent introduction allows us to move from any lines in a proof to a new line if the corresponding sequent is a substitution-instance of one we have already proved. The new line depends on all premises and assumptions which the lines we are using depend on. More precisely:

> **Rule of Sequent Introduction:** Suppose the sequent $r_1,...,r_n \vdash_{NK} s$ is a substitution-instance of the sequent $p_1,...,p_n \vdash_{NK} q$, that we have already proved the sequent $p_1,...,p_n \vdash_{NK} q$, and that the formulae $r_1,...,r_n$ occur at lines $j_1,...,j_n$ in a proof. Then we may infer s at line k, labeling the line '$j_1,...,j_n$ SI ⟨Identifier⟩' and writing on the left all the numbers which occur on the left of lines $j_1,...,j_n$. As a special case, when $n = 0$ and $\vdash_{NK} s$ is a substitution-instance of some theorem $\vdash_{NK} q$ which we have already proved, we may introduce a new line k into a proof with the formula s at it and no numbers on the left, labeling the line 'TI ⟨Identifier⟩'.

In this statement of the rule, 'TI' stands for 'Theorem Introduction', and the placeholder '⟨Identifier⟩' stands for a reference to the previously proved sequent or theorem which is being used.

We can now solve Examples 2 and 3 quite rapidly. As an identifier for the sequent A → B, ~B ⊢_{NK} ~A, we use its traditional Latin tag *Modus Tollens*, or MT for short. Hence:

Example 2: Show C → D, ~D ⊢_{NK} ~C.

1	(1)	C → D	Premise
2	(2)	~D	Premise
1,2	(3)	~C	1,2 SI (MT) ◆

We employ the identifier 'MT' to indicate which previously proved sequent we are using in deducing line 3; the lines we cite on the right are the lines where the premises of the sequent which is the substitution-instance of MT occur. This example is a rather special case, in that the lines which are the substitution-instances of the premises of MT, lines 1 and 2, are themselves the premises of the sequent we are trying to prove. More generally, we can use SI on any previous lines in a proof to produce a new line, not just on the premises; all that is required is that the lines in question be substitution-instances of the premises and conclusion of the sequent we use in applying SI. Thus:

Example 3: Show R → (V → (S ∨ T)), ~(S ∨ T) ⊢$_{NK}$ R → ~V.

1	(1)	R → (V → (S ∨ T))	Premise
2	(2)	~(S ∨ T)	Premise
3	(3)	R	Assumption
1,3	(4)	V → (S ∨ T)	1,3 →E
1,2,3	(5)	~V	4,2 SI (MT)
1,2	(6)	R → ~V	3,5 →I ◆

Though SI and TI allow the use of *any* previously proved sequent or theorem, the sequents and theorem listed below are the most useful (in this list, '*p* ⊣⊢$_{NK}$ *q*' abbreviates '*p* ⊢$_{NK}$ *q* and *q* ⊢$_{NK}$ *p*'):

(a)	A ∨ B, ~A ⊢$_{NK}$ B; or: A ∨ B, ~B ⊢$_{NK}$ A	(DS)
(b)	A → B, ~B ⊢$_{NK}$ ~A	(MT)
(c)	A ⊢$_{NK}$ B → A	(PMI)
(d)	~A ⊢$_{NK}$ A → B	(PMI)
(e)	A ⊢$_{NK}$ ~~A	(DN⁺)
(f)	~(A & B) ⊣⊢$_{NK}$ ~A ∨ ~B	(DeM)
(g)	~(A ∨ B) ⊣⊢$_{NK}$ ~A & ~B	(DeM)
(h)	~(~A ∨ ~B) ⊣⊢$_{NK}$ A & B	(DeM)
(i)	~(~A & ~B) ⊣⊢$_{NK}$ A ∨ B	(DeM)
(j)	A → B ⊣⊢$_{NK}$ ~A ∨ B	(Imp)
(k)	~(A → B) ⊣⊢$_{NK}$ A & ~B	(Neg-Imp)
(l)	A * B ⊢$_{NK}$ B * A	(Com)
(m)	A & (B ∨ C) ⊣⊢$_{NK}$ (A & B) ∨ (A & C)	(Dist)
(n)	A ∨ (B & C) ⊣⊢$_{NK}$ (A ∨ B) & (A ∨ C)	(Dist)
(p)	⊢$_{NK}$ A ∨ ~A	(LEM)
(q)	A * B ⊣⊢$_{NK}$ ~~A * ~~B; or: ~~A * B; or: A * ~~B	(SDN)
(r)	~(A * B) ⊣⊢$_{NK}$ ~(~~A * ~~B); or: ~(~~A * B); or: ~(A * ~~B)	(SDN)

As identifiers we can write out the sequents, or use the letters ordering this list, or else we can use the traditional names the sequents are known by, as indicated on the right. (a) is called Disjunctive Syllogism or Modus Tollendo Ponens; (b), as already remarked, is called Modus Tollens (traditionally, our rule →E is known as Modus Ponens); (c) and (d) are known as the Paradoxes of Material Implication (though they are not really paradoxical, by the account in §8 of Chapter Three); (e) is called Double Negation Addition; (f) through (i) are called De Morgan's Laws (after the mathematician who first investigated them); (j) is Implication and (k) Negation Implication; (l) is the Law of Commutation for any binary connective * *other than* '→'; (m) and (n) are the Laws of Distribution and

(p) is the Law of Excluded Middle. (q) and (r), Subformula Double Negation, require special comment. In (q), '*' stands for any binary connective, and so SDN allows adding or eliminating a '~~' prefix to one or other of the *main subformulae* of a formula—this is (q)—or the *main subformulae of the main subformula* of a negation—this is (r). By contrast, DN⁺ allows prefixing '~~' only to the whole formula. For example, applying DN⁺ to '(A & B)' would only produce '~~(A & B)', but with (q) we can, as it were, apply DN to the subformulae 'A' and 'B' without first using &E to get each by itself. So (q) could produce any of '(A & ~~B)', '(~~A & B)' and '(~~A & ~~B)'; and in the other direction, from any of these three SDN could produce '(A & B)'. Similarly, from '~(A ∨ ~~B)', (r) could produce '~(A ∨ B)' and vice versa.

Here is another example where we use SI, this time appealing to DS, to shorten a proof:

Example 4: Show R → ((S → T) ∨ V), ~(S → T), R → ~V ⊢_NK ~R.

1	(1)	R → ((S → T) ∨ V)	Premise
2	(2)	~(S → T)	Premise
3	(3)	R → ~V	Premise
4	(4)	R	Assumption
1,4	(5)	(S → T) ∨ V	1,4 →E
1,2,4	(6)	V	5,2 SI (DS)
3,4	(7)	~V	3,4 →E
1,2,3,4	(8)	⅄	7,6 ~E
1,2,3	(9)	~R	4,8 ~I ◆

The sequent (S → T) ∨ V, ~(S → T) ⊢_NK V is a substitution-instance of the sequent DS ((a) in the list on page 123) and we have already proved DS. Also, the premises of the sequent (S → T) ∨ V, ~(S → T) ⊢_NK V occur at lines 5 and 2 respectively in our proof. So at line 6 we cite these lines and say which sequent it is whose premises they instantiate. On the left at (6) we pool together all the premises and assumptions on which lines 5 and 2 depend.

It is never compulsory to use Sequent Introduction. By 'NK' we mean a particular collection of rules, so adding a new rule to NK by definition *extends* NK and gives us a new system. However, when we add a new rule to a system, it is an open question whether or not the new rule allows us to prove any *new sequents*, sequents which could not be proved without the new rule. If we extend a system by adding a rule which does *not* allow us to prove new sequents, then the extension is said to be a *conservative* extension. In adding SI to NK, we are only making a conservative extension of NK, since there is nothing we can prove using SI that we could not already prove—albeit at greater length—without. For we can eliminate any step of SI in a proof by inserting into the proof the relevant substitution-instance of the proof of the sequent used by the application of SI in question.

To illustrate Theorem Introduction, TI, which is the special case of SI with no premises, we give a proof which uses LEM. LEM is typically used in combination with ∨E.

Example 5: Show ⊢_NK (A → B) ∨ (B → A).

Intuitively, 'A' is either true or false; if 'A' is true, then '(B → A)' holds and so '(A → B) ∨ (B → A)' holds, while if 'A' is false, '(A → B)' holds and so once again '(A → B) ∨ (B → A)' holds. Either way, then, '(A → B) ∨ (B → A)' holds. This semantic argument uses the Principle of Bivalence, that 'A' is either true or false, and the table for material implication, that a true consequent or a false antecedent is sufficient for the truth of a conditional, to infer that ⊨ (A → B) ∨ (B → A). Its proof-theoretic counterpart is:

	(1)	A ∨ ~A	TI (LEM)
2	(2)	A	Assumption
2	(3)	B → A	2 SI (PMI)
2	(4)	(A → B) ∨ (B → A)	3 ∨I
5	(5)	~A	Assumption
5	(6)	A → B	5 SI (PMI)
5	(7)	(A → B) ∨ (B → A)	6 ∨I
	(8)	(A → B) ∨ (B → A)	1,2,4,5,7 ∨E ◆

One complaint about natural deduction that is sometimes made is that it often does not seem very *natural*, for example in proofs which require non-obvious assumptions and uses of DN. Those who make this complaint are contrasting formal proofs with ordinary reasoning. However, in ordinary reasoning, of which we have an exemplar in the argument in §1 of this chapter about who helped Smith open the safe, we use Sequent Introduction all the time. So some proofs earlier in this chapter, where SI was not available, seem unnatural by comparison. But when NK is equipped with SI we can produce natural proofs which exactly mirror the way one would reason in ordinary language. For example, here is such a proof of the argument about Smith's accomplice.

If the safe was opened, it must have been opened by Smith, with the assistance of Brown or Robinson. None of these three could have been involved unless he was absent from the meeting. But we know that either Smith or Brown was present at the meeting. So since the safe was opened, it must have been Robinson who helped open it.

We symbolize this argument using the following dictionary:

O: The safe was opened
S: Smith opened the safe
B: Brown assisted
R: Robinson assisted
X: Smith was absent from the meeting
Y: Brown was absent from the meeting
Z: Robinson was absent from the meeting

We obtain the following sequent:

O → (S & (B ∨ R)), (~S ∨ X) & ((~B ∨ Y) & (~R ∨ Z)), ~X ∨ ~Y, O ⊢$_{NK}$ R.

The following proof articulates exactly the principles of inference that the informal proof in §1 implicitly employs.

1	(1)	O → (S & (B ∨ R))	Premise
2	(2)	(~S ∨ X) & ((~B ∨ Y) & (~R ∨ Z))	Premise
3	(3)	~X ∨ ~Y	Premise
4	(4)	O	Premise
2	(5)	~S ∨ X	2 &E
2	(6)	(~B ∨ Y) & (~R ∨ Z)	2 &E
2	(7)	~B ∨ Y	6 &E
2	(8)	~R ∨ Z	6 &E
1,4	(9)	S & (B ∨ R)	1,4 →E
1,4	(10)	S	9 &E
1,4	(11)	B ∨ R	9 &E
1,4	(12)	~~S	10 SI (DN⁺)
1,2,4	(13)	X	5,12 SI (DS)
1,2,4	(14)	~~X	13 SI (DN⁺)
1,2,3,4	(15)	~Y	3,14 SI (DS)
1,2,3,4	(16)	~B	7,15 SI(DS)
1,2,3,4	(17)	R	11,16 SI (DS) ◆

Note carefully the need for SI at lines 12 and 14. In order to use the sequent DS, we need a disjunction and the *negation* of one of its disjuncts. (10) is not *the negation of* the first disjunct of (5), but (12) is.

As the example illustrates, with the addition of SI, we can use NK to mimic actual human reasoning, for in the latter an informal analogue of SI is used repeatedly. The system NK, then, provides us with not only a tool for executing proofs but also an idealized model of the human psychological faculty of deductive inference. If a machine with the same reasoning potential as humans is ever built, we can expect to find something like NK at the core of its program. On the other hand, to say that NK is an idealized model of the deductive reasoning faculty is not to say that people can reason deductively because rules like those of NK are encoded in their brains. The nature of the actual psychological mechanisms of deduction is an ongoing field of research in cognitive psychology; for further information, the reader should consult Johnson-Laird and Byrne. And obviously it is not being claimed that people *do* reason deductively in accordance with the rules of NK; we said, after all, that NK is an *idealized* model. There is a general distinction in cognitive psychology between *competence* and *performance*. A person's deductive competence is determined by the nature of the psychological mechanisms with which human beings are equipped. But one's individual performance is determined by many other factors, such as other aspects of one's psychological make-up, or just the kind of opportunities for practice which one has previously enjoyed. Though performance cannot exceed competence, it can fall well short of it for these kinds of reasons.

☐ Exercises

I   Some sequents exhibited below are substitution-instances of sequents in the list on page 123. For each such sequent below, identify the sequent in the list of which it is a substitution-instance. Justify your answer in every case by stating, for each sentence-letter π_i in the sequent in which substitution is made, which formula u_j has been substituted for π_i (refer to the definition of 'substitution-instance' on page 121).

(1) ~~(R & S) ∨ ~T, T ⊢$_{NK}$ ~(R & S)
(2) (A → B) → C, ~C ⊢$_{NK}$ ~(A → B)
(3) ~(~(R ∨ S) ∨ ~(~R ∨ ~S)) ⊢$_{NK}$ ~~(R ∨ S) & ~~(~R ∨ ~S)
(4) ((P → Q) ∨ R) & ((P → Q) ∨ S) ⊢$_{NK}$ (P → Q) ∨ (R & S)
*(5) ~(M ∨ N) ∨ (W & U) ⊢$_{NK}$ (M ∨ N) → (W & U)

II   Below there are two lists of sequents, and each sequent in the first list is a substitution-instance of a sequent in the second. Say which sequents are substitution-instances of which, justifying your answer in the same way as in I.

List 1:

(i)   ~(R & S) ∨ ~~(~T & S), ~W ∨ ~~T, ~(R & S) → ~~W ⊢$_{NK}$ (~T & S) → ~T
(ii)   ~~(R & S) ∨ ~(~T & S), ~~W ∨ ~T, ~~(R & S) → ~~~W
       ⊢$_{NK}$ ~(~T & S) → ~T
*(iii) ~~(R & S) ∨ ~(~T & S), ~~W ∨ ~~T, ~~(R & S)→ ~~~W
       ⊢$_{NK}$ ~(~T & S) → ~~T

List 2:

(a) A ∨ ~B, C ∨ ~D, A → ~C ⊢$_{NK}$ ~B → ~D
(b) ~~A ∨ B, ~~C ∨ D, ~~A → ~C ⊢$_{NK}$ B → D
(c) ~A ∨ ~~B, C ∨ ~~D, ~A → ~C ⊢$_{NK}$ B → ~D

III   Show the following. Wherever you apply Sequent Introduction, be sure to indicate which previously proved sequent you are using.

(1) ~A ⊢$_{NK}$ ~B → ~(A ∨ B)
(2) A → (B ∨ ~C), ~A → (B ∨ ~C), ~B ⊢$_{NK}$ ~C
(3) ⊢$_{NK}$ A ∨ (A → B)
*(4) ~B → A ⊢$_{NK}$ (B → A) → A
(5) ⊢$_{NK}$ (A → B) → [(A → ~B) → ~A]
(6) ~[A → (B ∨ C)] ⊢$_{NK}$ (B ∨ C) → A
(7) A → B, (~B → ~A) → (C → D), ~D ⊢$_{NK}$ ~C
*(8) (A ∨ B) → (A ∨ C) ⊢$_{NK}$ A ∨ (B → C)
(9) (A & B) ↔ C, ~(C ∨ ~A) ⊢$_{NK}$ ~B
(10) A → (B ∨ C) ⊣⊢$_{NK}$ (A → B) ∨ (A → C)

(11)   ~(A & ~B) ∨ ~(~D & ~E), ~(E ∨ B), C → (~E → (~D & A)) ⊢$_{NK}$ ~C
(12)   (A ∨ B) → (C & D), (~E ∨ C) → [(F ∨ G) → H],
       (~I → J) → [G & (H → ~K)] ⊢$_{NK}$ K → (~A ∨ ~I)
*(13) (A ↔ B) ↔ (C ↔ D) ⊢$_{NK}$ (A ↔ C) ↔ (B ↔ D)
(14)   (A ∨ B) & (C ∨ D) ⊢$_{NK}$ (B ∨ C) ∨ (A & D)
(15)   ~(A ↔ B), ~(B ↔ C), ~(C ↔ A) ⊢$_{NK}$ ⅄

IV   Symbolize the following arguments (state your dictionary explicitly). Then give proofs of the resulting argument-forms.

(1)   If God is omnipotent then He can do anything. So He can create a stone which is too heavy to be lifted. But that means *He* can't lift it, so there's something He can't do. Therefore, He isn't omnipotent.

(2)   If there is an empirical way of distinguishing between absolute rest and absolute motion, Newton was right to think that space is absolute, not relative. Also, if there is absolute space, there is a real difference between absolute rest and absolute motion—whether or not they are empirically distinguishable. So if, as some argue, there cannot really be a difference between absolute rest and absolute motion unless they are empirically distinguishable, an empirical way of distinguishing between absolute rest and absolute motion is necessary and sufficient for the existence of absolute space.

(3)   If God is willing to prevent evil but is unable to do so, He is impotent. If God is able to prevent evil but unwilling to do so, He is malevolent. If He is neither able nor willing, then he is both impotent and malevolent. Evil exists if and only if God is unwilling or unable to prevent it. God exists only if He is neither impotent nor malevolent. Therefore, if God exists evil does not.

9 Alternative formats for proofs

The format in which we have been setting out our proofs, taken from Lemmon, sits midway between two other formats, known respectively as *tree* format (this was Gentzen's original format) and *sequent-to-sequent* format. Lemmon format can be regarded either as a linearization of tree format or as a notational variant of sequent-to-sequent format. Since both other formats are revealing, we present them briefly here.

By contrast with parse trees and semantic tableaux, proof trees are not inverted: leaves are at the top and roots at the bottom. But like other trees, construction proceeds downward. A proof begins with a listing of premises and assumptions *across* the page, and an application of an inference rule extends a path or paths by adding a node below the current level and labeling it with the formula which that rule-application produces. In Lemmon format, the numbers of the premises and assumptions on which a formula occurrence ϕ depends are explicitly stated on its left. In tree format, what a formula occurrence ϕ depends on can be determined merely by tracing up through the proof from ϕ

and following all paths; the top nodes (with undischarged formulae) at which one arrives are the relevant premises and assumptions. For instance, the proof of Example 2.1 on page 89 can be arranged as a tree as follows:

Example 1: Show A & B, C & D, (A & D) → H ⊢$_{NK}$ H.

$$\cfrac{\cfrac{A\ \&\ B}{A}\ \&E \qquad \cfrac{C\ \&\ D}{D}\ \&E}{\cfrac{A\ \&\ D}{\qquad\qquad H\qquad\qquad}\ \&I \qquad\qquad (A\ \&\ D) \to H}\ \to E$$

We see that 'A & D' depends on 'A & B' and 'C & D', since these are the formulae at the leaves of the branches that lead upwards from the node labeled 'A & D'. More generally, in tree format the *root* formula (the conclusion) depends on whatever formulae there are on the leaves of the tree, except those which are discharged (none are discharged in this example). A *path* in a tree T is a sequence of formulae occurrences $\phi_1,...,\phi_n$ such that ϕ_1 is a leaf of T, ϕ_n is the root formula of T, and ϕ_{i+1} is the formula occurring in T below the line beneath ϕ_i. Thus the paths in Example 1 are ⟨A & B, A, A & D, H⟩, ⟨C & D, D, A & D, H⟩, and ⟨(A & D) → H, H⟩.

To discharge an assumption we draw a line over it. We indicate which step in the proof causes an assumption to be discharged by numbering the assumptions as we introduce them and using an assumption's number to label the rule-application which discharges it. Another feature of tree format is that if an assumption or premise is to be used twice, then we start *two* paths of the proof tree with that assumption or premise at the top. Both these features are illustrated by the following:

Example 2: Show A → (B & C) ⊢$_{NK}$ (A → B) & (A → C)

$$\cfrac{\cfrac{A \to (B\ \&\ C) \qquad \overline{\quad}^{(1)}\ A}{\cfrac{\cfrac{B\ \&\ C}{B}\ \&E}{A \to B}\ \to I\ (1)}\ \to E \qquad \cfrac{A \to (B\ \&\ C) \qquad \overline{\quad}^{(2)}\ A}{\cfrac{\cfrac{B\ \&\ C}{C}\ \&E}{A \to C}\ \to I\ (2)}\ \to E}{(A \to B)\ \&\ (A \to C)}\ \&I$$

In Lemmon format we would assume 'A' at line 2 and apply →I to it twice. In the corresponding proof in tree format, we put 'A' at the top of two paths, and since each of these formula occurrences will be used in an application of →E, we have to write the conditional premise 'A → (B & C)' at the top of two paths as well, one for each use of →E (hence the difference between a formula and its occurrences). The assumptions are numbered (1) and (2) respectively, and each is discharged by an application of →I. When we discharge an assumption, we indicate which assumption is being discharged by writing its number next to the rule-application that does the discharging. At that point, we draw a line over the assumption, indicating that it has been discharged. Undeniably, something of the dynamics of this process is lost in the final display.

One feature of tree format which Examples 1 and 2 illustrate is the way it factors out certain arbitrary features of a proof in Lemmon format. In Example 1 we have to apply &E to premises 1 and 2; it does not matter which application we make first, but in Lemmon format we must put one before the other. Similarly, in a Lemmon-format version of Example 2 we must decide which conjunct of the conclusion to derive first. But in tree format we can represent parts of the proof whose relative order is of no matter as proceeding in parallel.

The rule of ∨E is perspicuously represented as a schema in tree notation:

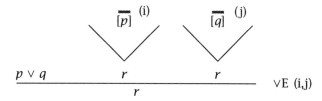

Here we represent a deduction of *r* from *p*, perhaps with the aid of other premises and assumptions (leaf formulae of paths which merge with the displayed *p*-paths before the first occurrence of *r*), and a deduction of *r* from *q*, again perhaps with aid, followed by an application of ∨E to discharge the assumptions *p* and *q*. The brackets indicate that the bracketed formula may be the leaf formula of more than one path and that all its occurrences are discharged. Here is an example which involves the negation rules as well as ∨E:

Example 3: Show A → (B ∨ C), ~B & ~C ⊢_NK ~A.

$$
\begin{array}{ccc}
& \overline{}\ (1) & \overline{}\ (2) \quad \dfrac{\sim B\ \&\ \sim C}{\sim B}\ \&E \qquad \overline{}\ (3) \quad \dfrac{\sim B\ \&\ \sim C}{\sim C}\ \&E \\[2ex]
\dfrac{A \to (B \lor C) \qquad A}{B \lor C}\ \to E & \dfrac{B \qquad \sim B}{\curlywedge}\ \sim E & \dfrac{C \qquad \sim C}{\curlywedge}\ \sim E
\end{array}
$$

$$
\dfrac{\curlywedge}{\dfrac{\curlywedge}{\sim A}\ \sim I\ (1)}\ \lor E\ (2,3)
$$

The undischarged leaf formulae are all premises of the sequent.

A formula can occur as leaf formula on two or more paths and have all its occurrences discharged by one rule-application when all the occurrences of the assumption lie on paths which merge at or before the discharging rule-application. In such a case, the occurrences of the assumption as leaf formulae may be given the same number. An example is the proof of the Law of Excluded Middle (compare the proof on page 116):

Example 4: Show $\vdash_{NK}$ A ∨ ~A.

$$
\frac{\dfrac{\rule{1cm}{1pt}\ (1)}{A}}{A \vee \text{~}A}\ \text{vI}
\qquad
\frac{\rule{3cm}{1pt}}{\text{~}(A \vee \text{~}A)}\ (2)
$$

$$
\frac{\qquad\qquad\qquad\qquad\qquad\qquad}{\curlywedge}\ \text{~E}
$$

$$
\frac{\dfrac{\rule{1cm}{1pt}}{\text{~}A}}{A \vee \text{~}A}\ \text{vI}\ \text{~I}\ (1)
\qquad
\frac{\rule{3cm}{1pt}}{\text{~}(A \vee \text{~}A)}\ (2)
$$

$$
\frac{\qquad\qquad\qquad\qquad\qquad}{\curlywedge}\ \text{~E}
$$

$$
\frac{\dfrac{\rule{1cm}{1pt}\ \text{~I}\ (2)}{\text{~~}(A \vee \text{~}A)}}{A \vee \text{~}A}\ \text{DN}
$$

In this tree, all leaf formulae are discharged, and so the root formula depends on nothing. We use '~(A ∨ ~A)' in two applications of ~E, but discharge it only once, with ~I as indicated. The one application of ~I discharges different occurrences of the assumption-formula because the paths which have their leaves labeled by those occurrences merge at the application of ~I in question.

The other format for proofs to which Lemmon format is related is sequent-to-sequent format. We presented rules of inference in previous sections as devices for deriving formulae from formulae while adjusting premise and assumption numbers on the left as a bookkeeping device. But it is quite revealing to realize that we can also regard our rules as devices for deriving *sequents* from *sequents*. This effect of the rules can be brought out by writing the relevant premise and assumption formulae on the left explicitly and putting a turnstile in the position where the line number goes in Lemmon format. Here is the proof of Example 2 above in Lemmon format (renumbered as Example 5), followed by the same proof in sequent-to-sequent format:

Example 5: Show A → (B & C) $\vdash_{NK}$ (A → B) & (A → C).

1	(1)	A → (B & C)	Premise
2	(2)	A	Assumption
1,2	(3)	B & C	1,2 →E
1,2	(4)	B	3 &E
1,2	(5)	C	3 &E
1	(6)	A → B	2,4 →I
1	(7)	A → C	2,5 →I
1	(8)	(A → B) & (A → C)	6,7 &I ◆

And now in sequent-to-sequent format:

(1)	A → (B & C) ⊢ A → (B & C)	Premise
(2)	A ⊢ A	Assumption
(3)	A → (B & C), A ⊢ B & C	1,2 →E
(4)	A → (B & C), A ⊢ B	3 &E
(5)	A → (B & C), A ⊢ C	3 &E
(6)	A → (B & C) ⊢ A → B	4 →I
(7)	A → (B & C) ⊢ A → C	5 →I
(8)	A → (B & C) ⊢ (A → B) & (A → C)	6,7 &I ◆

We see that each line in the Lemmon-format proof can be transcribed into the corresponding line in the sequent-to-sequent proof simply by replacing the numbers on the left in the Lemmon-format proof by the formulae for which they stand, and the line number by the turnstile. Notice that such a procedure always produces sequents of the form $p \vdash_{NK} p$ for premises and assumptions. If we think of Lemmon-format proofs as notational variants of sequent-to-sequent proofs, then the ultimate justification for the Rule of Assumptions is simply that in making an assumption, we are claiming no more than that a certain formula follows from itself.

Similarly, the justification for dropping an assumption-number on the left when applying →I is evident from inspection of lines 6 and 7 in the sequent-to-sequent proof above: →I moves a formula on which another depends across the turnstile and makes it the antecedent of a conditional, with the dependent formula as consequent. However, →I occasions a small divergence from simple transcription into sequent-to-sequent format, in that we need only cite a single line number on the right, the number of the sequent on which we are performing the move operation. In the same way, when applying ~I or ∨E in a sequent-to-sequent proof, one would only cite, respectively, one and three earlier lines, omitting the numbers of the assumption sequents.

Here are the sequent-to-sequent formulations of the rules of NK. We use uppercase Greek gamma, 'Γ' and uppercase Greek sigma, 'Σ', to stand for sets of LSL formulae, and 'Γ,Σ' to stand for the union of the two sets Γ and Σ (this is an alternative to the usual notation '$\Gamma \cup \Sigma$'). Recall that the union of two sets X and Y is the set whose members are all the members of X together with all the members of Y. So 'Γ,Σ' stands for the collection of all the formulae in Γ together with all those in Σ. For the set of formulae which results from removing p from Γ, if p is in Γ, we write 'Γ/p'; if p is not in Γ, then $\Gamma/p = \Gamma$.

> ***Rule of Assumptions:*** At any point in a proof, for any LSL formula p, we may introduce the sequent $p \vdash p$.
> ***Rule of &I:*** From sequents $\Gamma \vdash p$ and $\Sigma \vdash q$ we may infer the sequent $\Gamma,\Sigma \vdash p \& q$.
> ***Rule of &E:*** From the sequent $\Gamma \vdash p \& q$ we may infer the sequent $\Gamma \vdash p$ or the sequent $\Gamma \vdash q$.
> ***Rule of →I:*** From the sequent $\Gamma \vdash q$ we may infer the sequent $\Gamma/p \vdash p \to q$.

Rule of →E: From the sequents $\Gamma \vdash p \rightarrow q$ and $\Sigma \vdash p$ we may infer the sequent $\Gamma,\Sigma \vdash q$.

Rule of ~I: From the sequent $\Gamma \vdash \wedge$ we may infer the sequent $\Gamma/p \vdash \sim p$.

Rule of ~E: From the sequents $\Gamma \vdash q$, $\Sigma \vdash \sim q$ we may infer the sequent $\Gamma,\Sigma \vdash \wedge$.

Rule of DN: From the sequent $\Gamma \vdash \sim\sim p$ we may infer the sequent $\Gamma \vdash p$.

Rule of ∨I: From the sequent $\Gamma \vdash p$ we may infer either the sequent $\Gamma \vdash p \vee q$ or the sequent $\Gamma \vdash q \vee p$.

Rule of ∨E: From the sequents $\Gamma \vdash p \vee q$, $\Sigma \vdash r$ and $\Delta \vdash r$ we may infer the sequent $\Gamma,\Sigma/p,\Delta/q \vdash r$.

Rule of Df: The sequent $\Gamma \vdash p \leftrightarrow q$ may be expanded into the sequent $\Gamma \vdash (p \rightarrow q) \;\&\; (q \rightarrow p)$; and the sequent $\Gamma \vdash (p \rightarrow q) \;\&\; (q \rightarrow p)$ may be contracted into the sequent $\Gamma \vdash p \leftrightarrow q$.

❏ Exercises

*(1) Write out a schema for each rule of inference for use in tree format (see the example of ∨E in this section).

(2) Arrange the proofs of Examples 5.1–5.7 in tree format. [*(5.6)]

(3) Arrange the proofs of Examples 5.1–5.7 in sequent-to-sequent format.

10 Systems equivalent to NK

We remarked at the point of explaining the rule DN that it would be *redundant* to extend NK by adding a rule for prefixing '~~' to a wff also. Redundancy means that we would not be able to prove any *new* sequents, sequents not already provable without this rule; in the terminology of §8, the addition of such a rule would only yield a conservative extension of NK. However, not every change in NK which we might contemplate involves the *addition* of a rule: we might also consider *replacing* one of our rules by some other rule. To discuss the effects of both addition and replacement we need a more general concept than that of conservative extension, since replacement is a different kind of modification than extension.

> Two systems of natural deduction are *equivalent* if and only if every sequent provable in one is provable in the other, and vice versa.

We can show that a system S is equivalent to NK if we can show that the following two conditions hold:

- For every rule R of S which is not a rule of NK, there is some combination of rule-applications in NK which has the same effect as R.

- For every rule of R of NK which is not a rule of S, there is some combination of rule-applications in S which has the same effect as R.

We assume that every system S permits use of SI (SI is a conservative extension of any system). A combination of rule-applications in a system S has the same effect as a rule R in S' if and only if, whenever R is used in an S'-proof to infer a formula q depending on assumptions Σ, that combination of rule-applications could be used in S instead, resulting in the inference of the same q depending on the same assumptions Σ. The simplest cases involve rules which do not discharge assumptions. Suppose S is a system with no rule of &I. Then the most direct way of showing that there is a combination of rule-applications in S which has the same effect as &I in NK is to show that A, B $\vdash_S$ A & B. For then, wherever an NK-proof uses &I, we can employ SI in S using this sequent. The problem will be to show, for the particular S in question, that A, B $\vdash_S$ A & B.

If the rule that is not present in S involves discharge of assumptions, it may be less straightforward to explain how to get its effect in S. Suppose, for example, that S does not have the rule $\vee$E. What combination of rules would get the same effect as $\vee$E? A use of $\vee$E to infer r from $\ulcorner p \vee q \urcorner$ requires a derivation of r from p and then again from q. So if S contains the rule of $\rightarrow$I, then assuming that the derivation of r from p does not itself use $\vee$E (or any other rule missing from S), we can derive r from p in S and then add an extra application of $\rightarrow$I to obtain $\ulcorner p \rightarrow r \urcorner$. Similarly, we can get $\ulcorner q \rightarrow r \urcorner$ in S. So if we also have

$$\text{A} \vee \text{B}, \text{A} \rightarrow \text{C}, \text{B} \rightarrow \text{C} \vdash_S \text{C} \qquad (*)$$

we can then use SI in S to obtain r. Schematically, the S-proof will look like this:

$a_1,...,a_n$	(g)	$p \vee q$	
		$\vdots$	
h	(h)	p	Assumption
		$\vdots$	
$b_1,...,b_u$	(i)	r	
$b_1,...,b_u$/h	(i$'$)	$p \rightarrow r$	h,i $\rightarrow$I
		$\vdots$	
j	(j)	q	Assumption
		$\vdots$	
$c_1,...,c_w$	(k)	r	
$c_1,...,c_w$/j	(k$'$)	$q \rightarrow r$	j,k $\rightarrow$I
		$\vdots$	
X	(m)	r	g,i$'$,k$'$ SI (*)

where X is the set $\{a_1,...,a_n\} \cup \{b_1,...,b_u\}/h \cup \{c_1,...,c_w\}/j$. In sum, in a system S with $\rightarrow$I and the sequent (*), two applications of $\rightarrow$I and then SI using (*) yield the same result as an application of $\vee$E in NK (compare the schema for $\vee$E on page 112 with the schema just displayed). The problem will be to show, for the particular system S in question, that (*) is provable (see Exercise 10.2).

Returning now to our two conditions for a system S to be equivalent to NK, we see that the first condition guarantees that any sequent which has a proof in S has a proof in NK, for if a line in an S-proof is derived by a rule R of S, then we can infer the same line in NK using the appropriate combination of NK-rules. Conversely, the second condition guarantees that any line in an NK-proof inferred by a rule R of NK can be inferred in S, using the appropriate combination of S-rules. So the two conditions are exactly what is required for equivalence of the systems to hold.

Example 1: Let S be the system consisting in the rules of NK plus the rule DNI, from p to infer $\ulcorner{\sim}{\sim}p\urcorner$ depending on the same premises and assumptions. Show that S and NK are equivalent (hence DNI is redundant).

We have to show that the two conditions for equivalence of S and NK hold. Since every rule of NK is a rule of S, the second condition for S and NK to be equivalent is met trivially. To show that the first condition holds, we have to show that the effect of DNI can be obtained in NK by some combination of rules of NK. The simplest demonstration of this is to prove A $\vdash_{NK}$ ~~A, that is, DN⁺, since then wherever an S-proof uses DNI, an NK-proof can use SI (DN⁺). The proof of DN⁺ is straightforward:

1	(1)	A	Premise
2	(2)	~A	Assumption
1,2	(3)	⋏	1,2 ~E
1	(4)	~~A	2,3 ~I ◆

Another rule that sometimes features in Gentzen-style systems is the rule known as *ex falso quodlibet*, or *Absurdity*, that anything follows from an absurdity. This is the rule that if '⋏' has been inferred at line j, then at a later line k, any formula p may be inferred on the same assumptions as j, and the line labeled 'j EFQ'. Schematically,

$$a_1,...,a_n \quad (j) \quad \quad ⋏$$
$$\vdots$$
$$a_1,...,a_n \quad (k) \quad \quad p \quad \quad \text{j EFQ}$$

The system with EFQ *in place of* DN is known as NJ. NJ is a collection of rules for the alternative to classical logic which we have already mentioned, intuitionistic logic. NJ is not equivalent to NK, for there are many sequents provable in NK which are not provable in NJ. In particular, the sequent corresponding to DN is not provable in NJ. However, if we define S as the system resulting from *adding* EFQ to NK, S is equivalent to NK—no new sequents can be proved.

Example 2: Let S be the system which results from adding EFQ to NK. Show that S is equivalent to NK.

As before, the second condition holds trivially. The simplest way of showing that the first condition holds is again by appeal to SI, this time using the sequent ⋏ $\vdash_{NK}$ A, since then a use of EFQ in an S-proof can be mimicked by SI in a corresponding NK-proof. To show ⋏ $\vdash_{NK}$ A:

1	(1)	A	Premise
2	(2)	~A	Assumption
1	(3)	~~A	2,1 ~I
1	(4)	A	3 DN ◆

We also remarked earlier that the Law of Excluded Middle, $\vdash_{NK} A \vee {\sim}A$, like the rule DN, is characteristic of classical logic; indeed, intuitionistic logic is better known for rejecting the Law than it is for rejecting DN. The Law can also be formulated as a rule, namely, that at any line in a proof one may write down any substitution-instance of $\ulcorner p \vee {\sim}p \urcorner$, labeling the line 'LEM' and writing *no* numbers on its left (this is a special case of TI in NK). In view of the common fate of DN and the Law in intuitionistic logic, then, one might speculate that a system S equivalent to NK can be obtained by replacing DN in NK with the rule LEM. But this is not so. We have already seen how to prove the sequent corresponding to LEM in NK, but if S has only LEM in place of DN, there is no proof in S of the sequent corresponding to DN. Note in particular that it would be a mistake to offer the following proof:

1	(1)	~~A	Premise
	(2)	A ∨ ~A	LEM
1	(3)	A	1,2 SI (DS) ◆

The problem here is that our proof of the sequent DS (Example 5.3 on page 112) was an NK-proof, not an S-proof, and used the rule DN, which is not available in S. However, if instead we define S to be the result of replacing DN in NK with the *two* rules EFQ and LEM, then we do obtain a system equivalent to NK.

Example 3: Let S be the system which results from replacing DN in NK with the rules EFQ and LEM. Then S is equivalent to NK.

The first condition for equivalence is met since we have already given NK-proofs of the sequents corresponding to EFQ and LEM (both used DN), namely, Examples 5.7 and 10.2. To show that the effect of DN can be obtained in S using SI we give the following S-proof of the sequent corresponding to DN, $\sim\sim A \vdash_S A$:

1	(1)	~~A	Premise
	(2)	A ∨ ~A	LEM
3	(3)	A	Assumption
4	(4)	~A	Assumption
1,4	(5)	A	1,4 ~E
1,4	(6)	A	5 EFQ
1	(7)	A	2,3,3,4,6 ∨E ◆

This proof again illustrates the standard way of using LEM, that is, in combination with ∨E.

In this section, then, we have seen that there is more than one way of formulating a natural deduction system for classical logic (more are described in the exercises). But the system NK has a certain naturalness as a formulation of the logic of our five connectives: each connective has an introduction and an elimination rule, and there is one extra rule, DN, which captures an aspect of the meaning of negation which is missed by ~E and ~I. Furthermore, DN is the simplest way of capturing this extra aspect, for other rules or combinations of rules with the same effect seem to be in some sense *justified* by the fact that double negations collapse, rather than this fact being a consequence of something else to which those other rules make more direct appeal. For example, we proved LEM by showing that we could *reject* the *denial* of LEM, but it is only against the background of DN that rejecting the denial of LEM amounts to endorsing LEM. Similarly, EFQ's various possible justifications all depend on DN—LEM by itself is insufficient. Nor is it plausible that EFQ needs no justification (in the way that &E might be said to need no justification) until it is made plausible that our understanding of '⋏' does not involve our understanding of negation. On the other hand, it is hard to believe that our proof of $\sim\sim A \vdash_S A$ using LEM and EFQ articulates a more fundamental justification for identifying the meaning of the rejection of the denial of p with the meaning of the assertion of p.

There is also a question about our formulation of the disjunction rules. Some readers may be familiar with systems of logic in which reasoning from a disjunction is handled by the rule known variously as Modus Tollendo Ponens (MTP) or Disjunctive Syllogism, described in Exercise 2 below, in place of our own more complicated rule of ∨E. Why prefer the more complicated rule, granted the result of Exercise 2 that a system with MTP in place of ∨E, but otherwise like NK, is equivalent to NK? The reason is that MTP essentially involves negation: it says that from a disjunction and the *negation* of one of its disjuncts, we may infer the other disjunct. But the logic of disjunction does not involve negation *essentially*. There is no reason at all why we should have to use negation rules to prove, say, A ∨ B ⊢ B ∨ A, yet in a system S with MTP in place of ∨E, the negation rules would have to be used in proving this sequent (the reader may find it entertaining to construct such a proof). In fact, in NK any semantically correct sequent which contains no occurrences of negation, the conditional, or '⋏', can be proved without use of negation rules, whereas in the system S with MTP, many such sequents are unprovable without use of negation rules. This suggests that someone who regards a system with MTP in place of ∨E as providing the fundamental account of the logic of the sentential connectives is committed to the prediction that users of a language without negation would be unable to reason deductively from such premises as 'A ∨ B' to such conclusions as 'B ∨ A', or from '(A & B) ∨ (A & C)' to 'A'. But it seems highly implausible that this is true. Rather, the pattern of reasoning from a disjunction encapsulated in ∨E is the basic one, and so we prefer that rule to superficially simpler ones because of its fundamental nature.

❏ Exercises

(1) The rule of *Modus Tollendo Tollens* is the following:

> **Modus Tollendo Tollens:** For any formulae p and q, if $\ulcorner p \to q \urcorner$ occurs at a line j and $\ulcorner \sim q \urcorner$ occurs at a line k then $\ulcorner \sim p \urcorner$ may be inferred at line m, labeling the line 'j,k MTT' and writing on the left all numbers which appear on the left of line j and all which appear on the left of line k. We may have j > k or k > j. Schematically:

$$
\begin{array}{llll}
a_1,\ldots,a_n & \text{(j)} & p \to q & \\
& \vdots & & \\
b_1,\ldots,b_u & \text{(k)} & \sim q & \\
& \vdots & & \\
a_1,\ldots,a_n,b_1,\ldots,b_u & \text{(m)} & \sim p & \text{j,k MTT}
\end{array}
$$

Show that the system S defined as the result of replacing →E in NK with MTT is equivalent to NK.

(2) The rule of *Modus Tollendo Ponens* is the following:

> **Modus Tollendo Ponens:** for any formulae p and q, if $\ulcorner p \lor q \urcorner$ occurs at a line j and $\ulcorner \sim q \urcorner$ occurs at a line k then p may be inferred at line m, labeling the line 'j,k MTP' and writing on the left all numbers which appear on the left of line j and all which appear on the left of k. Alternatively, if $\ulcorner \sim p \urcorner$ occurs at k, q may be inferred at m with all the numbers from j and k on the left. Schematically:

$$
\begin{array}{llll}
a_1,\ldots,a_n & \text{(j)} & p \lor q & \\
& \vdots & & \\
b_1,\ldots,b_u & \text{(k)} & \sim q \ (\text{or} \sim p) & \\
& \vdots & & \\
a_1,\ldots,a_n,b_1,\ldots,b_u & \text{(m)} & p \ (\text{or } q) & \text{j,k MTP}
\end{array}
$$

Show that the system S defined as the result of replacing ∨E in NK with MTP is equivalent to NK. (Refer to the discussion of systems without ∨E on page 134 for guidance.)

*(3) The rule of *Classical Reductio* is the following:

> **Classical Reductio:** If 'λ' has been inferred at line k in a proof and $\{a_1,\ldots,a_n\}$ are the premise and assumption numbers λ depends upon, then if $\ulcorner \sim p \urcorner$ is the formula at line j, p may be inferred at line m, labeling the line 'j,k CR' and writing on its left the assumption numbers in $\{a_1,\ldots,a_n\}/\text{j}$. Schematically:

$$
\begin{array}{llll}
\text{j} & \text{(j)} & \sim p & \text{Assumption} \\
& & \vdots & \\
a_1,...,a_n & \text{(k)} & \curlywedge & \\
& & \vdots & \\
\{a_1,...,a_n\}/\text{j} & \text{(m)} & p & \text{j,k CR}
\end{array}
$$

As usual, j need not be in $\{a_1,...,a_n\}$. (The difference between CR and ~I, therefore, is that CR can be used to reduce the number of negation symbols prefixing an assumption.) Show that the system S defined as the result of replacing ~I and DN in NK with the rule of Classical Reductio is equivalent to NK.

(4) The rule of *Nonconstructive Dilemma* is the following:

Nonconstructive Dilemma: If p is assumed at line h, q is derived at line i, $\ulcorner\sim p\urcorner$ is assumed at line j and q is derived at line k, then at line m we may infer q, labeling the line 'h,i,j,k NCD' and writing on its left every number on the left of line i except h and every number on the left of k except j. Schematically:

$$
\begin{array}{llll}
\text{h} & \text{(h)} & p & \text{Assumption} \\
& & \vdots & \\
a_1,...,a_n & \text{(i)} & q & \\
& & \vdots & \\
\text{j} & \text{(j)} & \sim p & \text{Assumption} \\
& & \vdots & \\
b_1,...,b_u & \text{(k)} & q & \\
& & \vdots & \\
\text{X} & \text{(m)} & q & \text{h,i,j,k NCD}
\end{array}
$$

where X is the set $\{a_1,...,a_n\}/\text{h} \cup \{b_1,...,b_u\}/\text{j}$.

Show that the system S defined as the result of replacing DN in NK with the rules of Nonconstructive Dilemma and EFQ is equivalent to NK.

(5) The rule of *Negation-Definition* is the following:

Negation-Definition: If $\ulcorner(p \rightarrow \curlywedge)\urcorner$ occurs as the entire formula at a line j, then at line k we may write $\ulcorner\sim p\urcorner$, labeling the line 'j, Df~' and writing on its left the same numbers as on the left of j. Conversely, if $\ulcorner\sim p\urcorner$ occurs as the entire formula at a line j, then at line k we may write $\ulcorner(p \rightarrow \curlywedge)\urcorner$, labeling the line 'j, Df~' and writing on its left the same numbers as on the left of j.

Show that the system S in which Df~ replaces ~E and ~I (but not DN) is equivalent to NK.

*(6) The rule of *Generalized Negation-Definition* is the following:

> **Generalized Negation-Definition:** If $\ulcorner(p \rightarrow \wedge)\urcorner$ occurs as a subformula of a formula r at a line j, then at line k we may write the formula s, which is the same as r except that it has $\ulcorner\sim p\urcorner$ in place of that occurrence of $\ulcorner(p \rightarrow \wedge)\urcorner$, labeling the line 'j, Df$\sim_{Gen}$' and writing on its left the same numbers as on the left of j. Conversely, if $\ulcorner\sim p\urcorner$ occurs as a subformula of a formula r at a line j then at line k we may write the formula s, which is the same as r except that $\ulcorner(p \rightarrow \wedge)\urcorner$ replaces that occurrence of $\ulcorner\sim p\urcorner$, labeling the line 'j, Df$\sim_{Gen}$' and writing on its left the same numbers as on the left of j.

Show that the system S in which Df$\sim_{Gen}$ replaces $\sim$E and $\sim$I (but not DN) is equivalent to NK. (One part of this problem is a trivial consequence of the result of Problem 5. The other part is hard.)

11 Semantic and deductive consequence compared

We now have before us two very different ways of explaining what it is for the conclusion of an argument to follow from its premises. According to our first account, the conclusion of an argument follows from its premises when no interpretation of the argument's form makes the premises true and the conclusion false. In this situation, we write $p_1,...,p_n \vDash q$. According to our second account, the conclusion follows from the premises when the argument's form permits the derivation of the conclusion from the premises using the rules of inference of NK. In this situation, we write $p_1,...,p_n \vdash_{NK} q$. The second account says nothing about truth or interpretations, while the first says nothing about derivation or rules of inference. So it is not at all obvious how our two accounts are related.

What *is* obvious, however, is how we would *like* them to be related: we would like the two accounts to agree. In other words, (i) whenever the conclusion of a sequent can be derived from the sequent's premises in NK, we would like that conclusion also to be a semantic consequence of those premises. Conversely, (ii), whenever the conclusion of a sequent is a semantic consequence of the premises, we would also like to be able to derive it from those premises in NK. The desirability of both conditions can be brought out by an analogy. Suppose we have a lie detector which is designed to indicate when a subject is lying by flashing a green light to indicate truth-telling. Such a lie detector would be inadequate if *only* the following is true:

> (1) If the green light flashes, then the subject is telling the truth

since (1) is consistent with the green light *failing* to flash when the subject is telling the truth, and so misleadingly implying a lie. But equally, the lie detector is inadequate if only the following is true:

(2) If the subject is telling the truth, then the green light flashes

since (2) is consistent with the green light flashing when the subject is lying—indeed, (2) is true if the green light constantly flashes, even when no one is attached to the machine, and (1) is true if the green light never flashes, no matter what the subject is saying. Evidently, what we require of a lie detector is that (1) and (2) *both* be true.

Analogously, it would be unsatisfactory if only the following is true:

(3) If $p_1,...,p_n \vdash_{NK} q$ then $p_1,...,p_n \vDash q$

since this is consistent with there being many valid arguments which we *cannot* prove in NK. But equally, it would be unsatisfactory if only the following is true:

(4) If $p_1,...,p_n \vDash q$ then $p_1,...,p_n \vdash_{NK} q$

since this is consistent with there being many invalid arguments which are provable. Ideally, what we require of our system of inference NK is that both (3) and (4) be true.

One way of looking at (3) and (4) is as expressing claims about the strength of the rules of NK. According to (4), the rules are *strong enough*, in that they allow us to prove every valid argument-form, while according to (3), the rules are *not too strong*: reading (3) contrapositively, it says that the rules will not permit a proof to be given of an invalid argument-form. The two properties of *sufficient* yet *not excessive* strength in a collection of rules are known respectively as the *completeness* and *soundness* of the rules. We have the following definitions:

> A system S of rules of inference is said to be *sound* if and only if every S-provable argument-form is valid, that is, if and only if whenever $p_1,...,p_n \vdash_S q$, then $p_1,...,p_n \vDash q$.

> A system S of rules of inference is said to be *complete* if and only if every valid argument-form is S-provable, that is, if and only if whenever $p_1,...,p_n \vDash q$, then $p_1,...,p_n \vdash_S q$.

Our analogy with the lie detector suggests that the definition of semantic consequence is the basic elucidation of the intuitive idea of a conclusion following from premises, and it is then up to us to find some collection S of rules of inference (some machine) that is in agreement with the semantic criterion. But it should be emphasized that this is not the only possible perspective. There is an alternative view on which the rules of inference embody the basic way of capturing the meanings of the connectives, and it is then up to us to find a notion of semantic consequence that is in agreement with the rules (this was Gentzen's own perspective). If we adopt this perspective, the lie-detector analogy is misleading, for from this point of view it is the derivability of a conclusion from certain premises that *makes* the argument correct, but we would not want to say that it is the green light flashing that *makes* it the case that the subject is telling the truth.

But no matter which perspective we adopt, we require answers to the questions *is* NK sound, and *is* NK complete? The answer to both questions is yes, and there exist rigorous proofs of this, though ones beyond the scope of this book. However, we can see informally that it is quite plausible that NK is sound, since each rule by itself is perfectly acceptable. For example, if $\ulcorner p \;\&\; q \urcorner$ is a semantic consequence of some premises Γ, it is clear from the truth-table for '&' that each of the formulae we can infer from $\ulcorner p \;\&\; q \urcorner$ by &E is also a semantic consequence of Γ. Rules such as $\vee$E are more complex, but in the light of the truth-tables for the connectives, reflection indicates that they are semantically unobjectionable. Consequently, they could not be used to prove an invalid sequent. In this respect, the system NK contrasts with the system S defined as the result of adding to NK the rule of (for example) Disjunction Reduction, which is the fallacious rule that given a disjunction $\ulcorner p \vee q \urcorner$ one may infer either disjunct, depending on the same premises and assumptions as $\ulcorner p \vee q \urcorner$. Obviously, there are invalid sequents which have S-proofs. The simplest example is that we can show A $\vee$ B $\vdash_S$ A though of course A $\vee$ B $\nvDash$ A. Referring to the definition of 'sound', we conclude that S as defined is an *unsound* system. And it is because none of NK's rules have the fallacious character of Disjunction Reduction that the soundness of NK seems evident.

Completeness is another matter, however. To say that NK is complete is to say that every valid argument-form can be proved in it. But while we can tell by inspecting the list of NK's rules that none are objectionable, we cannot tell by inspecting the list that we have *every rule we need* in order to provide proofs for all valid arguments. In other words, NK might be incomplete because there are certain valid arguments whose proof requires a rule we have forgotten about, and we cannot tell by inspecting the list of rules that this is not the case: the presence of a bad rule is perceptible, but not the absence of a good one.

However, NK is indeed complete (see Hodges), and we can use this fact to give examples of incomplete systems, exploiting the results of the previous section. First, the system S which consists in NK without DN is incomplete, since although $\sim\sim$A $\vDash$ A, we cannot show $\sim\sim$A $\vdash_S$ A, since S does not have DN or an equivalent rule such as Classical Reductio or a combination of rules such as EFQ/LEM. Similarly, replacing DN in NK with just EFQ or just LEM yields an incomplete system S, since we will still be unable to show $\sim\sim$A $\vdash_S$ A. In general, then, if S is a system which NK extends *non*conservatively, S is incomplete.

Here are two final comments about completeness. First, we have previously mentioned a kind of logic called intuitionistic logic, for which a system of rules is obtained by replacing DN with EFQ in NK; this collection of rules, called 'NJ', does not suffice to prove the sequents corresponding to LEM and DN. This means that NJ is incomplete, and since we have said that completeness is a *desideratum* in a system of sentential logic, the reader may wonder what interest there can be in intuitionistic logic. The answer is that its interest lies in a perspective from which NJ is *not* incomplete. For intuitionists reject the classical definition of '$\vDash$' which we are implicitly invoking when we say NJ is incomplete, and which is based on the Principle of Bivalence. They have their own, completely different, way of defining '⊨', and on *their* definition, NJ is complete—in particular, the sequents corresponding to LEM and DN are invalid (see Chapter 10). Just as we have been subscripting the single turnstile to indicate what system of rules is in question, so, according to intuitionists, we should subscript the *double* turnstile to indicate what definition of semantic consequence is in question. Suppose we use '⊨ᵢ' for the intuitionistic definition (whose details we do not need to know for the purposes of this discussion) and '⊨_C' for the classical definition. Then although NJ is *classically incomplete*, since ~~A ⊨_C A while ~~A ⊬_NJ A, it is *intuitionistically complete* (hence ~~A ⊭ᵢ A). Similarly, though NK is classically sound, it is intuitionistically unsound, since ~~A ⊢_NK A but ~~A ⊭ᵢA. So the notion of semantic consequence also has a kind of relativity in it, and the main area of dispute between classical and intuitionistic logic concerns the relative merits of the two definitions of semantic consequence (see Dummett). However, in most of this book we are only concerned with the classical account, so we will not bother to continue subscripting the double-turnstile.

The second point about completeness is that although it is a desirable property, it is less important than soundness: an unsound system is epistemically dangerous, since it could cause us to accept falsehoods on the basis of truths or even to accept implicit contradictions. When it is *possible* to formulate a sound and complete system of rules for a notion of semantic consequence, an incomplete system would be unsatisfactory. But if we had to choose *between* soundness and completeness, it would be better to choose soundness. And this choice is not hypothetical. Sentential logic is the simplest form of classical logic. The next simplest, to which the rest of this book is devoted, is classical first-order logic, for which we can also give sound and complete rules. But beyond first-order logic there lie classical 'higher-order' logics, for which it is provable that no set of rules can be both sound and complete (see Van Benthem and Doets). For the higher-order logics we have to content ourselves with sound systems of natural deduction only.

❏ Exercises

(1) Give your own example of a system *S* which is complete but unsound. Demonstrate its unsoundness by exhibiting a proof of an argument-form which you show to be invalid. (Compare our discussion of the fallacious rule of Disjunction Reduction.)

(2) Say that a set of sentences Σ is *NK-consistent* if and only if $\Sigma \nvdash_{NK} \wedge$. Which of the following sets of sentences are consistent, which inconsistent? For each set that you judge inconsistent, give a proof which demonstrates its inconsistency.

    (a) {A ∨ B, ~A → ~B, ~A}   (b) {A → (B → C), ~A, ~B, ~C}
    (c) {A ∨ B, ~A ∨ ~B, A ↔ B}

(3) Say that a set of sentences Σ is *satisfiable* if and only if there is an interpretation $\mathcal{I}$ on which every member of Σ is true. Which of the following are satisfiable, which unsatisfiable? Explain your reasoning, and for each set that you judge to be satisfiable, give an interpretation which establishes its satisfiability.

    (a) $\{\sim A \vee \sim B, B \rightarrow A\}$
    (b) $\{(A \rightarrow A) \rightarrow B, \sim B\}$
    (c) $\{\sim(A \rightarrow \sim A), \sim(B \rightarrow \sim B), \sim(A \& B)\}$

*(4) Let Σ be a set of LSL sentences. Argue informally that the following is correct:

    (a) $\Sigma \vdash_{NK} A$ if and only if $\Sigma, \sim A$ is inconsistent.

Here '$\Sigma, \sim A$' stands for the set of sentences whose members are '$\sim A$' and all the members of Σ. (Hint: for the left-to-right direction, show that if $\Sigma \vdash_{NK} A$ then $\Sigma, \sim A \vdash_{NK} \curlywedge$. To establish this conditional, assume that you have an NK-proof of A from premises in Σ. Describe how you would construct an NK-proof of '$\curlywedge$' from '$\sim A$' and premises in Σ. Alternatively, using natural-deduction rules in sequent-to-sequent versions, there is a three-line proof whose premise is '$\Sigma \vdash_{NK}$ A' and whose conclusion is '$\Sigma, \sim A \vdash_{NK} \curlywedge$'. This is the left-to-right direction of (a). Now establish the right-to-left direction.)

(5) Let Σ be a set of LSL sentences. Argue informally that the following is correct:

    (a) $\Sigma \vDash A$ if and only if $\Sigma, \sim A$ is unsatisfiable.

(6) The completeness of NK is usually stated this way:

    (Comp) If $\Sigma \vDash A$ then $\Sigma \vdash_{NK} A$.

Using Exercises 4 and 5 restate Comp in terms of consistency and satisfiability, simplifying the restatement as much as possible. (It is in this restated form that completeness results are normally proved, using a method discovered by Leon Henkin.)

12 Summary

- Every connective is governed by an introduction rule and an elimination rule, and there is an extra rule, known as DN, to remove double negations.

- A proof can be regarded either as a progression from formulae to formulae or as a progression from sequents to sequents.
- Sequents which have already been proved can be used again to shorten proofs of new sequents, using an abbreviatory technique called Sequent Introduction.
- There are other collections of rules which define systems that are equivalent to NK, in the sense that they have the same provable sequents as NK. There are also collections which define systems that are weaker than NK, in the sense that their provable sequents are a proper subset of NK's.
- NK is sound and complete with respect to the classical definition of '⊨'.

PART II

MONADIC PREDICATE LOGIC

5 Predication and Quantification in English

1 A different type of argument

The system of logic developed in Part I of this book has considerable power, yet it falls far short of the goal we set ourselves at the start. That goal was to explain the difference between arguments whose conclusions follow from their premises and arguments whose conclusions do not, as this distinction applies to any English argument whatsoever. However, there are many English arguments where the conclusion clearly follows from the premises, and yet this fact cannot be represented within sentential logic. Such arguments are *intuitively* valid, but *sententially* invalid. Therefore there had better be more to logic than sentential logic, if we are to meet our goal.

Here are two examples of sententially invalid arguments which are intuitively valid:

> A: (1) Socrates is wise
> (2) ∴ Someone is wise
>
> B: (1) Everyone is happy
> (2) ∴ Plato is happy

To translate these arguments into LSL, the best we can do is

> C: P
> ∴ Q

since we cannot use the same sentence-letter for both the premise and the conclusion—after all, in both A and B the premise means something quite different from the conclusion. C is of course an invalid argument-form of LSL, as is shown by the assignment of ⊤ to 'P' and ⊥ to 'Q'. But in both A and B there is a clear sense in which the conclusion *follows* from the premise, and we want a symbolization which reflects this.

It is not difficult to see the broad outlines of the apparatus we need. The feeble symbolization C is most obviously deficient in two respects. First, it fails to represent the fact that in each of A and B, the premise and conclusion have something in common, the phrase 'is wise' in the case of A and the phrase 'is happy' in the case of B. The repetition of the phrases is crucial to the intuitive validity of each argument, so a symbolization would require an element which recurs in both premise and conclusion. The other kind of connection between the premise and conclusion in each argument which is crucial to its intuitive validity is the relationship between the meaning of the first word of the premise and the first word of the conclusion. If we interchange the 'someone' in A with the 'everyone' in B, we obtain the arguments

> D: (1) Socrates is wise
>   (2) ∴ Everyone is wise
>
> E: (1) Someone is happy
>   (2) ∴ Plato is happy

which, intuitively, are invalid: the truth of (1) in D does not preclude the existence of fools, and the truth of (1) in E does not preclude Plato's being ridden with *angst*. Consequently, adequate symbolizations of these arguments must not only represent the common element in the premise and conclusion but also have distinguishing symbols for 'someone', 'everyone' and words like 'Plato' and 'Socrates'.

We will use the following terminology for the categories of expression which we have just distinguished. Expressions like 'is wise' and 'is happy' are called *predicates* (or, more accurately, *monadic* predicates, but the point of the 'monadic' will not be apparent until later). A predicate is something which *applies to* an object, or is *true of* an object, or which an object *satisfies*.[1] Thus, according to (1) in D, 'is wise' applies to, or is true of, Socrates; or we can say that Socrates satisfies 'is wise'. Predicates are sometimes said to stand for *properties*, and if a predicate applies to an object then the object *possesses* the corresponding property. So according to (1) in D, Socrates possesses the property of wisdom. Make careful note of the correct use of these terms: possession concerns two things, an object and a property, but application or satisfaction concern a *thing* and a *word* or phrase. It would be wrong to say that 'is wise' applies to 'Socrates'; it is not words which are wise, but people.

Words like 'Plato' and 'Socrates' are *proper names*. Proper names are words or phrases which *stand for*, or *refer to*, or *denote*, specific persons or things; thus 'London', 'the Holy Roman Empire', and 'Hurricane Camille' are all proper names. However, the thing which a proper name denotes is fixed, so we do not count 'I' or 'that building' or 'tomorrow' as proper names, since each of these words changes its denotation according to who is speaking, or what the speaker is pointing at, or when the speaker is speaking. Sometimes the reference of a proper name changes according to context as well: there is Socrates the philosopher and Socrates the Brazilian soccer star (or perhaps these are two dif-

---

[1] A monadic ('one-place', 'unary') predicate is one which applies to things one at a time, by contrast with dyadic ('two-place', 'binary') predicates, which apply to things two at a time. 'is wise' is monadic because it is single individuals who satisfy or fail to satisfy it. 'is wiser than' is dyadic, because it is (ordered) pairs of individuals which satisfy or fail to satisfy it.

ferent names which just happen to be spelled the same way). However, we shall idealize slightly and speak as if no proper name has more than one reference.

Words like 'someone' and 'everyone' are *quantifier phrases*, so called because they specify a quantity: 'someone' means '*at least one* person' and 'everyone' means '*all* people'. 'At least one person' and 'all people' are themselves examples of quantifier phrases, as are 'everything', 'all objects', 'each thing', 'there is a person/thing such that' and 'anything' or 'anyone'. We will call these quantifier phrases the *classical* quantifiers, by contrast with such quantifiers as 'there are infinitely many things such that', 'few', and 'most'. That is, the classical quantifier phrases are the ones which connote the quantities 'at least one' and 'all' and others definable in their terms (*nonclassical* quantifier phrases are discussed briefly in §8 of Chapter 8). Note that quantifier phrases are quite different from proper names: there is no sense in which 'someone' stands for a specific person in the fixed way a proper name does. This contrast should be respected by whatever formal notation we come up with to symbolize arguments like A, B, D and E above.

When we use a quantifier phrase in ordinary conversation, there is usually some restriction on it implicit in the context. For instance, a person who smells cigarette smoke in a room full of people but cannot see who is producing it may say 'someone is smoking'. This is naturally understood to mean 'someone in this room is smoking'; in other words, the quantifier phrase 'someone' is *restricted* or *relativized* to the people in the room, and what is said is true only if *one of them* is smoking. If it was demonstrated that no one in the room is smoking, the statement could not be defended by pointing out that probably someone is smoking in another room, or in another country, or on another planet! It is only the people in the room who are relevant to the truth or falsity of the statement.

The collection of people or objects to which the quantifiers in a statement are relativized is called the *domain of discourse* of the statement. In the previous example, then, the domain of discourse is the collection, or set, of people in the room. As this example indicates, the domain of discourse of a statement is usually given by the circumstances in which the statement is made. But in this book we will be considering sentences in isolation from contexts in which they are used to make statements, so implicit domains of discourse will not be evident. However, there are very general kinds of domains of discourse which we find explicitly signaled in quantifier phrases themselves. The difference between 'someone' and 'something' illustrates this. We can regard these two words as consisting in the common component 'some', a quantifier *word*, suffixed with the domain of discourse indicators 'one' and 'thing' respectively, to produce the quantifier *phrases* 'someone' and 'something'. The indicator 'one' indicates that the domain of discourse is restricted to people, while the indicator 'thing' indicates that the domain of discourse is not restricted at all. Other indicator words in English are (i) 'where', which can be suffixed to 'some' and 'every', and (ii) 'ways', which can be suffixed to 'all' (dropping the second 'l'). (i) indicates restriction of the domain of discourse to places, and (ii) to times. In the next section we will exploit domains of discourse to simplify translations.

2 Further steps in symbolization: the existential quantifier

We are now going to develop a formal language with predicates, names and quantifiers. The idea is quite simple. The sentence-letter is the basic unit of sentential logic, but the discussion in the previous section of the shortcomings of the sentential form C has shown that sentential logic is too crude to represent some logical relationships. Whereas in this logic we can only break up compound sentences into their basic *sentential* constituents, that is, atomic sentences, in monadic predicate logic we can break up atomic sentences (sentences with no connectives) themselves. The category of atomic sentence is now taken to include expressions with predicates and proper names as constituents, and we distinguish such constituents by using different types of symbols for each. In English, proper names are capitalized and typically precede predicates, as in

(1) Socrates is wise.

Perhaps perversely, in the symbolization of (1) we capitalize the element corresponding to 'is wise', use lowercase for the proper name, and reverse the order (this will come to seem natural). By convention, proper names are symbolized by lowercase letters (excluding any after 's' in the alphabet), while for predicates we can just use the English word. Symbolized this way, it is useful to surround the name with parentheses to distinguish it from the predicate. Thus for (1) using the letter 's' for 'Socrates', we could write

(2) Wise(s).

However, it is quicker to abbreviate the predicate by some letter, usually its initial letter, and then there is no need to use the parentheses to mark off the symbol for the name. For (1) this gives us our preferred symbolization

(1.s) Ws.

As with sentential symbolizations, we will think of (1.s) as being a sentence in a formal language. In this language, uppercase letters are called *predicate-letters* and lowercase letters 'a'–'s' *individual constants*. We will call the language 'LMPL' for 'the language of monadic predicate logic'. Thus (1.s) is the translation of (1) into LMPL; English predicates are translated by the predicate-letters of LMPL, and English proper names are translated by the individual constants of LMPL. LMPL is an *extension* of LSL, so we can use the sentential connectives of LSL to form compound sentences out of the atomic sentences of LMPL, and, if it is ever necessary, sentence-letters as constituents of LMPL sentences. Since we would symbolize 'Plato is happy' by analogy with (1.s) as 'Hp', we can then symbolize such sentences as (3) by (3.s):

(3) Socrates is happy unless Plato is unhappy

(3.s) Hs ∨ ~Hp.

But quantifier phrases require more apparatus. The sentence 'someone is wise' contains structure that is not apparent to the naked eye, structure which is brought out by two paraphrases:

(4) Someone is wise
(5) There exists at least one person who is wise
(6) There exists at least one person such that he/she is wise.

(4), (5) and (6) are listed in order from most natural to most awkward, but they all mean the same. And it is (6) that is most useful for our purposes, since it suggests that we can regard 'someone is wise' as being assembled from two constituents, a phrase 'he/she is wise', which syntactically appears to have a structure close to the structure of atomic sentences like 'Socrates is wise', and a prefixed quantifier phrase 'there exists at least one person such that'. Notice that the 'he/she' of (6) is not being used in the way that 'I', 'that building' or 'tomorrow' is used, to pick out specific objects on a particular occasion of use. This 'he/she' is not the one which accompanies pointing, as in '*He* is next'. Instead, the 'he/she' refers *back* to the phrase 'one person' (linguists call this phenomenon 'anaphora' and the pronouns which exhibit it *anaphoric pronouns*). We can make this back-reference completely explicit, and also eliminate the inelegance of 'he/she', by replacing the English pronoun with the letter 'x' and inserting 'x' at an appropriate point in the quantifier phrase as well. This gives us:

(7) There exists at least one person x such that x is wise.

Since 'Socrates is wise' is symbolized 'Ws', we can symbolize 'x is wise' by 'Wx'. It is also useful to abbreviate the quantifier prefix 'there exists at least one person x such that' by symbols, for which we use '($\exists$x)'. As a result, we obtain for (7) the symbolization

(8) ($\exists$x)Wx.

However, though we are using '($\exists$x)' for 'there exists at least one person x such that', strictly speaking the '$\exists$' corresponds just to 'there exists at least one _ such that'. So '($\exists$x)' really only means 'there exists at least one x such that'; the domain of discourse indicator 'person' is not represented in the symbolism. This is acceptable because we will treat each translation into LMPL as *relativized to* a domain of discourse. Thus '($\exists$x)Wx' is an adequate translation of (4) with '$\exists$' relativized to the domain of people, but not with it relativized to the domain of things. Relativized to the domain of things, '($\exists$x)Wx' just means 'something is wise' and one somehow has to add a constituent corresponding to 'person'. How then do we know what the domain for a particular translation into LMPL should be? There are two rules which settle most cases (but see the discussion following (12a) below). First, when a sentence's quantifier phrases all invoke the same domain, that domain is the one to which all the quantifiers are relativized: the domain is people if the quantifier phrases are 'someone', 'somebody', 'at least one person' and so on, places if the quantifiers are 'somewhere' and 'everywhere', and so on. Second, if a sentence has quantifier phrases with indicators for different domains, as in 'someone has been everywhere', the domain of relativization is the most general domain of all, the domain of things. In this case we must represent the indicators explicitly in the symbolization, using predicate-letters for 'person' and 'place'. As a consequence of these rules, we recognize only the four domains for which there are explicit indicators in English quantifiers: the domains of things, people, places and times. Notice that by places and times we mean places and times *literally*. For example, we distinguish between places and the things located at them. Thus a city is a thing, not a place (consider the fact that Troy no longer exists, but the place at which it was located still exists).

There are a number of other points to observe:

- The quantifier in (8) is called the *existential* quantifier, because, as (7) makes explicit, it asserts the *existence* of something. A sentence such as (8) is called an *existential sentence*. Existential sentences, sentences which assert the existence of something, are said to be *existentially committing*, because if you sincerely assert one, you are committing yourself to the truth of what you say and therefore to the existence of an item of the appropriate kind. Thus 'there are ghosts in this house' is existentially committing to ghosts, because anyone who sincerely asserts it commits himself or herself to the existence of ghosts.

- (7) is not a sentence of English, since in English we use pronouns, not letters like 'x'. However, in translating English sentences into LMPL, the language of monadic predicate logic, it is often helpful to work through some intermediate steps in which English and LMPL are mixed. Sentences in which English is mixed with symbols, such as (7), are sometimes said to be sentences of *Loglish*.

- The letter 'x' in (7) and (8) is called an *individual variable*. By convention, individual variables are single letters chosen from the end of the alphabet (to distinguish them from individual constants). In (7) and (8) we could just as easily have used 'y' or 'z' instead of 'x'. Thus the formulae '$(\exists y)Wy$' and '$(\exists z)Wz$' are also correct symbolizations of (4), (5) and (6). When the only difference between two formulae is that the first has a particular individual variable v in exactly the positions that the second has a different individual variable v', the formulae are said to be *single-variable alphabetic variants* of each other. Thus '$(\exists x)Wx$', '$(\exists y)Wy$' and '$(\exists z)Wz$' are all single-variable alphabetic variants of each other.

- The existential quantifier in (8) is said to *bind* the occurrences of the individual variable which follow it after the quantifier prefix. Thus the 'x' of 'x is wise' in (7) and of 'Wx' in (8) is a *bound* variable. If a variable occurs in a formula and is not bound by any quantifier, the variable is said to be *free*. Thus in the expression 'Wy', 'y' occurs free, or is a *free variable*, while in the expression '$(\exists y)Wy$', the second 'y' is bound by the quantifier; in every case, for occurrences of a variable v to be bound by a quantifier, there must be an occurrence of v immediately to the right of the quantifier.

- Expressions with free variables cannot be true or false. For instance, 'x is wise' or 'Wx' cannot be true or false, since 'x' does not stand for any particular person or thing. So individual variables are different from individual constants. When we need a word for both, we use 'term': both individual variables and individual constants are terms. An expression with free individual variables is called an *open sentence*, and is said to be *atomic* if it contains no connectives or quantifiers ('Wx' is atomic). A *closed sentence*, or a *sentence*, for short, contains no free variables. So it is only closed sentences which can be true or false. In this book, 'sentence' usually means 'closed sentence'; for a word to include both open and closed sentences we use 'formula'.

- Since '∃' is 'E' written backward, it may be pronounced by pronouncing 'E' backward.

It is important to master this terminology, since proper discussion of predicate logic is impossible without it.

Like atomic sentences, existentially quantified sentences can be combined with the connectives of LSL. However, there are two rather different ways of making compound sentences using LSL connectives in LMPL. Comparing the English sentences

    (9) Someone is happy and someone is wise

and

    (10) Someone is happy and wise

a difference in meaning is immediately evident, for (9) could be true even if no single person has both the property of happiness and the property of wisdom, while (10) would be false in this situation. Consequently, the symbolizations will differ: (9) is a conjunction since its main connective is '&', while (10) is an existential sentence, since its main connective is 'someone' (we will define 'main connective' for LMPL in §4 below). We can therefore symbolize (9), relativizing '∃' to the domain of people, by symbolizing each conjunct separately, obtaining

    (9.s) (∃x)Hx & (∃x)Wx

or an alphabetic variant such as '(∃y)Hy & (∃y)Wy'. Notice that there is no problem about using the same individual variable 'x' or 'y' in the two conjuncts. This does *not* carry the implication that there is a happy person and there is a wise person and they are the *same* person, because the 'x' of 'Hx' in (9) is bound by the first existential quantifier and the 'x' of 'Wx' is bound by the second. (9) can be paraphrased long-windedly in English as:

> (11)   There is at least one person such that he/she is happy and there is
>   at least one person such that he/she is wise.

In (11) we use 'he/she' twice, but again there is no implication that it is the *same* person who is in question, since the anaphoric reference is to different quantifier phrases.

To symbolize (10) we first formulate it in Loglish. The initial 'someone' becomes, say, 'there is at least one z such that'. But how should we continue? We want to say that there is someone with both properties, so we should assert that z is happy and wise. However, we do not conjoin the predicate-letters 'H' and 'W' into a complex predicate-letter 'H & W', for this would not correspond to our procedure for names. 'John is happy and wise' has as its full sentential form 'John is happy and John is wise', so 'z is happy and wise' should expand into 'z is happy and z is wise'. We can also call this a full sentential form provided we note that 'sentential form' now includes forms for open sentences. Thus the new role for the sentential connectives, as exhibited in (10) but not in (9), is to combine open sentences as well as closed. The Loglish for (10), relativizing 'there is at least one x such that' to the domain of people, is therefore

> (10.a)   There is at least one x such that x is happy and x is wise

which goes into LMPL as

> (10.s) $(\exists x)(Hx \ \& \ Wx)$.

Parentheses are required to distinguish (10.s) from the incorrect '$(\exists x)Hx \ \& \ Wx$', which is a conjunction with '$(\exists x)Hx$' as first conjunct and the open sentence 'Wx' as second.

In symbolizing English we do not distinguish between singular and plural. For

> (12)   There are economists who are wealthy and economists who are not

we use 'at least one' again rather than find some way to express 'at least two'. That is, we shall regard (12) as true even if only one of the many economists is wealthy. As for predicate-letters, we will continue the practice of beginning a symbolization problem with a dictionary. For (12) the dictionary is:

> $E_{\_}$: _ is an economist
> $L_{\_}$: _ is wealthy

where the underline marks the position to be filled by an individual variable. As an intermediate step we state the Loglish for (12). In full, 'and economists who are not' means 'and there are economists who are not wealthy', so this gives us

(12.a) There is at least one y such that y is an economist and y is wealthy
and there is at least one z such that z is an economist and z is not
wealthy.

Here we use 'y' and then 'z' simply for variety. As far as the adequacy of (12.a)
as a symbolization is concerned, the domain to which the quantifiers are relativized can be either people or things, though neither is explicitly signaled in
the English. But if asked, for definiteness, to specify a domain, it would be more
natural to say that the domain is that of people, since 'there are economists
who are wealthy' is more naturally paraphrased as 'someone who is an economist is wealthy' rather than 'something which is an economist is wealthy'. The
final step in symbolizing (12) is to transcribe from (12.a) into LMPL using the
dictionary:

(12.s) $(\exists y)(Ey \ \& \ Ly) \ \& \ (\exists z)(Ez \ \& \ {\sim}Lz)$.

The existential quantifier, in conjunction with negation, can be used to
express other quantifiers. For example,

(13) No unwise person is happy

asserts that it is not the case that some unwise person is happy. 'Some unwise
person is happy' is '$(\exists x)({\sim}Wx \ \& \ Hx)$', so for (13) we have

(13.s) ${\sim}(\exists x)({\sim}Wx \ \& \ Hx)$.

Returning now to the arguments of the previous section, we see that A and
E are to be formalized in LMPL (relative to the domain of people) respectively as

A.s Wa
 ∴ $(\exists x)Wx$

and

E.s $(\exists x)Hx$
 ∴ Hb

Since A was intuitively valid and E intuitively invalid, we would expect A.s to be
a valid argument-form of LMPL, and E.s to be invalid. However, we have yet to
explain what is meant by 'valid' and 'invalid' in monadic predicate logic, and
this will be one of the topics of the next chapter.

❏ Exercises

Symbolize each of the following sentences using names, predicates and the existential quantifier, as appropriate. State your dictionary and say what domain your quantifiers are relativized to. Show at least one intermediate step in Loglish for each example.

   (1)   Some mathematicians are famous.
   (2)   Some mathematicians are not famous.
 *(3)   There is no mathematician who is famous.
   (4)   Some Germans are famous mathematicians.
   (5)   Gödel was a famous German mathematician.
   (6)   If Fermat was a French mathematician, then he was famous.
   (7)   Ada Lovelace was a brilliant English mathematician but she was not famous.
   (8)   Some famous mathematicians are neither German nor French.
   (9)   New Orleans is polluted but not smoggy.
 (10)   Some cities are smoggy and polluted.
 (11)   Some polluted cities are smoggy.
*(12)   Some polluted cities are smoggy and some aren't.
 (¹3)   No smoggy city is unpolluted.
 (14)   No city is smoggy if it is unpolluted.
*(15)   If a wealthy economist exists so does a famous mathematician.
 (16)   If no wealthy economist exists then no famous mathematician exists.
 (17)   Vampires don't exist.
 (18)   Nothing is both a ghost and a vampire.
 (19)   There aren't any ghosts, nor vampires either.
 (20)   If ghosts and vampires don't exist then nothing can be a ghost without being a vampire.

3 More symbolizations: the universal quantifier

The other arguments of §1, B and D, contain the quantifier 'everyone'. In Loglish, 'every' becomes 'for every _' and so the premise of B, 'everyone is happy', is rendered

    (1.a) For every x, x is happy

relativizing 'for every _' to the domain of people. In place of 'every' we may have 'each', or 'any', or 'all'. To turn (1.a) into a sentence of LMPL, we need a symbol for 'for every_'; the symbol we use is '∀', and so we obtain

    (1.s) (∀x)Hx.

'∀' is called the *universal* quantifier, and a sentence like (1.s) with '∀' as its main connective is called a *universal sentence.*
    The following two examples parallel (2.9) and (2.10) at the syntactic level:

    (2) Everyone is happy and everyone is wise.

(3) Everyone is happy and wise.

But by contrast with (2.9) and (2.10), there are no situations in which the truth-value of (2) is different from that of (3): both sentences require that everyone have the two properties of happiness and wisdom. However, we symbolize them differently to reflect the syntactic differences between them:

(2.a) For every x, x is happy, and for every x, x is wise.
(2.s) ($\forall$x)Hx & ($\forall$x)Wx.

(3.a) For every x, x is happy and x is wise.
(3.s) ($\forall$x)(Hx & Wx).

(2.s) and (3.s) are correct relativized to the domain of people. And when we define logical equivalence for LMPL, they will prove to be logically equivalent.

While (2) and (3) cannot differ in truth-value in the same situation, they can differ in truth-value together from

(4) Everyone who is happy is wise

since (4) does not entail that everyone is happy, while (2) and (3) do. (4) says merely that *if* a person is happy, then that person is wise, which leaves the question of the existence and virtues of unhappy people open. This paraphrase indicates that the conditional is the relevant connective for (4):

(4.a) For every x, if x is happy then x is wise.
(4.s) ($\forall$x)(Hx $\rightarrow$ Wx)

So (4.s) is also the correct LMPL translation of the paraphrase itself:

(5) If a person is happy then that person is wise

all this, of course, relativized to the domain of people. (4) and (5) say the same thing, so they can be symbolized with the same formula. Yet (5) appears to be a conditional with antecedent 'a person is happy' and consequent 'that person is wise', so one might expect that there should also be a symbolization of it with the conditional as *main* connective. But in fact this is not possible. (5) does not mean

(6) If everyone is happy then everyone is wise

since the falsity of (6) requires that everyone be happy (true antecedent) and not everyone be wise (false consequent), while (5) can be false even though there are unhappy people (making (6) true), so long as there are *some* happy ones, and one of them is unwise. And (5) does not mean

(7) If someone is happy then someone is wise

since the falsity of (7) requires that no one be wise (false consequent), while (5) could be false even if there are some wise people, so long as there are some unwise people as well and at least one of them is happy. (5) does mean the same as

  (8) If someone is happy then he/she is wise

but this also cannot be straightforwardly symbolized. The formula

  (9) $[(\exists x)Hx] \rightarrow Wx$

appears closest to (8) since it has the right antecedent 'someone is happy', but (8) is a closed sentence which is either true or false, while (9) is an open sentence, since its last 'x' is bound by no quantifier. So (9) cannot be true or false. If we solve this problem by moving the parentheses, as in

  (10) $(\exists x)(Hx \rightarrow Wx)$

we obtain the symbolization of

  (11) There is someone such that if he/she is happy then he/she is wise

which does not say that *all* happy people are wise, while (8), (5) and (4) do say this. The point is that in English we can form conditionals to the general effect

  (12) If an arbitrary person/thing is F, then he/she/it is G

which make universal claims about all Fs, but we cannot translate these conditionals into LMPL as conditionals—we have to symbolize them explicitly as universal sentences, as we did in (4.s).

There are other locutions which are also symbolized by universal sentences where the quantifier has a conditional within its scope, for example

  (13) Only the wise are happy

and

  (14) No one who is unhappy is wise.

Evidently, (13) does not mean the same as 'all the wise are happy', which says that wisdom is sufficient for happiness. (13) says instead that it is necessary, so that everyone who is happy must be wise. Consequently, relativizing '$\forall$' to the domain of people, a symbolization is

  (13.s) $(\forall x)(Hx \rightarrow Wx)$.[2]

(14) is also a universal sentence, or rather, can be paraphrased by a universal

sentence, namely, 'Everyone who is wise is happy'. So we get:

(14.s) $(\forall x)(Wx \rightarrow Hx)$.

It might be thought that if we have special symbols for 'every' and 'some' we should have a special symbol for 'no' as well, and indeed there is no particularly deep reason to have symbols for just two of these words. But traditionally, 'every' and 'some' have symbols, and other quantifier expressions are handled in terms of these two.

However, as we already saw in connection with §2.13, there is another equally natural way of approaching (14). (14) *denies the existence* of a person who combines unhappiness with wisdom, and we can symbolize existence-denials with the combination $\ulcorner\sim(\exists \nu)\urcorner$ where ν is an individual variable of LMPL. In other words, (14) has the paraphrase

(15) It is not the case that there exists a person who is unhappy and wise

for which the Loglish, relativizing 'there is at least one' to the domain of people, is

(15.a) Not: there is at least one x such that x is unhappy and x is wise

and the LMPL translation is

(15.s) $\sim(\exists x)(\sim Hx \,\&\, Wx)$.

The possibility of symbolizing (14) both as (14.s) and (15.s) reflects an underlying relationship between the universal and the existential quantifier, which can be brought out by the simple examples

(2.4) Someone is wise

and

(1) Everyone is wise.

To assert that someone is wise is to deny that everyone is unwise, so we can symbolize (2.4), relativizing '$\forall$' to the domain of people, by

(16) $\sim(\forall x)\sim Wx$

as well as by '$(\exists x)Wx$'; and to assert that everyone is wise is to deny that some-one is unwise, so we can symbolize (1) by

---

[2] However, it is arguable that (13.s) does not take account of the 'the' in (13). It might be held that while (13.s) is correct for 'only wise people are happy', if we use 'the wise' we imply that there are wise people. On this view, (13.s) needs '$\&\,(\exists x)Wx$' to be attached. It is not appropriate to be dogmatic about this. The reader may choose a policy, but the author prefers to remain with the simpler formula.

(17) $\sim(\exists x)\sim Wx$

as well as by '$(\forall x)Wx$'. The two quantifiers '$\forall$' and '$\exists$' are said to be *interdefinable*, a notion we will explain more precisely later.

Another common English quantifier word is 'any', as in 'anyone' and 'anything'. The meaning of 'any' changes from context to context, and it is important to ask whether 'any' means 'at least one' or 'every' before symbolizing it. Here are three examples:

> (18) Anyone who is wise is happy.
> (19) If anyone is wealthy then he/she is happy.
> (20) If anyone is wealthy then economists are.

(18) is a universal sentence: 'any' means 'every'. (19) is also universal, since it is the kind of sentence characterized by (12): 'anyone' means 'an arbitrary person'. So for (18) and (19), relativizing '$\forall$' to the domain of people, we have respectively

(18.s) $(\forall x)(Wx \rightarrow Hx)$

and

(19.s) $(\forall x)(Lx \rightarrow Hx)$.

But (20) is different. The natural reading is brought out by inserting 'at all' after 'anyone' to obtain 'if anyone at all is wealthy then economists are'. 'Anyone at all' means 'at least one person', and since there is no pronoun in the consequent of (20) which refers back to the antecedent, we can symbolize it as a conditional without encountering the problems raised by (8). The antecedent is 'at least one person is wealthy' and the consequent is 'economists are'. The plural indicates the presence of a quantifier, which in (20) is clearly universal. Hence 'economists are' is really 'every economist is wealthy', so the final formula is

(20.s) $(\exists x)Lx \rightarrow (\forall x)(Ex \rightarrow Lx)$.

As with (12), we say that the domain of relativization for the quantifiers in (20.s) is the domain of people. Admittedly, there is also a less natural reading of (20) on which 'anyone' is read as 'everyone' (perhaps even 'anyone at all' in the context of (20) can be heard this way). This reading is symbolized as

(20.s') $(\forall x)Lx \rightarrow (\forall x)(Ex \rightarrow Lx)$.

What makes (20.s') the *less* natural reading is that (20) is hardly worth asserting if (20.s') is what is meant: if the antecedent (whose English translation is) 'everyone is wealthy' is true, then *trivially* the consequent 'everyone who is an economist is wealthy' is true as well. (20.s') is an example of a formula which is guaranteed to be true for reasons similar to those which guarantee the truth of tautologies in LSL. But (20.s) is a much more substantial assertion: according to it, all it takes for every economist to be wealthy is *one* wealthy person.

Most of the English sentences we have considered to this point have involved only two predicates, but we can symbolize more complicated statements as well. For example, we can form complex open sentences to make statements about subgroups of groups to which a single predicate applies, for example, wise economists instead of just economists. Consider the sentence

(21) Every wealthy logician is happy.

A wealthy logician is someone who is *both* wealthy *and* a logician, so the kind of combination which 'wealthy logician' and 'wise economist' exemplify corresponds to conjunction in LMPL. Relativizing 'every' to the domain of people, we can put (21) into Loglish as

(21.a) For every x, if x is wealthy and x is a logician, then x is happy

and so, using 'G_' for '_ is a logician', its LMPL translation is

(21.s) $(\forall x)((Lx \ \& \ Gx) \rightarrow Hx)$.

Similarly,

(22) No wealthy economists are happy

in its negative existential paraphrase 'there does not exist a wealthy economist who is happy', becomes

(22.a) Not: there is at least one x such that x is wealthy & x is an economist & x is happy

in Loglish, which translates into

(22.s) $\sim(\exists x)((Lx \ \& \ Ex) \ \& \ Hx)$.

The inner parentheses in (22.s) are not needed for disambiguation, but when we come to apply rules of inference to formulae of LMPL, the syntax will have to determine a main connective for each line in a proof; therefore, we will set up the syntactic rules to supply enough parentheses to ensure this.

We have yet to consider an example where a domain of discourse indicator such as 'place' or 'person' must be used explicitly because the sentence cannot be translated into LMPL relativizing all its quantifiers to that domain. Here is such a case:

(23) A fetus is a person but an embryo is not.

The domain of relativization should not be taken to be people, since the sentence concerns some entities, embryos, which it explicitly says are not in that domain. Regardless of the sentence's truth-value, its intended domain, when it is used sincerely, must therefore be wider than the domain of people. So the domain to which the quantifiers in (23) are relativized will be the most general domain, that of things. This means that the word 'person' occurs in (23) as a predicate like 'logician' and 'economist' in earlier examples, and so we need to give it its own predicate-letter. With this in mind, (23) becomes

(23.a) For every x, if x is a fetus then x is a person and for every x, if x is an embryo then x is not a person.

Using the dictionary

P_: _ is a person
F_: _ is a fetus
R_: _ is an embryo

we transcribe (23.a) into the formula

(23.s) $(\forall x)(Fx \rightarrow Px)$ & $(\forall x)(Rx \rightarrow \sim Px)$.

The example also illustrates the use of the indefinite article 'a' or 'an' for the universal quantifier—(23) is not saying merely that at least one fetus is a person and at least one embryo is not. In fact, the repeated quantifier in (23.s) is not strictly necessary, since the formula

(23.s′) $(\forall x)((Fx \rightarrow Px)$ & $(Rx \rightarrow \sim Px))$

is logically equivalent to (23.s) in the way that (2) and (3) are logically equivalent. However, in keeping with our general policy of following English syntax closely, (23.s) is the preferred symbolization.[3]

Finally, the reader should notice that we made frequent informal use of predicate logic in Part I of this book, in stating rules of inference, grammatical rules and rules for evaluating kinds of formulae. For instance, in the rule (f-$\sim$), 'for any symbol string p, if p is a wff then $\ulcorner \sim p \urcorner$ is a wff', the phrase 'for any symbol string p' abbreviates 'for any p, if p is a symbol string then...', and the letter 'p' is being used as an individual variable.

---

[3] As the reader may have noted, our discussion of 'no one' conflicts with our working principle that English syntax determines a unique preferred symbolization for each sentence, since we have not insisted that one of our quantifiers is to be preferred to the other for symbolizing 'no one' or 'none'. In fact, there is a case for preferring the negative existential analysis of 'none', since 'none' derives from 'not one', which is in turn elliptical for 'not at least one'. But it will often be more convenient to translate 'none' using the universal quantifier, and we shall not forbid this. Thus for a given English sentence there may be no single preferred translation, but rather two or more, all of which will be logically equivalent. So when we speak of 'the' translation of an English sentence we really mean any of the preferred translations, if no specific translation is given in the context.

❏ Exercises

I   Translate each of the following sentences in LMPL using names, predicates and quantifiers as appropriate. State your dictionary and the domain to which your quantifiers are relativized. Show at least one intermediate step in Loglish for each example.

(1)   All donkeys are stubborn.
(2)   Every inflationary economy is faltering. ('I_', 'E_', 'F_')
*(3)   Only *private* universities are expensive. ('P_', 'U_', 'E_')
(4)   Whales are mammals.
(5)   If it rains, only the killjoys will be happy. ('A': it rains)
(6)   It's always the men who are overpaid.
*(7)   All that glitters is not gold. ('G_': _ glitters; 'O_': _ is gold)
(8)   If a woman is elected, someone will be happy.
(9)   If a woman is elected, she will be happy.
(10)   If Mary is elected, the directors will all resign.
(11)   No corrupt politicians were elected.
*(12)   None but corrupt politicians were elected.
(13)   If any elected politician is corrupt, no voter will be satisfied.
(14)   If an elected politician is corrupt, he will not be re-elected.
(15)   No voter will be satisfied unless some politician who is elected is incorrupt.
*(16)   Invariably, a wealthy logician is a textbook author.
(17)   Any logician who is a textbook author is wealthy.
(18)   Occasionally, a logician who is a textbook author is wealthy.
(19)   Among the wealthy, the only logicians are textbook authors.
(20)   Except for textbook authors, no logicians are wealthy.

II   With the same symbols as in (7), (2) and (5) respectively, translate the following formulae into idiomatic sentences of ordinary English.

(a) $(\forall x)(Gx \rightarrow \sim Ox)$ (b) $\sim(\exists x)((Fx \ \& \ Ex) \ \& \ \sim Ix)$ (c) $A \rightarrow (\forall x)(\sim Hx \rightarrow \sim Kx)$

4 The syntax of LMPL

We take the same perspective on the language of monadic predicate logic as we took on the language of sentential logic: LMPL is a language in its own right, with its own lexicon and its own formation rules for well-formed formulae. The lexicon includes the new categories of symbol which have been introduced in the previous examples:

The lexicon of LMPL:

All items in the lexicon of LSL, now including 'λ'; an unlimited supply of individual variables 'x', 'y', 'z', 'x'', 'y'', 'z''...; an unlimited supply of individual constants 'a', 'b', 'c', 'a'', 'b'', 'c''...; an unlimited supply of monadic predicate-letters 'A_1', 'B_1', 'C_1',...,'A_1'', 'B_1'', 'C_1'',...; quantifier symbols '$\forall$' and '$\exists$'.

Since we are including the vocabulary of LSL in LMPL, we have to distinguish between the use of, say, 'F', as a sentence-letter and as a predicate-letter, so in the official definition of the lexicon we subscript predicate-letters. But in actual practice we omit the subscripts, since context always shows which use of the letter is intended; thus in 'F & G', 'F' occurs as a sentence-letter, while in '(∃x)Fx' it occurs as a monadic predicate-letter. The point of allowing ourselves an *unlimited* supply of predicate-letters, variables, constants and so on, is so that sentences of any finite length can be constructed without repetition of any predicate-letter or variable or constant.

Since LMPL is an extension of LSL, every well-formed formula of LSL will also be a wff of LMPL. So the construction rules for formulae in LMPL include those of LSL. Recall that a formula is said to be an *open* sentence if it contains a variable not bound by any quantifier. Such a variable is said to be *free*. The new cases we have to cover are formation rules (i) for structured atomic formulae, that is, open or closed atomic sentences formed from a predicate-letter and a term, and (ii) for universal and existential sentences, again open or closed. For example, the closed sentence '(∃x)(∃y)(Fx & Gy)' is arrived at by combining the variable 'x' with the predicate-letter 'F' and 'y' with 'G', applying conjunction-formation to obtain the open sentence '(Fx & Gy)' (in which 'x' and 'y' are free), then prefixing '(∃y)' to obtain the open sentence '(∃y)(Fx & Gy)' (in which only 'x' is free) and finally prefixing '(∃x)' to obtain the closed '(∃x)(∃y)(Fx & Gy)'. The formation rules which allow us to accomplish such a construction are:

The formation rules of LMPL:

(*f-at*): For any sentence-letter p, variable v, individual constant s and predicate-letter λ (all from the lexicon of LMPL), p, λv and λs are (atomic) wffs.

(*f-con*): If ϕ and ψ are wffs (atomic or otherwise) so are $\ulcorner \sim\!\phi \urcorner$, $\ulcorner (\phi \mathbin{\&} \psi) \urcorner$, $\ulcorner (\phi \lor \psi) \urcorner$, $\ulcorner (\phi \to \psi) \urcorner$ and $\ulcorner (\phi \leftrightarrow \psi) \urcorner$.

(*f-q*): If ϕv is an open wff with all occurrences of v free, then $\ulcorner (\exists v)\phi v \urcorner$ and $\ulcorner (\forall v)\phi v \urcorner$ are wffs.

(*f!*): Nothing is a wff unless it is certified as such by the previous rules.

Here we use Greek letters phi, 'ϕ', and psi, 'ψ', as metalanguage variables for formulae of LMPL, and lambda, 'λ', as a metalanguage variable for predicate-letters. 'v' is a metalanguage variable for object language variables ('x', 'y', etc.). The rule (*f-at*) allows the formation of atomic formulae; it could be abbreviated to cover the cases of variables and individual constants simultaneously, using 'term': 'if p is a sentence-letter, t is a term and λ is a predicate-letter, p and λt are atomic wffs'. By including sentence-letters in the atomic formulae, our rules allow the construction of such formulae as '(∀x)(Fx → A)'. (*f-con*) is a single rule covering all the connectives of LSL. Recall that (*f!*) is the *closure* condition, as explained on page 37.

(*f-q*) is a rule which allows the formation of quantified formulae, and has two features of note:

- The use of 'ϕv' in (f-q) rather than simply 'ϕ' is meant to indicate that the expression ϕ contains at least one free occurrence of v. By requiring that a formula to which $\ulcorner(\exists v)\urcorner$ or $\ulcorner(\forall v)\urcorner$ is prefixed has free occurrences of v, we rule out so-called *redundant* quantification, as in '$(\exists x)(\exists y)Fy$'. We will not have any use for redundant quantification.
- By requiring that *every* occurrence of v in ϕv must be free before we can prefix $\ulcorner(\exists v)\urcorner$ or $\ulcorner(\forall v)\urcorner$, we rule out *double binding*, as in '$(\exists x)(Fx \,\&\, (\exists x)Gx)$'. Such formulae would be unambiguous on the understanding that a variable v is bound by the *closest* v-quantifier within whose scope it lies, but we lose nothing by excluding them from the category of wffs. (For a variable v to be double bound in a formula ϕ, it must be within the scope of at least two v-quantifiers in ϕ. See the definition of scope on page 168.)

As with LSL, we can draw parse trees for LMPL formulae which illustrate their construction and which can be used to define important syntactic concepts, such as 'scope' and 'main connective', for LMPL. At the bottom nodes of the trees we select the required lexical elements, and at each subsequent node we apply a formation rule to one or two previous nodes.

Example 1: Give the parse tree for '$(\exists x)(Fx \,\&\, (\forall y)(Gy \rightarrow \sim Fx))$'.

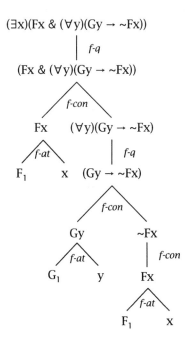

We have labeled each node of the tree with the formation rule which is used at that node. Notice that all the well-formed formulae displayed at nodes in the tree are *open* sentences except for the formula at the very top node. Do not confuse the open/closed distinction with the non-well-formed/well-formed distinction: an open sentence such as 'Gy → ~Fx' is perfectly well formed. (The invisibility convention is in force in LMPL as well as LSL, so you cannot see the outer parentheses on 'Gy → ~Fx'.)

We can use the formation rules to distinguish between strings of symbols which are well formed and those which are not. Apart from obvious examples of gibberish such as '∃F&xxyG', problematic expressions are likely to involve double-binding, redundant quantification or improper parentheses. For instance, '((∃x)(Fx & Gx)) & A' is not well formed because the rule (*f-q*) does not introduce outer parentheses around the formula which is formed when '(∃x)' is prefixed to '(Fx & Gx)', so the outer parentheses on the first conjunct of this formula are illegal.

The tree apparatus allows us to define the notions of subformula, main connective, scope, and being within the scope of, as before.

- The *subformulae* of a formula ϕ of LMPL are all the formulae which appear at nodes in the parse tree for ϕ (since predicate-letters and terms are not formulae, there are no subformulae at the bottom nodes of the parse tree for Example 1).
- The *main connective* of a formula ϕ is the connective introduced at the top of ϕ's parse tree.
- The *scope* of an occurrence of a binary sentential connective in a formula ϕ is the subformula at the node in the parse tree for ϕ at which that occurrence of the connective is introduced, while the scope of an occurrence of '~' or a quantifier is the subformula at the immediately preceding node. For instance, in our example '(∃x)(Fx & (∀y)(Gy → ~Fx))' the scope of '(∀y)' is the formula at the fourth top node on the right of its tree, '(Gy → ~Fx)', which is also the scope of the '→'.
- A lexical item in ϕ or a subformula of ϕ is *within the scope* of a connective or quantifier c in ϕ if and only if that lexical item or subformula occurs in the formula which is the scope of c.

For the purposes of defining 'main connective' we count quantifier symbols as connectives. Sentential connectives and quantifiers are collectively known as *logical constants*, for reasons which will transpire when we see how to extend the semantics for LSL to LMPL. This is our next topic.

❏ Exercises

I   Which of the following are non-wffs? In each case explain where the problem lies.

  (1) ((∃x)(Fx & ~Fx))
  (2) (∀x)(Fx → ((Gx) & (∃y)Hy))
  (3) (∀x)Fx ↔ (∀x)(Gx & Hy)
  (4) (∀x)(Fx → (∀x)Gx)
 *(5) (∃x)(~(∃y)(Fx & ~Gy))
  (6) (∀x)(∃y)(∀z)(∃w)(∃u)((Fx & Gu) → (Hw & (Tz ∨ Tw)))

II   Draw the parse tree for the formula '(∀x)(Fx → (∃y)(~Gy & Hy))'.

  (1) List all the subformulae of this formula.
  (2) What is the scope of '∃' in this formula?
 *(3) Is the '&' within the scope of '~' or vice versa?

5 Summary

- There are many arguments in English which are easily seen to be valid or invalid but which cannot be formalized adequately in sentential logic.
- In order to symbolize such arguments properly, we need symbols for proper names, predicates and quantifier phrases.
- The quantifiers we use are called the *universal* and the *existential* quantifier, and mean respectively 'for every _' and 'there is at least one _ such that'.
- With the aid of these two quantifiers we can express many quantifier locutions such as 'only the', 'none' and 'no one'.
- In fact, one of our two quantifiers is all we need since we can express each of them using the other and negation.
- The formal language of monadic predicate logic is called LMPL, and its grammatical rules are those of LSL together with extra rules to allow the formation of atomic formulae from predicates and variables or individual constants, and of existential and universal sentences from open sentences with at least one free variable.

6 Validity and Provability in Monadic Predicate Logic

1 Semantics for the quantifiers

We now wish to give a completely precise characterization of the difference between valid and invalid argument-forms in LMPL, that is, of the difference between the case where the conclusion of an LMPL argument-form is a semantic consequence of its premises and the case where it is not. To make our way into this topic, we consider two of the simplest invalid argument-forms:

> A: (∃x)Fx & (∃x)Gx
> ∴ (∃x)(Fx & Gx)

and

> B: (∀x)(Fx ∨ Gx)
> ∴ (∀x)Fx ∨ (∀x)Gx.

It is easy to see that these LMPL forms are invalid, since we can find English instantiations of each which have actually true premises and actually false conclusions. For example, with quantifiers relativized to the domain of people, A is the translation of

> A_E: Someone is male and someone is female
> ∴ Someone is both male and female

while B is the translation of

> B_E: Everyone is either male or female
> ∴ Either everyone is male or everyone is female

The problem with A, then, is that its premise does not require that it be the *same* thing which satisfies 'F' and 'G', while its conclusion does require this; and the problem with B is that its premise does not require every object to satisfy 'F' and does not require every object to satisfy 'G', while its conclusion requires one or the other of these conditions to hold.

The procedure illustrated by A_E and B_E is one way of establishing the invalidity of an argument-form in LMPL: we find an English argument with true premises and false conclusion which has the form in question. But this procedure rapidly becomes useless as argument-forms become more complex. Consequently, we will develop a method of showing that LMPL argument-forms such as A and B are invalid, one which does not depend on finding English arguments with true premises and false conclusions which have those LMPL forms.

The idea behind the method is to describe in an abstract way the essential structure of any situation in which the premises of a given LMPL argument-form are true and its conclusion is false. For instance, returning to A, we can see that if its premise '(∃x)Fx & (∃x)Gx' is to be true and its conclusion '(∃x)(Fx & Gx)' false, there would have to be an object which satisfies 'F' and an object which satisfies 'G', but at the same time there would have to be no object which satisfies both. The *minimum* number of objects needed to realize such a situation is two; there could be more, but the extra objects would be inessential, and we are trying to describe what is *essential* to any situation where the premise of A holds and the conclusion does not. What is essential, then, is that there be two objects, one satisfying 'F' and the other satisfying 'G', thereby verifying the two conjuncts of A's premise; in addition, the object which satisfies 'F' should not satisfy 'G', and the object which satisfies 'G' should not satisfy 'F', thereby falsifying A's conclusion.

What we are calling a 'situation' we shall refer to more formally as an *interpretation* of the argument-form. The collection of objects which occur in the situation, which must include at least one object, is called the *domain of discourse* or *universe of discourse* of the interpretation (since we are dealing with formal arguments in LMPL, we can think of these objects as being of any kind); and when we say which objects satisfy 'F' and which satisfy 'G' we are specifying the *extensions* of the predicates 'F' and 'G', that is, we are saying what things they apply to or 'extend over'. There is a standard way of describing an interpretation, which we can illustrate in connection with A. We use lowercase Greek letters as names of objects in the domain and set-braces '{' and '}' to specify the sets which constitute the domain and the extensions of the predicates. Then the interpretation which shows A to be invalid is the following.

Interpretation: $D = \{\alpha, \beta\}$
$Ext(F) = \{\alpha\}$
$Ext(G) = \{\beta\}$

'D' abbreviates 'domain of discourse' and 'Ext(F)' abbreviates "the extension of 'F'" (so the parentheses do duty for the single quotes around 'F'). There is also another way in which the same information can be represented which is perhaps more revealing, at least initially, and that is in the form of a matrix, where we write the domain on the side and the predicate-letters along the top, and indicate by the entries in the matrix which objects satisfy which predicates. The matrix representation of the interpretation just given is displayed at the top of page 172. A '+' at a position in the matrix indicates that the object on the row in question satisfies the predicate of the column in question, a '−' that it does

$$
\begin{array}{c|cc}
 & F & G \\
\hline
\alpha & + & - \\
\\
\beta & - & + \\
\end{array}
$$

not satisfy it (the technique is borrowed from Gustason and Ulrich, Ch. 5.1).

We can do something similar for argument-form B; we could use the very same interpretation to show that B is invalid, or else this slightly different one:

Interpretation: $D = \{\alpha, \beta\}$
$Ext(F) = \{\beta\}$
$Ext(G) = \{\alpha\}$

The premise of B is true because each object satisfies 'F' or satisfies 'G', but the conclusion is false since both its disjuncts are false: it is not true that both objects satisfy 'F', nor that both satisfy 'G'.

Establishing the invalidity of an argument-form in monadic predicate logic is a two-stage process. First we state the interpretation which shows the argument-form to be invalid, and secondly we *explain why* the interpretation establishes invalidity. Since anyone can fill in a matrix with pluses and minuses, the second stage is required to show that we understand why our proposed interpretation does what we claim it does. In the examples we have been discussing, we have only roughly explained why the interpretations work, whereas what we need is some reliable format in which we can give complete explanations, no matter how complex the example.

Underlying our intuitive grasp of why the two preceding interpretations demonstrate the invalidity of A and B are implicit principles about when quantified sentences are true or false on interpretations. We will now make these principles explicit, and the rest of this section will be devoted to understanding how they are applied—we will return to the construction of 'counterexamples' to other invalid argument-forms in the next section. First we need to define the notion of a sentence's being an *instance* of a quantified sentence *in* an interpretation. In this chapter the domains we consider will all be nonempty finite sets, but we will not presuppose finiteness in our definitions. So let $\ulcorner(Qv)\phi v\urcorner$ be a (closed) quantified sentence, where Q is either '$\forall$' or '$\exists$', v is an individual variable, ϕv is a wff with all and only occurrences of v free, and Q is the main connective of $\ulcorner(Qv)\phi v\urcorner$. Then if I is an interpretation with domain $D = \{\alpha, \beta, \gamma,...\}$, we form instances of $\ulcorner(Qv)\phi v\urcorner$ by choosing individual constants t_i from the lexicon of LMPL, a different one for each element of the domain of I, and form the sentence $\phi(t_i/v)$. That is, each instance of $\ulcorner(Qv)\phi v\urcorner$ for a given interpretation I is obtained by deleting the quantifier prefix $\ulcorner(Qv)\urcorner$ and replacing every occurrence of v in ϕv by an LMPL name, using the *same* name throughout, so that we end up with one instance of $\ulcorner(Qv)\phi v\urcorner$ for each element of the domain of I. Conventionally, we use 'a', 'b', 'c',... as LMPL individual constants for α, β, γ,... That is, we match the English and Greek letters alphabetically. This is called the *alphabetic convention.* So in the preceding interpretations,

  (i)   the instances of '(∃x)Fx' and '(∀x)Fx' are 'Fa' and 'Fb';
  (ii) the instances of '(∃x)Gx' and '(∀x)Gx' are 'Ga' and 'Gb';
  (iii) the instances of '(∀x)(Fx ∨ Gx)' are 'Fa ∨ Ga' and 'Fb ∨ Gb';
  (iv) the instances of '(∃x)(Fx & Gx)' are 'Fa & Ga' and 'Fb & Gb'.

Thus the number of instances of a quantified sentence in an interpretation *I* is determined by the number of objects in the domain of *I*.

Let us be clear about the difference between the roman and Greek letters. When we specify the domain of an interpretation, we use the Greek letters as if they are metalanguage (English) names of specific objects—indeed, instead of using Greek letters we could just use actual English names, such as numerals or city names or names of rivers, to specify domains. By contrast, the letters 'a', 'b', 'c' and so on are taken to be names in the object language, LMPL. Unlike 'Socrates', '3' or (we are supposing) 'α', the individual constants of LMPL are uninterpreted: in advance of stipulating some interpretation we cannot say what they stand for, and we can stipulate any reference we like. This difference is a little obscured by the alphabetic convention and the use of artificial names in the metalanguage to specify domains, but these are simply matters of convenience and the reader who wishes to can use numerals to specify domains and assign LMPL names to numbers any way he or she pleases.

We can now state the rules which determine the truth-values of quantified sentences in interpretations. What we are doing here for the quantifiers is analogous to giving truth-tables for the sentential connectives, with two differences. A truth-table for a connective tells us how on an interpretation *I* the truth-value of a compound sentence with that connective as main connective is determined by the truth-values on *I* of the main subsentences. In predicate logic, an interpretation is not merely an assignment of truth-values to sentence-letters, but also includes a specification of a nonempty domain of discourse and an assignment of extensions to predicates. Moreover, the truth-value of ⌜(Qv)φv⌝ on an interpretation *I* is not determined by the *syntactic* component to which ⌜(Qv)⌝ connects, since that component, φv, does not have a truth-value, being an open sentence. But we can think of the instances as components in a semantic sense, so the dissimilarity with our procedure in sentential logic is not too great (we give a more precise description of the difference in §1 of Chapter 8, using the notion of *full compositionality*).

The semantic rules for quantifiers in predicate logic are these:

(∀⊤) A universal sentence is *true* on an interpretation *I* if *all* its instances in *I* are true.

(∀⊥) A universal sentence is *false* on an interpretation *I* if *at least one* of its instances in *I* is false.

(∃⊤) An existential sentence is *true* on an interpretation *I* if *at least one* of its instances in *I* is true.

(∃⊥) An existential sentence is *false* on an interpretation *I* if *all* its instances in *I* are false.

Each rule states a sufficient condition for a quantified sentence to have a particular truth-value, but granted the intuitive meaning of 'all' and 'some', the stated conditions are clearly necessary as well as sufficient. Indeed, in view of the Principle of Bivalence, each pair of rules could be expressed in a single rule: a universal sentence is true if and only if all its instances are true, an existential sentence is true if and only if at least one of its instances is true. However, in constructing and reasoning about interpretations, it is useful to have an explicit, separate formulation of the truth condition and the falsity condition for each kind of quantified sentence. The reader should pay careful attention to the rule (∃⊥). The most common mistake in constructing an interpretation with a view to making an existential sentence false is forgetting that this requires *all* the instances to be false. To see the justification for this requirement, note that the falsity of 'At least one U.S. president was a spy' requires that it be false that Bush was a spy, that Reagan was a spy, that Carter was a spy and so on back to Washington.

In principle, we should supplement the four quantifier rules with rules for each of the other syntactic categories of sentence of LMPL: atomic sentences, negations, conjunctions, disjunctions, conditionals and biconditionals. However, the rules for these five kinds of sentence are the same as in sentential logic, where we embodied them in truth-tables. For example, a conjunction is true on an interpretation $\mathit{1}$ if both of its conjuncts are true on $\mathit{1}$, and false on $\mathit{1}$ if at least one conjunct is false on $\mathit{1}$; a biconditional is true on $\mathit{1}$ if both of its sides have the same truth-value on $\mathit{1}$, false if they do not. And if an atomic sentence is a sentence-letter, it is true or false on $\mathit{1}$ according to whether $\mathit{1}$ assigns it ⊤ or ⊥ (remember that every interpretation $\mathit{1}$ assigns ⊥ to 'λ'). Since there is nothing new in this, we will not bother to recapitulate the principles for the connectives in separate clauses. However, atomic sentences built out of predicates and individual constants do represent something new, so we should make explicit how the truth-value on an interpretation $\mathit{1}$ of such an atomic sentence relates to what $\mathit{1}$ says about the extension of the relevant predicate. The rule, which is simple, is stated so as to include the sentential case:

> (A⊤) An atomic sentence λ*t* is true on an interpretation $\mathit{1}$ if the object referred to by the individual constant *t* *belongs* to the extension in $\mathit{1}$ of λ; a sentence-letter *p* is true on $\mathit{1}$ if $\mathit{1}$ assigns ⊤ to *p*.
>
> (A⊥) An atomic sentence λ*t* is false on an interpretation $\mathit{1}$ if the object named by the individual constant *t* does *not* belong to the extension in $\mathit{1}$ of λ; a sentence-letter *p* is false on $\mathit{1}$ if $\mathit{1}$ assigns ⊥ to *p*; every $\mathit{1}$ assigns ⊥ to 'λ'.

For example, 'Socrates is wise' is true on an interpretation $\mathit{1}$ if the object named by 'Socrates' is in the extension in $\mathit{1}$ of 'is wise'; if $\mathit{1}$ is the real world, this just means that Socrates is one of those who are wise.

We can now use these six rules to give a complete demonstration of the invalidity of arguments A and B. For both arguments, we will use our first interpretation: D = {α,β}, Ext(F) = {α}, Ext(G) = {β}. We have to show that in each case, the premise of the argument-form is true and the conclusion is false.

Example 1: Show that the LMPL argument-form A is invalid.

> A: (∃x)Fx & (∃x)Gx
> ∴ (∃x)(Fx & Gx)

Interpretation: D = {α,β}, Ext(F) = {α}, Ext(*G*) = {β}, or as a matrix, below; 'a' refers to α and 'b' refers to β. *Explanation:* The premise is true because both

	F	G
α	+	−
β	−	+

conjuncts are true. '(∃x)Fx' is true because 'Fa' is true (rule ∃⊤). '(∃x)Gx' is true because 'Gb' is true (rule ∃⊤). 'Fa' is true because 'a' denotes α and α belongs to Ext(F), and 'Gb' is true because 'b' denotes β and β belongs to Ext(G) (rule A⊤). '(∃x)(Fx & Gx)' is false because 'Fa & Ga' is false *and* 'Fb & Gb' is false (rule ∃⊥). 'Fa & Ga' is false because 'Ga' is false, and 'Fb & Gb' is false because 'Fb' is false (we assume knowledge of the truth-table for '&'). Finally, 'Ga' is false because 'a' denotes α and α does not belong to Ext(G) and 'Fb' is false because 'b' denotes β and β does not belong to Ext(F).

Since this is our first example of a demonstration of invalidity for an argument-form in LMPL, we have spelled it out in complete detail. But in future, steps in the explanation which appeal to truth-tables to account for the truth-values of sentential combinations of atomic formulae (e.g., 'Fb & Gb') will be omitted. As abbreviations, we will use the standard mathematical symbols '∈' and '∉' for 'belongs to' and 'does not belong to' respectively. And we will not explicitly state the part of the interpretation that is covered by the alphabetic convention, in this case that 'a' denotes α and 'b' denotes β, unless the individual constant of LMPL in question actually occurs in the argument-form for which an interpretation is being given. If it does so occur, we state its reference in the form ⌜Ref(*t*) = x⌝, which abbreviates ⌜*t* refers to x⌝. And in the subsequent reasoning, we will not cite which of the quantifier rules we are appealing to when we use them. All this enables us to give a more succinct demonstration of B's invalidity.

Example 2: Show that the LMPL argument-form B is invalid.

> B: (∀x)(Fx ∨ Gx)
> ∴ (∀x)Fx ∨ (∀x)Gx.

Interpretation: As in Example 1. *Explanation:* The premise is true because both its instances '(Fa ∨ Ga)' and '(Fb ∨ Gb)' are true, since α ∈ Ext(F) and β ∈ Ext(G). The conclusion is false because both its disjuncts are: '(∀x)Fx' is false since 'Fb' is false (β ∉ Ext(F)), and '(∀x)Gx' is false since 'Ga' is false (α ∉ Ext(G)).

The most important points to notice about these examples are how we explain the falsity of the conclusion in **A** and the truth of the premise in **B**. In explaining why the conclusion of **A**, '(∃x)(Fx & Gx)', is false, we cite the falsity of *both* instances. It would not be sufficient to say that '(∃x)(Fx & Gx)' is false because 'Fa & Ga' is false, since that is not sufficient for the falsity of '(∃x)(Fx & Gx)' in a domain which contains other objects (recall the example 'at least one U.S. president was a spy'). Similarly, it would not be sufficient to explain the truth of '(∀x)(Fx ∨ Gx)' to cite merely the truth of one instance: the truth of *all* instances has to be cited.

As noted earlier, in forming the instances of a quantified sentence we use LMPL individual constants 'a', 'b' and so on, whose interpretation we stipulate. In **A** and **B** none of these names occur in the arguments themselves, but in §1 of Chapter 5 (page 149) we considered the intuitively invalid English arguments

> D: (1) Socrates is wise
> (2) ∴ Everyone is wise

and

> E: (1) Someone is happy
> (2) ∴ Plato is happy.

We now translate these into LMPL and demonstrate their invalidity.

Example 3: Show that the LMPL argument-form **D.s** is invalid.

> D.s: Wa
> ∴ (∀x)Wx

Interpretation: D = {α,β}, Ext(W) = {α}, Ref(a) = α. *Explanation:* '(∀x)Wx' is false because 'Wb' is false (since β ∉ Ext(W)), while 'Wa' is true since Ref(a) ∈ Ext(W).

Example 4: Show that the LMPL argument-form **E.s** is invalid.

> E.s (∃x)Hx
> ∴ Hb

Interpretation: D = {α,β}, Ext(H) = {α}, Ref(b) = β. *Explanation:* 'Hb' is false since β ∉ Ext(H), while '(∃x)Hx' is true since 'Ha' is true, since α ∈ Ext(W).

In these symbolizations of the English arguments, the interpreted metalanguage proper names 'Socrates' and 'Plato' are rendered by the uninterpreted LMPL individual constants 'a' and 'b' and then an interpretation is specified for the resulting argument-forms in which 'a' and 'b' denote α and β (whatever *they* are!) rather than Socrates and Plato. This further emphasizes the fact that the validity or invalidity of the original English arguments turns on their forms, not their subject matter.

We can now explicitly state the conception of interpretation and validity underlying the four demonstrations of invalidity just given. Since the object language LMPL includes LSL, the account of interpretation has to include the kind of interpretation appropriate for LSL as well as the kind appropriate for the new apparatus.

> An *interpretation* of an argument-form in LMPL consists in a specification of a nonempty domain of discourse together with an assignment of extensions to the predicate-letters, if any, in the argument-form, of references to the individual constants, if any, in the argument-form, and of truth-values to the sentence-letters, if any, in the argument-form. An extension for a predicate in an interpretation 7 is a subset of the domain of 7. 'λ' is always assigned $\bot$.

It should be noted that since the empty set is a subset of every set, this definition allows us to assign an empty extension to a predicate-letter; in other words, we can have predicate-letters which are true of no objects, just as we can have English predicates which are true of no objects (consider 'is a Martian').[1] A standard symbol for the empty set is '$\varnothing$'.

> An argument-form in LMPL is *valid* if and only if there is no interpretation of it on which all its premises are true and its conclusion is false.

> An English argument is *monadically* valid if and only if it translates into a valid argument-form of LMPL.

As in the sentential case, a monadically valid English argument is valid absolutely. A monadically invalid English argument may be absolutely invalid, or it may translate into a valid argument-form of a more powerful kind of logic. D and E have just been shown to be monadically invalid. Although we cannot *prove* that they are invalid absolutely, it seems plausible that they are, since their translations into LMPL appear to capture all relevant aspects of the structure of the English sentences.

The definition of LMPL validity is like the definition of LSL validity; what has changed is just the notion of interpretation in the two definitions. Another definition which carries over from LSL is that of logical equivalence:

> If p and q are sentences of LMPL, p and q are said to be *logically equivalent* if and only if, for each interpretation 7, the truth-value of p on 7 is the same as the truth-value of q on 7.

---

[1] Why is the empty set a subset of every set? To say that X is a subset of Y is to say that every member of X is a member of Y, or in symbols, '$(\forall x)(x \in X \to x \in Y)$'. If X is the empty set, then no matter what the domain of discourse, every instance of '$(\forall x)(x \in X \to x \in Y)$' will be true since the antecedent of every instance will be false. Consequently, no matter what set Y is, if X is empty, X is a subset of Y.

We now wish to apply these techniques to demonstrate particular LMPL arguments to be invalid. But before beginning on this, we need to familiarize ourselves further with the truth and falsity conditions of sentences of LMPL. Here is an interpretation, chosen completely at random, which is rather more complex than the two we have considered so far:

	F	G	H	I	J
α	+	+	−	+	−
β	−	−	−	+	+
γ	+	−	−	−	+

Thus D = $\{\alpha,\beta,\gamma\}$, Ext(F) = $\{\alpha,\gamma\}$, Ext(G) = $\{\alpha\}$, Ext(H) = $\varnothing$, Ext(I) = $\{\alpha,\beta\}$, Ext(J) = $\{\beta,\gamma\}$. The following six sentences are also chosen at random, and we have to determine their truth-values in this interpretation.

> (1) ~Ja
> (2) Fc → Ic
> (3) (∃x)(Jx ↔ Hx)
> (4) (∀x)(Jx → (Gx ∨ Fx))
> (5) (∃x)Gx → (∀y)(Fy ∨ Gy)
> (6) (∃y)(∀x)(Gy & (Jx → (Ix ∨ Fx)))

(1) '~Ja' is true because 'Ja' is false, since $\alpha \notin$ Ext(J).
(2) 'Fc → Ic' is false because $\gamma \in$ Ext(F) and $\gamma \notin$ Ext(I).
(3) '(∃x)(Jx ↔ Hx)' is true because 'Ja ↔ Ha' is true, since $\alpha \notin$ Ext(J) and $\alpha \notin$ Ext(H).
(4) '(∀x)(Jx → (Gx ∨ Fx))' is false because 'Jb → (Gb ∨ Fb)' is false, since $\beta \in$ Ext(J) but $\beta \notin$ Ext(G) and $\beta \notin$ Ext(F).
(5) '(∃x)Gx → (∀y)(Fy ∨ Gy)' is false because '(∃x)Gx' is true and '(∀y)(Fy ∨ Gy)' is false. '(∃x)Gx' is true because 'Ga' is true, since $\alpha \in$ Ext(G). '(∀y)(Fy ∨ Gy)' is false because 'Fb ∨ Gb' is false, since $\beta \notin$ Ext(F) and $\beta \notin$ Ext(G).
(6) '(∃y)(∀x)(Gy & (Jx → (Ix ∨ Fx)))' is true because '(∀x)(Ga & (Jx → (Ix ∨ Fx)))' is true, in turn because '(Ga & (Ja → (Ia ∨ Fa)))', '(Ga & (Jb → (Ib ∨ Fb)))' and '(Ga & (Jc → (Ic ∨ Fc)))' are all true. '(Ga & (Ja → (Ia ∨ Fa)))' is true because $\alpha \in$ Ext(G) and $\alpha \notin$ Ext(J). '(Ga & (Jb → (Ib ∨ Fb)))' is true because $\alpha \in$ Ext(G) and $\beta \in$ Ext(I). And '(Ga & (Jc → (Ic ∨ Fc)))' is true because $\alpha \in$ Ext(G) and $\gamma \in$ Ext(F).

In all our evaluations we follow the syntactic structure of the formula: we begin with the main subformulae and then work down through their main subformulae, and so on until we arrive at atomic formulae. (5) should be noted. (5) is a conditional, *not* a quantified sentence, and it would be a mistake to begin its evaluation by applying a quantifier rule. Rather, since (5) is a conditional, we calculate the truth-values of its antecedent and consequent separately, and then use the truth-table for '→'.

Special attention should be paid to (6), in which two quantifiers prefix the body of the formula. This makes the search process longer. Since (6) is existential, we can find out whether it is true or false by finding out whether or not it has a true instance. The instances of (6) are

(6a) (∀x)(Ga & (Jx → (Ix ∨ Fx)))
(6b) (∀x)(Gb & (Jx → (Ix ∨ Fx)))
(6c) (∀x)(Gc & (Jx → (Ix ∨ Fx))).

Each of these instances is a universal sentence, and each of them in turn has three instances. For example, the instances of (6b) are:

(6b$_1$) Gb & (Ja → (Ia ∨ Fa))
(6b$_2$) Gb & (Jb → (Ib ∨ Fb))
(6b$_3$) Gb & (Jc → (Ic ∨ Fc))

Had (6) been false, therefore, we would have had to consider a total of nine quantifier-free sentences to confirm this. Fortunately, (6) is true, and this is shown by its instance (6a).[2]

❑ Exercise

Evaluate the numbered formulae in the displayed interpretation. Explain your reasoning in the same way as in (1)–(6) above, accounting for the truth-values of quantified sentences in terms of the truth-values of their instances.

	F	G	H	I	J
α	+	−	+	−	+
β	+	−	+	−	−
γ	+	−	−	+	+

(1) (Ha ∨ Hc) → Ib
*(3) (∃x)(Fx & Gx)
(5) (∃x)(Ix → Hx)
(7) (∀x)((Fx & Hx) → Jx)
(9) (∀x)(∃y)(Fx → (Hx ∨ Jy))
(11) (∃x)(Ix → (∀y)(Jy → Iy))
(12) (∀x)(∀y)((Fx ↔ Gy) ↔ (∃w)(∃z)(Hw & Jz))

(2) (Ha & Hc) ∨ (Ja & Jc)
(4) ~(∃x)Gx
(6) (∀x)((Hx ∨ Ix) → Fx)
*(8) (∀x)(Hx → (∃y)(Jx & Iy))
(10) (∃x)Ix → (∀x)(Jx → Ix)

---

[2] The rules of this section explain the term 'logical constant' mentioned earlier, which is applied both to quantifiers and to sentential connectives. Domains and extensions of predicates *vary* from interpretation to interpretation, but the evaluation rules for connectives and quantifiers are *constant* across all interpretations. Any expression which has a constant evaluation rule is called a logical constant.

2 Constructing counterexamples

Now that we understand how sentences of LMPL are evaluated in interpreta-tions, we turn to the question of how to find interpretations which show invalid LMPL argument-forms to be invalid. An interpretation which shows an LMPL argument-form to be invalid is called a *counterexample* to the argument-form. The question, then, is how to go about constructing counterexamples. We will illustrate the techniques in connection with a number of examples. But first we reintroduce the double-turnstile notation, which we are going to use to express semantic consequence exactly as we did for LSL in §4 of Chapter 3:

- For any sentences $p_1,...,p_n$ and q of LMPL, we write $p_1,...,p_n \vDash q$ to mean that q is a semantic consequence of $p_1,...,p_n$, that is, that no interpretation of $p_1,...,p_n$ and q makes all of $p_1,...,p_n$ true and q false.
- For any sentences $p_1,...,p_n$ and q of LMPL, we write $p_1,...,p_n \nvDash q$ to mean that q is *not* a semantic consequence of $p_1,...,p_n$, that is, that *some* interpretation of $p_1,...,p_n$ and q makes all of $p_1,...,p_n$ true and q false.
- For any sentence q of LMPL, we write $\vDash q$ to mean that there is no interpretation that makes q false, or in other words, that every interpretation makes q true. Such a q is said to be *logically true*. An example: $\vDash (\forall x)((Fx \ \& \ Gx) \rightarrow Fx)$.

Logical truth is the special case $n = 0$ of semantic consequence, in that a logical truth is a semantic consequence of the empty set of premises.

To give a counterexample to an argument-form with premises $p_1,...,p_n$ and conclusion q is to show that $p_1,...,p_n \nvDash q$, and this will be our preferred way of expressing our goal.

Example 1: Show $(\forall x)(Fx \rightarrow Gx), (\forall x)(Fx \rightarrow Hx) \nvDash (\forall x)(Gx \rightarrow Hx)$.

Provisionally, we begin by setting up a domain $D = \{\alpha\}$. No matter what the problem, this can always be the first step, since every interpretation must have a nonempty domain. We say that the specification of D at this stage is provi-sional because in the course of making the conclusion false and the premises true, it may be necessary to add further objects to the domain.

When the conclusion formula is a universal sentence, it is clear what we must do: we must arrange that some object in the domain provides a false instance of the universal sentence. Such an object is also known as a counterex-ample—not a counterexample to the argument-form but to the universal sen-tence. In this particular case, what is required is an object in the extension of 'G' which is not in the extension of 'H'. So we set Ext(G) = $\{\alpha\}$, Ext(H) = $\varnothing$; again, this is merely provisional, since it may later be necessary to add to the exten-sions of 'G' and 'H'. Our interpretation now makes the conclusion false, but in order to evaluate the premises in it we have to specify an extension for 'F'. To make both premises true, what we must avoid is having an object in the extension of 'F' which is not in the extension of 'G' or not in the extension of 'H'. In the current setup, the simplest way of avoiding such an object is to let the extension of 'F' be empty. So we do not need to add to the domain or to the extensions of the other two predicates. The interpretation we arrive at is: D = {α}, Ext(F) = ∅, Ext(G) = {α} and Ext(H) = ∅, with the matrix displayed below. To

$$
\begin{array}{c|ccc}
 & F & G & H \\
\hline
\alpha & - & + & - \\
\end{array}
$$

complete the solution to the problem, we explain why this interpretation is a counterexample. '(∀x)(Gx → Hx)' is false because 'Ga → Ha' is false, since α ∈ Ext(G) and α ∉ Ext(H). '(∀x)(Fx → Gx)' and '(∀x)(Fx → Hx)' are both true because 'Fa → Ga' and 'Fa → Ha' are both true, since α ∉ Ext(F).

Example 2: Show (∃x)(Fx & Gx), (∃x)(Fx & Hx), (∀x)(Gx → ~Hx) ⊭
    (∀x)(Fx ↔ (Gx ∨ Hx)).

We begin as before with {α} as provisional domain. A counterexample to '(∀x)(Fx ↔ (Gx ∨ Hx))' requires either (i) an object in Ext(F) which is in neither Ext(G) nor Ext(H), or else (ii) an object in at least one of Ext(G) and Ext(H) which is not in Ext(F). Since premise 1 will require an object in Ext(F) anyway, we start with (i) (if we can find no way of making all the premises true, we will have to come back to (ii)). Provisionally, then, we set Ext(F) = {α} and put nothing into the extensions of 'G' and 'H'; this gives us the false conclusion-instance 'Fa ↔ (Ga ∨ Ha)'. Turning to the premises, we see that to make all three true we have to set up Ext(F), Ext(G) and Ext(H) so that true instances of the first two premises are provided and at the same time no object is in both Ext(G) and Ext(H) (otherwise premise 3 would be false). We could obtain a true instance of premise 1 by adding α to Ext(G), but this would defeat what we have already done to ensure the falsity of the conclusion. Consequently, we have to add a second object to the domain to provide a true instance of premise 1. Thus we now put D = {α,β}, Ext(F) = {α,β}, Ext(G) = {β} (making 'Fb & Gb' true) and add nothing to the extension of 'H', as in the matrix below. This means that we do

$$
\begin{array}{c|ccc}
 & F & G & H \\
\hline
\alpha & + & - & - \\
\beta & + & + & - \\
\end{array}
$$

not yet have a true instance for premise 2. If we set Ext(H) = {α} we have a true instance for premise 2, but we will have made 'Fa ↔ (Ga ∨ Ha)' true, and as 'Fb ↔ (Gb ∨ Hb)' is also true, we would have made the conclusion true. And if we set Ext(H) = {β}, we have an object in both Ext(G) and Ext(H), which is exactly

$$
\begin{array}{c|ccc}
 & \text{F} & \text{G} & \text{H} \\
\hline
\alpha & + & - & - \\
\\
\beta & + & + & - \\
\\
\gamma & + & - & + \\
\end{array}
$$

what we must avoid in order to verify premise 3. Consequently, we must add a third object to D to provide a true instance for premise 2 that does not refute premise 3. The interpretation at which we arrive, therefore, is D = {α,β,γ}, Ext(F) = {α,β,γ}, Ext(G) = {β}, Ext(H) = {γ}, displayed above, and we confirm that this is right with the following explanation: '($\forall$x)(Fx ↔ (Gx ∨ Hx))' is false because 'Fa ↔ (Ga ∨ Ha)' is false, since α ∈ Ext(F), α ∉ Ext(G) and α ∉ Ext(H). '($\exists$x)(Fx & Gx)' is true because 'Fb & Gb' is true, since β ∈ Ext(F) and β ∈ Ext(G). '($\exists$x)(Fx & Hx)' is true because 'Fc & Hc' is true, since γ ∈ Ext(F) and γ ∈ Ext(H). '($\forall$x)(Gx → ~Hx)' is true because all three of 'Ga → ~Ha', 'Gb → ~Hb' and 'Gc → ~Hc' are true. First, 'Ga → ~Ha' is true because α ∉ Ext(G); next, 'Gb → ~Hb' is true because β ∈ Ext(G) and β ∉ Ext(H); last, 'Gc → ~Hc' is true because γ ∉ Ext(G).

Example 2 illustrates the usual reason for increasing the size of the domain: a number of existential premises require different objects to provide true instances, since using the same object for all those premises would make another premise false or the conclusion true.

The next example illustrates how we handle arguments which contain sentence-letters as well as predicates and quantifiers.

Example 3: Show ($\exists$x)(Fx → A) ⊭ ($\exists$x)Fx → A.

It is important to note the different forms of the premise and conclusion: the premise is an existential sentence in which '→' is within the scope of '$\exists$', while the conclusion is a conditional in which '$\exists$' is within the scope of '→'. Since the conclusion has antecedent '($\exists$x)Fx' and consequent 'A', we require an interpretation in which '($\exists$x)Fx' is true and 'A' is false. However, the interpretation $\mathcal{I}$ with D = {α}, Ext(F) = {α}, and ⊥ assigned to 'A' also makes the premise false, since there is only the instance 'Fa → A'. To make the premise true, we have to provide another instance, which means adding an object to the domain. But since 'A' is false and we want the new object to provide a true instance of the premise, we should *not* add the new object to the extension of 'F'. So the interpretation at which we arrive is: D = {α,β}, Ext(F) = {α}, ⊥ assigned to 'A', as displayed below. *Explanation:* '($\exists$x)Fx → A' is false because 'A' is false and '($\exists$x)Fx'

$$
\begin{array}{c|c}
 & \text{F} \\
\hline
\alpha & + \\
\\
\beta & - \\
\end{array}
$$

'A' is false

is true; '(∃x)Fx' is true because α ∈ Ext(F). And '(∃x)(Fx → A)' is true because 'Fb → A' is true, since β ∉ Ext(F).

Finally we give an example involving successive quantifiers.

Example 4: Show (∃x)(∀y)(Fx → Gy) ⊭ (∀x)(∃y)(Fx → Gy).

To make the conclusion false, we must have at least one of its instances false. On any interpretation, the instances of '(∀x)(∃y)(Fx → Gy)' are existential sentences of the form '(∃y)(F_ → Gy)', where the blank is filled by an individual constant. One of these instances must be false for '(∀x)(∃y)(Fx → Gy)' to be false. Since our domain will contain α anyway, we may as well begin by making '(∃y)(Fa → Gy)' false. This means, by (∃⊥), that all *its* instances have to be false, so in particular '(Fa → Ga)' must be false. Hence α ∈ Ext(F), α ∉ Ext(G). This gives us the interpretation immediately below. However, as things stand in this

	F	G
α	+	-

interpretation, the premise of Example 4 is false. The premise is an existential sentence, and so only requires one true instance for it to be true itself. But with just α in the domain, the premise has only one instance, '(∀y)(Fa → Gy)', and this universal sentence is false because it has the false instance '(Fa → Ga)'. Since we do not want to alter any entries we have already made in the interpretation (they were required to make the conclusion false) it follows that to make the premise '(∃x)(∀y)(Fx → Gy)' true, we should provide it with another instance '(∀y)(Fb → Gy)' and ensure that this instance is true.

There are now two constraints to satisfy simultaneously: we have to make '(∀y)(Fb → Gy)' true, which means making both its instances 'Fb → Ga' and 'Fb → Gb' true, and at the same time we have to avoid doing anything that would make the conclusion true. Since 'Ga' is false, our only option for making 'Fb → Ga' true is to make 'Fb' false as well. Consequently, we set β ∉ Ext(F), and then 'Fb → Ga' and 'Fb → Gb' are both true, so '(∀y)(Fb → Gy)' is true as desired, which in turn makes the premise '(∃x)(∀y)(Fx → Gy)' true. As for the conclusion, the important thing is to keep its previously false instance '(∃y)(Fa → Gy)' still false. The new instance of this existential sentence is 'Fa → Gb', and since 'Fa' is true, we have to set β ∉ Ext(G) to make this conditional false. So the final interpretation is D = {α,β}, Ext(F) = {α}, Ext(G) = ∅, as exhibited below:

	F	G
α	+	-
β	-	-

To summarize: '(∃x)(∀y)(Fx → Gy)' has two instances, (i) '(∀y)(Fa → Gy)' and (ii) '(∀y)(Fb → Gy)', and it is true because (ii) is true. (ii) is true because it has two instances, 'Fb → Ga' and 'Fb → Gb' and both are true since both have false antecedents. On the other hand, '(∀x)(∃y)(Fx → Gy)' is false. It has two instances, (iii) '(∃y)(Fa → Gy)' and (iv) '(∃y)(Fb → Gy)', and (iii) is false. (iii) is false because both its instances, '(Fa → Ga)' and 'Fa → Gb', are false, since both have true antecedent and false consequent.

It is noticeable that all our problems of showing failure of semantic consequence have been solved with small domains, whereas in Chapter 5, the domains with respect to which our symbolizations are relativized are large: people, places, things. But counterexamples with small domains to argument-forms derived from symbolizations of English relativized to large domains are not irrelevant to English arguments, for if the argument-form can be shown to be invalid by an interpretation with a small domain, then it *is* shown to be invalid, and if it is the form of an English argument, it follows that that English argument is monadically invalid. Moreover, a counterexample with a small domain can be 'blown up' into one with a large domain by a duplication process (see Exercise II.2), so our preference for simplicity does not entail irrelevance.

❑ Exercises

I   Show the following, with explanations:

    (1) (∀x)(Fx → Gx) ⊭ (∀x)(Gx → Fx)
    (2) (∀x)(Fx ∨ Gx), (∀x)(Fx ∨ Hx) ⊭ (∀x)(Gx ∨ Hx)
    (3) (∀x)(Fx → ~Gx), (∀x)(Gx → Hx) ⊭ (∀x)(Fx → ~Hx)
 *(4) (∀x)((Fx & Gx) → Hx) ⊭ (∀x)(Fx ∨ Gx) ∨ (∀x)(Fx ∨ Hx)
    (5) (∃x)(Fx & ~Hx), (∃x)(Gx & ~Hx) ⊭ (∃x)(Fx & Gx)
    (6) (∃x)(Fx ↔ Gx) ⊭ (∃x)(Fx ∨ Gx)
    (7) (∃x)(Fx & Gx), (∀x)(Gx → Hx) ⊭ (∀x)(Fx → Hx)
    (8) (∀x)Fx → (∃x)Gx ⊭ (∀x)(Fx → Gx)
    (9) (∃x)(Fx ∨ Gx), (∀x)(Fx → ~Hx), (∃x)Hx ⊭ (∃x)Gx
  (10) (∀x)(Fx → Gx) ⊭ ~(∀x)(Fx → ~Gx)
  (11) (∃x)~Fx ⊭ ~(∃x)Fx
  (12) ~(∀x)Fx ⊭ (∀x)~Fx
 *(13) (∀x)(Fx → Gx) → (∀x)(Hx → Jx) ⊭ (∃x)(Fx & Gx) → (∀x)(Hx → Jx)
  (14) (∃x)(Fx → A), (∃x)(A → Fx) ⊭ (∀x)(A ↔ Fx)
  (15) ~(A → (∀x)Fx) ⊭ (∀x)(A → ~Fx)
  (16) (∀x)Fx ↔ A ⊭ (∀x)(Fx ↔ A)
  (17) (∀x)Fx → (∀x)Gx ⊭ Fa → (∀x)Gx
  (18) Fa → (∃x)Gx ⊭ (∃x)Fx → (∃x)Gx
  (19) (∀x)Fx ↔ (∀x)Gx ⊭ (∃x)(Fx ↔ Gx)
 *(20) (∀x)Fx → (∃y)Gy ⊭ (∀x)(Fx → (∃y)Gy)
  (21) (∃x)(Fx → (∀y)Gy) ⊭ (∃x)Fx → (∀y)Gy
  (22) ~(∃x)Fx ∨ ~(∃x)Gx ⊭ ~(∃x)(Fx ∨ Gx)

(23) $(\exists x)(Fx \leftrightarrow Gx), (\forall x)(Gx \rightarrow (Hx \rightarrow Jx)) \nVdash (\exists x)Jx \lor \sim(\exists x)Fx$
(24) $(\forall x)(\exists y)(Fy \rightarrow Gx) \nVdash (\forall x)(\exists y)(Gy \rightarrow Fx)$
(25) $(\exists x)(Fx \rightarrow (\exists y)Gy) \nVdash (\exists x)(\forall y)(Fx \rightarrow Gy)$
(26) $(\forall x)(\exists y)(Fx \rightarrow Gy) \nVdash (\exists x)(\forall y)(Fx \rightarrow Gy)$
*(27) $(\forall x)(\exists y)(Gy \rightarrow Fx) \nVdash (\forall x)[(\exists y)Gy \rightarrow Fx]$
(28) $(\forall x)[(\forall y)Gy \rightarrow Fx] \nVdash (\forall x)(\forall y)(Gy \rightarrow Fx)$
(29) $(\forall x)(Fx \rightarrow (\exists y)Gy) \nVdash (\forall x)(\forall y)(Fx \rightarrow Gy)$
(30) $(\exists x)(\forall y)(Fx \rightarrow Gy) \nVdash (\exists y)(\forall x)(Fx \rightarrow Gy)$

II   Show $(\exists x)(Fx \& Gx) \& (\exists x)(Fx \& \sim Gx) \& (\exists x)(\sim Fx \& Gx) \nVdash (\forall x)(Fx \lor Gx)$. Then evaluate the following two statements as true or false. Explain your answer.

> (1) If a sentence is true on at least one interpretation whose domain has n members ($n \geqslant 2$), it is true on at least one interpretation whose domain has $n - 1$ members.
>
> *(2) If a sentence is true on at least one interpretation whose domain has n members ($n \geqslant 2$), it is true on at least one interpretation whose domain has $n + 1$ members. (Hint: think of how you could define the notion of two objects being *indistinguishable* in an interpretation.)

3 Deductive consequence: quantifiers in NK

It is perfectly natural to respond to an English argument by saying that the conclusion does not follow from the premises and in support of this to describe a possible situation in which the premises of the argument would be true and the conclusion false. Our abstract model of this procedure, as described in the previous section, is therefore quite realistic. However, we do not usually *advance* an argument by stating our premises and conclusion and defying an opponent to describe a situation in which the premises would be true and the conclusion false (the strategy is not totally unnatural: it is embodied in the rhetorical question 'How could it *fail* to follow?'). In advancing an argument it is much more common to try to reason from the premises to the conclusion. So we now wish to extend our formal model of this procedure, the natural-deduction system NK, to those arguments which involve quantification and predication. Actual arguments in natural language are taking place in an interpreted language, of course. But the reasoning principles involved in giving an argument are independent of interpretation and can be stated for an uninterpreted formal language like LMPL.

The new logical constants are the two quantifiers, so at the very least we shall have to add two I and two E rules for these symbols to NK. However, we shall continue to call the system 'NK' rather than give it a new name to indicate the presence of new rules; whenever we want to contrast NK before and after the new rules are added we will speak of sentential NK versus quantificational NK. This section introduces quantificational NK through three of the four new rules. The fourth, being more complicated, is postponed until the next section. Then we conclude the discussion of natural deduction with some considerations about how Sequent Introduction can be extended into quantificational NK.

Each quantifier must be provided with an introduction and an elimination rule. The introduction rule for '∃' and the elimination rule for '∀' are the most straightforward, and arguments A and B from the beginning of Chapter 5 consist essentially in one application of ∃I and ∀E respectively:

> A: (1) Socrates is wise
> (2) ∴ Someone is wise

> B: (1) Everyone is happy
> (2) ∴ Plato is happy

Intuitively, if we have established a statement ϕt which contains occurrences of an individual constant t then we can infer a related existential sentence: whatever is true of a particular object is true of *something*. However, if there is more than one occurrence of t in ϕt, more than one existential sentence may be inferred. For example, from 'Wa & Ha' we can infer '(∃x)(Wx & Ha)', '(∃x)(Wa & Hx)' and '(∃x)(Wx & Hx)'. By contrast, ∀E involves using a particular individual constant to replace *every* occurrence of the variable bound by the quantifier being eliminated. Thus from '(∀x)(Fx → Gx)' we can infer 'Fa → Ga', 'Fb → Gb', 'Fc → Gc', and so on, but not 'Fa → Gx', which is an open sentence, and not 'Fb → Gc', since we cannot change individual constants in midstream. (Indeed, it is easy to see that (∀x)(Fx→ Gx) ⊭ Fb → Gc; consider the interpretation D = {β,γ}, Ext(F) = {β}, Ext(G) = {β}. Here we have '(∀x)(Fx → Gx)' true since 'Fb → Gb' and 'Fc → Gc' are both true, while 'Fb → Gc' is false.) Intuitively, ∀E expresses the idea that whatever is true of everything is true of any single thing we care to mention.

The new rules are simply to be added to NK as it stands; no changes are being made in the layout of proofs. We therefore state the rules in the same format as the rules of sentential NK:

> ***Rule of ∃-Introduction:*** For any sentence ϕt, if ϕt has been inferred at a line j in a proof, then at line k we may infer $\ulcorner(\exists v)\phi v\urcorner$, labeling the line 'j ∃I' and writing on its left the same numbers as occur on the left of line j. Schematically,

$$
\begin{array}{lll}
a_1,...,a_n & (j) & \phi t \\
& \vdots & \\
a_1,...,a_n & (k) & (\exists v)\phi v \qquad j\ \exists I
\end{array}
$$

$\ulcorner(\exists v)\phi v\urcorner$ is obtained syntactically from ϕt by replacing one or more occurrences of t in ϕt by an individual variable v, *which must not already occur in ϕt*, and then by prefixing $\ulcorner(\exists v)\urcorner$.

The point of the proviso that v must not already occur in ϕt is to avoid double-binding. For example, we can apply ∃I to the formula '(∃x)(Wa & Hx)', where t = 'a' and ϕ = '(∃x)(W_ & Hx)', to obtain '(∃y)(∃x)(Wy & Hx)', but we cannot put 'x' for 'a' to obtain '(∃x)(∃x)(Wx & Hx)', since 'x' already occurs in '(∃x)(Wa & Hx)'. The rule also prevents us from inferring a sentence that contains a vacuous quantifier. From '(∃x)(Wx & Hx)' we cannot infer '(∃y)(∃x)(Wx & Hx)' since this requires taking ϕ to be '(∃x)(Wx & Hx)', so that in this purported application of ∃I there is no t such that *one or more* occurrences of t are replaced by 'y'.

>**Rule of ∀-Elimination:** For any sentence $\ulcorner(\forall v)\phi v\urcorner$ and individual constant t, if $\ulcorner(\forall v)\phi v\urcorner$ has been inferred at a line j in a proof, then at line k we may infer ϕt, labeling the line 'j ∀E' and writing on its left the same numbers as appear on the left at line j. Schematically,

$$a_1,...,a_n \quad \text{(j)} \qquad (\forall v)\phi v$$
$$\vdots$$
$$a_1,...,a_n \quad \text{(k)} \qquad \phi t \qquad\qquad \text{j ∀E}$$

ϕt is obtained syntactically from $\ulcorner(\forall v)\phi v\urcorner$ by deleting the quantifier prefix $\ulcorner(\forall v)\urcorner$ and then replacing *every* occurrence of v in the open sentence ϕv by *one and the same* individual constant t. Since '∀' means literally *everything*, there are no constraints on which individual constant may be chosen.

It is extremely important to remember that the connective to which an elimination rule is applied at a line must be the *main* connective of the formula on that line. It would be entirely incorrect to infer 'Gb → A' from '(∀x)Gx → A' by a purported application of ∀E. '(∀x)Gx → A' is a conditional and its main connective is '→'; consequently, the only E-rule which can be applied to it is →E. Intuitively, 'if everyone works hard then the project will be completed on time' does not entail 'if Bill works hard then the project will be completed on time', since if Bill works hard while everyone else takes it easy, the project will *not* be completed on time. It is simple to construct a formal counterexample to the inference, which shows that (∀x)Gx → A ⊭ Gb → A. Also, Gb → A ⊭ (∃x)Gx → A: intuitively, it may be something special about b which would make its possession of G-ness sufficient for the truth of 'A'. For example, even if it is true that if Mary works hard then the project will be completed on time, it does not follow that if someone (or other) works hard then the project will be completed on time—perhaps it has to be *Mary*. Therefore '(∃x)Gx → A' does not follow from 'Gb → A'. And while it is indeed the case that A → Gb ⊨ A → (∃x)Gx (if the project will be completed on time only if Bill works hard, then it follows that it will be completed on time only if someone works hard), we cannot make this inference in one line citing ∃I, since ∃I always *prefixes* a quantifier. Thus from 'A → Gb', using ∃I we can only obtain '(∃x)(A → Gx)' or an alphabetic variant such as '(∃z)(A → Gz)'. To circumvent this difficulty, we derive 'A → (∃x)Gx' using →I, and this will be our first example of a proof in quantificational NK.

As before, we write '$p_1,...,p_n \vdash_{NK} q$', to mean that $p_1,...,p_n$ deductively entail q in NK; the definition is word-for-word identical to the one on page 100.

Example 1: Show A → Gb $\vdash_{NK}$ A → (∃x)Gx.

Since the conclusion is a conditional we should use →I to derive it, so we have the subgoal of obtaining '(∃x)Gx' from the premise 'A → Gb' and the assumption 'A', and that, as we see below, is not difficult.

1	(1)	A → Gb	Premise
2	(2)	A	Assumption
1,2	(3)	Gb	1,2 →E
1,2	(4)	(∃x)Gx	3 ∃I
1	(5)	A → (∃x)Gx	2,4 →I ◆

When the conclusion has a quantifier as *main* connective, then the last line of the proof will often be obtained by a quantifier introduction rule, as in this example:

Example 2: Show (∀x)(Fx → Gx), Fa $\vdash_{NK}$ (∃x)(~Gx → Hx).

Since the conclusion is an existential formula, we should expect to infer it with ∃I. The previous line will therefore be a formula $\ulcorner$~Gt → Ht$\urcorner$ for some individual constant t, and we use →I to obtain this formula:

1	(1)	(∀x)(Fx → Gx)	Premise
2	(2)	Fa	Premise
3	(3)	~Ga	Assumption
4	(4)	~Ha	Assumption
1	(5)	Fa → Ga	1 ∀E
1,2	(6)	Ga	2,5 →E
1,2,3	(7)	⋏	3,6 ~E
1,2,3	(8)	~~Ha	4,7 ~I
1,2,3	(9)	Ha	8 DN
1,2	(10)	~Ga → Ha	3,9 →I
1,2	(11)	(∃x)(~Gx → Hx)	10 ∃I ◆

This example illustrates a standard pattern in quantificational proofs: quantifiers are removed from the premises using elimination rules, sentential rules are applied, and then quantifiers are reintroduced using introduction rules to obtain the conclusion.

We turn now to the rule of Universal Introduction. Suppose we are asked to prove the following valid sequent:

Example 3: Show (∀x)(Fx → Gx), (∀x)Fx $\vdash_{NK}$ (∀x)Gx.

Since the conclusion is a universal sentence, we would expect to obtain the last line of a proof of it by a rule of ∀-Introduction. We might try the following:

1	(1)	(∀x)(Fx → Gx)	Premise
2	(2)	(∀x)Fx	Premise
1	(3)	Fb → Gb	1 ∀E
2	(4)	Fb	2 ∀E
1,2	(5)	Gb	3,4 →E
1,2	(6)	(∀x)Gx	5 ∀I ◆

But the last line of this proof looks suspicious. At line 5 we have derived 'Gb'. But 'b' stands for just *one* object: how can (5) justify (6), which says that *every* object has the property G? To emphasize the apparently objectionable nature of the move from (5) to (6) we note that what seems to be the same step would allow us to prove an *invalid* argument-form. We have (∀x)(Fx → Gx), Fb ⊭ (∀x)Gx, as the interpretation D = {α,β}, Ext(F) = {β}, Ext(G) = {β}, shows: 'Fb' is true, and '(∀x)(Fx → Gx)' is true because 'Fa → Ga' and 'Fb → Gb' are both true, while '(∀x)Gx' is false because 'Ga' is false. But if the proof in Example 3 is correct, what would be wrong with the following?

Example 4: Show (∀x)(Fx → Gx), Fb ⊢NK (∀x)Gx (?)

1	(1)	(∀x)(Fx → Gx)	Premise
2	(2)	Fb	Premise
1	(3)	Fb → Gb	1 ∀E
1,2	(4)	Gb	2,3 →E
1,2	(5)	(∀x)Gx	4 ∀I (no!) ◆

However, even though the step from (4) to (5) in this 'proof' looks no worse than the step from (5) to (6) in Example 3, there is actually a significant difference, which we can bring out using our semantic account of the universal quantifier. A universal sentence is true in an interpretation *7* if and only if all its instances in *7* are true. This means that in each interpretation *7*, a universal sentence is true if and only if the *conjunction* of all its instances in *7* is true. Indeed, one could regard a universal statement such as '(∀x)Fx' as simply a convenient way of asserting a conjunction 'Fa & (Fb & (Fc &...))' which continues with one conjunct for each object which exists, a conjunction that is infinitely long if there are infinitely many objects.[3]

This perspective on universal sentences implies a constraint on when we can infer a universal sentence in an interpreted language: a universal sentence can be inferred in the same circumstances as the conjunction of its instances given by the domain. To infer a conjunction, we have to be able to infer every conjunct, so the same condition should apply to universal sentences. But in the

---

[3] The analogy between universal quantification and conjunction has its limitations, since the conjunction of instances does not say that *all* objects have been mentioned. Any conjunction of instances is consistent with the addition of one more conjunct using a name that has not already appeared, a conjunct which is the *negation* of an instance; for example, '(Fa & Fb) & ~Fc'. But the universal quantification '(∀x)Fx', of course, is not consistent with this extra conjunct.

course of ordinary reasoning in English, where the background domain of discourse may be very large (or infinite, if we are reasoning about numbers), we could not literally be required to infer every conjunct which is an instance of the universal sentence at which we are aiming, since there would be far too many conjuncts for this to be practical. Fortunately, the ability to infer every conjunct can be reduced to the ability to infer a single conjunct *in a particular way*. What is required is that we should infer a single conjunct *by a sequence of rule-applications that would equally well serve to effect the derivation of any other conjunct from the same premises and assumptions*. And since the notion of two conjuncts being different is purely syntactic, this principle about reasoning with '∀' can sensibly be applied to an uninterpreted language like LMPL.

Examples 3 and 4 differ precisely in respect of derivability of other conjuncts by the same steps. In Example 4 the conjunct 'Gb' we derive at line 4 is obtained by reasoning that cannot be used to infer other conjuncts, other instances of '(∀x)Gx'. For example, if we try to derive 'Gd' by inferring 'Fd → Gd' at line 3 using ∀E, we will be unable to make any progress, since we need a conditional with 'Fb' as its antecedent in order to be able to use premise 2. And 'Fb→Gd' is of course *not* obtainable from '(∀x)(Fx→Gx)' by ∀E, since we cannot put different individual constants for the same variable. The alleged 'proof' of Example 4 contrasts with the proof of Example 3, where we *can* use the same rules in the same order to derive any other conjunct; for we can use any name we like in applying ∀E to the two premises, since both are universal sentences, and so we can obtain ⌜Gt⌝ for any individual constant *t* whatsoever. It is for this reason that the final introduction of a universal quantifier is justified in Example 3 but not in Example 4.

When the reasoning used in a proof to derive a sentence ϕt could also be used to derive any other sentence $\phi t'$ which is the same as ϕt except for containing t' in exactly the positions where ϕt contains t, we say the reasoning is *generalizable*. In Example 3, then, the sequence of steps /1,∀E/2,∀E/3,4,→E/ is generalizable reasoning, while in Example 4 the step-sequence /1,∀E/2,3,→E/ is not generalizable reasoning. The rule of Universal Introduction should say that from ϕt we can infer ⌜(∀v)ϕv⌝ if and only if the reasoning that leads to ϕt is generalizable.

It is fairly easy to see that in Example 3 the reasoning used to arrive at 'Gb' is generalizable while in Example 4 it is not. However, in more complicated proofs it may not be so easy to determine if a particular sequence of steps is generalizable, so it would be helpful if our way of writing out proofs provided some simple test which settles when ∀I may be applied correctly. It is one of the features of Gentzen's approach that it does provide a simple test: we only have to check that the name *t* which we are replacing in the line to which we are applying ∀I *does not occur in any premise or assumption on which that line depends*. This means that line 6 in Example 3 is correct, since the name being replaced in line 5, 'b', does not occur in any line on which line 5 depends, that is, it does not occur in lines 1 or 2; but line 5 in Example 4 is incorrect, since 'b' occurs in line 2 and line 2 is one of the lines on which line 4 depends. With these motivating remarks, then, we can state the rule of ∀I in a way that precisely embodies the idea of generalizability:

Rule of ∀ -Introduction: For any sentence ϕt, if ϕt has been inferred at a line j in a proof, then *provided t does not occur in any premise or assumption whose line number is on the left at line j,* we may infer $\ulcorner(\forall v)\phi v\urcorner$ at line k, labeling the line 'j ∀I' and writing on its left the same numbers as occur on the left of line j. Schematically,

$$a_1,...,a_n \quad (j) \qquad \phi t$$
$$\vdots$$
$$a_1,...,a_n \quad (k) \qquad (\forall v)\phi v \qquad j\ \forall I$$

where t is not in any of the formulae on lines $a_1,...,a_n$ and v is not in ϕt. $\ulcorner(\forall v)\phi v\urcorner$ is obtained by replacing *every* occurrence of t in ϕt with v and then prefixing $\ulcorner(\forall v)\urcorner$.

The main contrast between ∀I and ∃I is therefore the restriction on when ∀I may be applied, a restriction with no counterpart for ∃I. However, there is also a contrast in the *way* the two rules may be applied. With ∀I we have to replace *every* occurrence of t in ϕt, while with ∃I we need only replace some occurrence. This condition, though important, is easiest to motivate in the context of predicate logic with identity (see §3 of Chapter 8). For motivation at this point, suppose we have derived 'Fa → Ga' by steps which could also lead to 'Fb → Gb', 'Fc → Gc' and so on. Then we are entitled to conclude that everything which is F is G, '(∀x)(Fx → Gx)', replacing both occurrences of 'a' in 'Fa → Ga' with 'x'. But we are not entitled to conclude '(∀x)(Fa → Gx)', replacing just one occurrence, since this says that for any object, a's being F is sufficient for that object's being G, while what our proof shows is that *that thing's* being F is sufficient for its being G.

In the light of our statement of the rule ∀I, our earlier proof of Example 3 is seen to be a correct example of a proof involving ∀I, and one which follows the pattern mentioned earlier: quantifiers are removed from the premises using elimination rules, sentential rules are applied, and then quantifiers are reintroduced using introduction rules to obtain the conclusion. In other proofs there may be a minor deviation from this procedure, for example, if a quantified sentence is needed at an intermediate point for ~E:

Example 5: Show ~(∃x)(Fx & Gx) ⊢$_{NK}$ (∀x)(Fx → ~Gx).

To derive '(∀x)(Fx → ~Gx)' we need a formula of the form $\ulcorner Ft → \sim Gt\urcorner$ for some individual constant t, say 'a'. To derive 'Fa → ~Ga' we proceed as in sentential logic, assuming 'Fa' to obtain '~Ga'. To obtain the latter, we assume 'Ga' to derive '⋏' by ~E. Since premise 1 is a negation, it is likely that it is one of the two formulae needed for ~E, so the other, which we have to derive, is '(∃x)(Fx & Gx)'. We can obtain '(∃x)(Fx & Gx)' by ∃I from a formula of the form $\ulcorner Ft \& Gt\urcorner$ and such a formula will itself be derived by &I. So the problem reduces to deriving the required conjuncts of the forms Ft and Gt. But it is immediately evident how to do this, given our assumptions for →I and ~I. The proof we construct following these tactics is:

1	(1)	~(∃x)(Fx & Gx)	Premise
2	(2)	Fa	Assumption
3	(3)	Ga	Assumption
2,3	(4)	Fa & Ga	2,3 &I
2,3	(5)	(∃x)(Fx & Gx)	4 ∃I
1,2,3	(6)	⅄	1,5 ~E
1,2	(7)	~Ga	3,6 ~I
1	(8)	Fa→ ~Ga	2,7 →I
1	(9)	(∀x)(Fx → ~Gx)	8 ∀I ◆

We have at various points in the foregoing noted the expressibility of one quantifier using negation and the other. One aspect of this is captured in two sequents often known as the *quantifier shift* sequents, and which we shall label 'QS':

QS: (∀x)~Fx ⊣⊢$_{NK}$ ~(∃x)Fx
         (∃x)~Fx ⊣⊢$_{NK}$ ~(∀x)Fx

When a sequent and its converse are both provable in a system S, the two formulae involved are said to be *provably equivalent in S* or *S-equivalent*. With the quantifier rules we have developed so far, we are in a position to prove the right-to-left directions of each of these equivalences. The right-to-left direction of the first is easy, and is an exercise. Here is the right-to-left direction of the (much harder) second:

Example 6: Show ~(∀x)Fx ⊢$_{NK}$ (∃x)~Fx.

A first thought here is that we should attempt to derive, say, '~Fa', and then obtain '(∃x)~Fx' by ∃I. To derive '~Fa' we should presumably assume 'Fa' and try to derive '⅄'. But how would we obtain contradictory formulae from 'Fa' and '~(∀x)Fx'? We could not apply ∀I to 'Fa', since 'Fa' is an assumption and so by the restriction on ∀I, we cannot apply ∀I to any line which depends on it, in particular, itself, if it is 'a' we are replacing. Thus our first thought does not lead to a workable strategy. However, if the direct way of deriving '(∃x)~Fx', by using ∃I, breaks down, we can try assuming the negation of what we want and use ~E, ~I and DN; that is, if we can obtain contradictory formulae from the premise and the assumption '~(∃x)~Fx', we would get '~~(∃x)~Fx' and hence '(∃x)~Fx', as required, by DN. So at the outset we have the premise '~(∀x)Fx' and the assumption '~(∃x)~Fx' to work with. But since *both* of these are negative formulae, the role of each is most likely to be to figure as one of two mutually contradictory formulae in an application of ~E. Hence we expect two uses of the ~E-~I sequence. The second is, as discussed, the one which involves the assumption '~(∃x)~Fx' and yields '~~(∃x)~Fx'. So the first is the one which will involve the premise '~(∀x)Fx'. Hence we have to derive '(∀x)Fx', which we can obtain from 'Fa' by ∀I provided 'Fa' rests on no assumptions involving 'a'. To derive 'Fa', we use the familiar strategy of assuming its negation, deriving '⅄' and applying DN. This leads to the following proof:

1	(1)	~(∀x)Fx	Premise
2	(2)	~(∃x)~Fx	Assumption
3	(3)	~Fa	Assumption
3	(4)	(∃x)~Fx	3 ∃I
2,3	(5)	⋏	2,4 ~E
2	(6)	~~Fa	3,5 ~I
2	(7)	Fa	6 DN
2	(8)	(∀x)Fx	7 ∀I
1,2	(9)	⋏	1,8 ~E
1	(10)	~~(∃x)~Fx	2,9 ~I
1	(11)	(∃x)~Fx	10 DN ◆

The problem is to satisfy the condition for legal use of ∀I, and the trick which solves the problem is at line 6, where we discharge the assumption (3), thereby legitimizing use of ∀I at line 8. For (7) rests only on (2), and (2) does not contain the individual constant 'a'.[4]

As a final illustration of the rules for the universal quantifier, we give an example with consecutive quantifiers. In this example it is important to note that each quantifier gets its own line for its elimination or introduction: we do not try to do two at once. Later we shall relax this stricture for applying ∀E (only!) to consecutive universal quantifiers, but at the moment we lack the familiarity with the system that would justify taking such shortcuts.

Example 7: Show (∀x)(Fx → (∀y)Gy) ⊢NK (∀x)(∀y)(Fx → Gy).

1	(1)	(∀x)(Fx → (∀y)Gy)	Premise
2	(2)	Fa	Assumption
1	(3)	Fa → (∀y)Gy	1 ∀E
1,2	(4)	(∀y)Gy	3,2 →E
1,2	(5)	Gb	4 ∀E
1	(6)	Fa → Gb	2,5 →E
1	(7)	(∀y)(Fa → Gy)	6 ∀I
1	(8)	(∀x)(∀y)(Fx → Gy)	7 ∀I ◆

Note in this proof that use of a new individual constant at line 5 is essential. If we were to use 'a' again this would lead to 'Fa → Ga' at line 6, and ∀I applied to this formula will produce '(∀y)(Fy → Gy)', since ∀I replaces *every* occurrence of the individual constant being removed. We would then be unable to reach the conclusion of the sequent by another step of ∀I.

Last, we note that we can employ Sequent Introduction in monadic predicate logic in the same way as we used it in sentential logic. It may already have occurred to the reader that our eleven-line proof of Example 2, (∀x)(Fx → Gx), Fa ⊢NK (∃x)(~Gx → Hx), can easily be shortened using SI, as follows:

---

[4] As one would anticipate from the uses of DN in this proof, the sequent is not derivable in intuitionistic logic.

$$
\begin{array}{llll}
1 & (1) & (\forall x)(Fx \rightarrow Gx) & \text{Premise} \\
2 & (2) & Fa & \text{Premise} \\
1 & (3) & Fa \rightarrow Ga & 1\ \forall E \\
1,2 & (4) & Ga & 3,4 \rightarrow E \\
1,2 & (5) & {\sim}{\sim}Ga & 4\ \text{SI (DN}^+) \\
1,2 & (6) & {\sim}Ga \rightarrow Ha & 5\ \text{SI (PMI)} \\
1,2 & (7) & (\exists x)({\sim}Gx \rightarrow Hx) & 6\ \exists I \quad \blacklozenge
\end{array}
$$

This proof illustrates the standard way in which SI is used in quantificational NK: the instructions for applying the rule do not change at all, but more sequents are available as substitution instances, since we can put sentences of LMPL for the sentence-letters of the LSL sequents we use for SI. But it must be *closed* sentences of LMPL which are substituted for sentence-letters. For example, we do not permit the move from '$(\forall x)Fx$' to '$(\forall x){\sim}{\sim}Fx$' in one line by SI, citing DN or SDN, since this involves substituting the open sentence 'Fx' for 'A' in $A \vdash_{NK} {\sim}{\sim}A$. To move from '$(\forall x)Fx$' to '$(\forall x){\sim}{\sim}Fx$' we must instead apply $\forall E$ to '$(\forall x)Fx$' to obtain, say, 'Fa', *then* use SI (DN$^+$), which yields '${\sim}{\sim}Fa$', and then obtain '$(\forall x){\sim}{\sim}Fx$' by $\forall I$. On the other hand, we *are* permitted to move from '$(\forall x)Fx$' to '${\sim}{\sim}(\forall x)Fx$' in one line by SI, citing DN, since this just involves a straightforward substitution of the closed sentence '$(\forall x)Fx$' for 'A' in $A \vdash_{NK} {\sim}{\sim}A$. And of course we can also use SDN to move in one line from, say, '$(\forall x)Fx \rightarrow (\forall x)Gx$' to '$(\forall x)Fx \rightarrow {\sim}{\sim}(\forall x)Gx$', since again this just involves replacing a sentence-letter with the closed sentence '$(\forall x)Gx$'.

The rationale for restricting substitutions to closed sentences is twofold. First, while it is in fact possible to formulate a version of SI which allows replacement by open sentences, stating this version is rather involved. Secondly, there are not many occasions when the lack of this more complicated version of SI is sorely missed. But there are some, and so in the interests of keeping the frustration level down, we will later make two entirely *ad hoc* extensions to Sequent Introduction that will prove very convenient.

❑ Exercises

I   Show the following.

(1) $(\forall x)Fx\ \&\ (\forall x)Gx \dashv\vdash_{NK} (\forall x)(Fx\ \&\ Gx)$

(2) $(\forall x){\sim}Fx \vdash_{NK} (\exists x)(Fx \rightarrow Gx)$

*(3) $(\forall x)(Fx \rightarrow Gx) \vdash_{NK} (\forall x)Fx \rightarrow (\forall x)Gx$

(4) $(\forall x)Fx \lor (\forall x)Gx \vdash_{NK} (\forall x)(Fx \lor Gx)$

(5) ${\sim}(\exists x)Fx \vdash_{NK} (\forall x){\sim}Fx$

(6) $(\exists x)Fx \rightarrow (\forall x)Gx \vdash_{NK} (\forall x)(Fx \rightarrow Gx)$

*(7) $(\exists x)Fx \rightarrow Ga \vdash_{NK} (\exists x)(Fx \rightarrow Gx)$

(8) $(\forall x)({\sim}Fx \rightarrow {\sim}Kx) \vdash_{NK} (\exists x)((Fx\ \&\ Kx) \lor {\sim}Kx)$

(9) $(\forall x)(A\ \&\ Fx) \dashv\vdash_{NK} A\ \&\ (\forall x)Fx$

(10) $(\forall x)(A \rightarrow Fx) \dashv\vdash_{NK} A \rightarrow (\forall x)Fx$

(11) $(\forall x)(\forall y)(Fx \rightarrow Gy) \vdash_{NK} (\forall x)(Fx \rightarrow (\forall y)Gy)$

(12) $(\forall x)(\forall y)(Gy \rightarrow Fx) \vdash_{NK} (\forall x)((\forall y)Gy \rightarrow Fx)$

II Symbolize the following English arguments (quantifiers in (1)-(3) are relativized to the domain of things, in (4) to the domain of persons) and then prove their forms. State your dictionary for each example.

(1) All elephants are large. Jumbo is not large. Therefore Jumbo is not an elephant.
*(2) All tigers are fierce. No antelope is fierce. Therefore no antelope is a tiger.
(3) 2 is a number. No number is both prime and composite. Every number is either prime or composite. Therefore 2 is either prime and not composite or composite and not prime. (Use 'a' as a name of the number 2.)
(4) All Glaswegians are Scots. All Britons are Europeans. Therefore, if every Scot is British then every Glaswegian is European. (A Glaswegian is a citizen of Glasgow, not a Norwegian glass-blower.)

4 The rule of Existential Elimination

The remaining rule of inference which we have yet to explain is the rule which allows us to reason *from* an existential sentence, the rule of Existential Elimination ($\exists$E). What we are looking for is a rule which will allow us to prove, for example, '$(\exists x)(Fx \& Gx) \vdash_{NK} (\exists x)Fx \& (\exists x)Gx$', but will *not* allow us to prove the converse, since as we established in §1, $(\exists x)Fx \& (\exists x)Gx \nvdash (\exists x)(Fx \& Gx)$.

We can arrive at the correct form of the rule by reflecting on our semantic account of the existential quantifier. An existential sentence is true on an interpretation 1 if and only if at least one of its instances in 1 is true. On any given 1, then, there is a *disjunction* whose truth is necessary and sufficient for the truth of an existential sentence, namely, the disjunction of all the instances of the existential sentence: to say that at least one instance is true is to say that the first instance is true *or* the second one is true *or* the third one is true, and so on, for as many disjuncts as there are objects in the domain of 1. For example, in the interpretation D = $\{\alpha,\beta,\gamma\}$, Ext(F) = $\{\alpha,\beta,\gamma\}$, Ext(G) = $\{\beta\}$, Ext(H) = $\varnothing$,

	F	G	H
α	+	-	-
β	+	+	-
γ	+	-	-

it is necessary and sufficient for the truth of '$(\exists x)Fx$' that 'Fa $\lor$ (Fb $\lor$ Fc)' is true, it is necessary and sufficient for the truth of '$(\exists x)Gx$' that 'Ga $\lor$ (Gb $\lor$ Gc)' is true, and it is necessary and sufficient for the truth of '$(\exists x)(Gx \& Hx)$' that '(Ga & Ha) $\lor$ ((Gb & Hb) $\lor$ (Gc & Hc))' is true; thus the first two of these existential sentences is true in the displayed interpretation, since at least one disjunct in its associated disjunction is true, while the third is false, since all the disjuncts of its associated disjunction are false.

Consequently, the requirement for inferring a statement ψ from an existential sentence $\ulcorner(\exists v)\phi v\urcorner$ in an interpreted language is that we should be able to infer ψ from the disjunction which is associated with $\ulcorner(\exists v)\phi v\urcorner$ by whatever the background domain of discourse is. The rule of $\vee$-Elimination indicates that to infer a target formula ψ from a two-disjunct disjunction, we must infer it from the first disjunct and then from the second; so to infer ψ from a many-disjunct disjunction, we have to be able to infer it from each disjunct in turn until we have inferred it from all of them. However, this is impractical in ordinary real-world reasoning, where the background domain of discourse may be very large, for instance, all people, since there are then far too many disjuncts. The solution to the difficulty, echoing that to the corresponding problem over $\forall$I, is to show not that we can infer ψ from *every* disjunct, but that we can infer it from *one* disjunct *in a way which would work for every other disjunct as well.* Suppose ϕt is an instance of $\ulcorner(\exists v)\phi v\urcorner$, and so a possible disjunct in a disjunction associated with $\ulcorner(\exists v)\phi v\urcorner$ by some interpretation. Let $\phi t'$ be the same as ϕt except for containing t' in exactly the positions where ϕt contains t. Then if the sequence of steps by which ψ is derived from ϕt would also suffice for derivation of ψ from $\phi t'$, we say that that step-sequence is *generalizable.* The rule of Existential Elimination, $\exists$E, should say that ψ follows from $\ulcorner(\exists v)\phi v\urcorner$ if ψ follows from ϕt by a generalizable sequence of steps.

Given the closeness in sense between an existential statement and a disjunction,[5] it is not surprising that we implement Existential Elimination much as we implemented $\vee$E: we assume an instance (a disjunct), and once the target formula ψ has been inferred from it by generalizable reasoning, we restate ψ, adjusting numbers on the left to reflect the fact that we have now shown that ψ follows from the original existential sentence.

Example 1: Show $(\exists x)(Fx \,\&\, Gx) \vdash_{NK} (\exists x)Fx$.

1	(1)	$(\exists x)(Fx \,\&\, Gx)$	Premise
2	(2)	$Fa \,\&\, Ga$	Assumption
2	(3)	Fa	2 &E
2	(4)	$(\exists x)Fx$	3 $\exists$I
1	(5)	$(\exists x)Fx$	1,2,4 $\exists$E ◆

In order to derive line 5 from line 1, we assume an instance of (1), namely, (2), and derive the target formula '$(\exists x)Fx$' from (2) at line 4. Having arrived at the target formula on the basis of (2), we can then assert that this formula follows from the original existential sentence whose instance we have employed. So on the left of (5) we replace the number of the instance with the numbers which appear on the left of the line where the existential sentence occurs, '1'; and on the right, we label this maneuver with the line numbers of (a) the existential sentence, (b) the instance and (c) the target sentence where it is derived from

---

[5] This is not to say that an existential sentence means the *same* as any particular disjunction. For example, in the interpretation displayed on page 195, '$(\exists x)Hx$' and 'Ha $\vee$ (Hb $\vee$ Hc)' are both false, but we could add an object to the domain and put it in the extension of 'H', so that '$(\exists x)Hx$' would be true while 'Ha $\vee$ (Hb $\vee$ Hc)' is still false. The point is analogous to the one about '$\forall$' and '&' in note 3.

the instance (the target sentence may have other, earlier, occurrences, for example, if ∨E was used in deriving it from the instance). In this example, it is clear that our reasoning is generalizable, since even if we had picked another instance of (1), say 'Fd & Gd', exactly the same steps of &E and ∃I would lead us to (5). So (5) follows from each instance of (1), and hence (5) follows from (1) itself.

Actually, we have already used the notion of a certain sequence of steps being generalizable, in formulating the conditions under which ∀I is legal. The present use is slightly different. For ∀I, generalizable reasoning is reasoning which can lead indifferently *to any target formula* in a certain class, while for ∃E, generalizable reasoning is reasoning which can lead indifferently *from any assumption* in a certain class to a fixed target formula. Context usually makes it clear which kind of generalizability is in question—are we talking about an application of ∀I or an application of ∃E?—but it may be useful to draw the distinction terminologically as well. We can say that ∀I is legitimate when the reasoning that leads to the instance of $\ulcorner(\forall v)\phi v\urcorner$ is *generalizable with respect to goal formula*, while ∃E is legitimate when the reasoning that leads from the instance of $\ulcorner(\exists v)\phi v\urcorner$ is *generalizable with respect to assumed instance*.

It will not always be as obvious as it is in Example 1 that the reasoning is generalizable with respect to assumed instance, and thus we have a need for a precise criterion for when an application of ∃E is correct. We approach this by considering some 'proofs' of invalid sequents which work by violating the intuitive idea of generalizability, the constraint that the target formula should be deducible in the same way from any other instance of the existential sentence besides the one actually chosen. For example, the following sequent is invalid, and the so-called proof of it displayed is wrong:

Example 2: Show (∃x)Fx, Ga ⊢$_{NK}$ (∃x)(Fx & Gx) (?)

1	(1)	(∃x)Fx	Premise
2	(2)	Ga	Premise
3	(3)	Fa	Assumption
2,3	(4)	Fa & Ga	2,3 &I
2,3	(5)	(∃x)(Fx & Gx)	4 ∃I
1,2	(6)	(∃x)(Fx & Gx)	1,3,5 ∃E (no!) ◆

To see the invalidity, consider the interpretation D = {α,β}, Ext(F) = {β}, Ext(G) = {α}, which establishes that (∃x)Fx, Ga ⊭ (∃x)(Fx & Gx). So the earlier attempt to prove the sequent must contain a mistake. An instance of (1) is chosen at (3), and the target formula is inferred at (5) using this instance and premise 2; then we assert at (6) that the target formula therefore follows from the existential premise, (1), and premise 2; we make this assertion by replacing the assumption number of the instance (3) which is on the left at (5) with the number of the existential premise (1) (the reader should postpone worrying about how the numbers on the left of a step of ∃E are determined; at the moment we are just trying to work out when ∃E is legitimate at all).

Reading through this description of the proof, it is not difficult to see that the reasoning is not generalizable with respect to assumed instance: at (3) we were careful to pick an instance of (1) with the *same* individual constant as occurs in (2); for it is only if the names are the same in (4) that ∃I can be applied correctly to get (5). So if we had chosen a different instance of (1) at (3), the 'proof' would have ground to a halt after line (4). The idea that at (5) we have done enough to show that '(∃x)(Fx & Gx)' follows from *every* disjunct in the disjunction corresponding to (1) is therefore incorrect, and so intuitively, the claim of line 6 that '(∃x)(Fx & Gx)' follows from lines 1 and 2 is wrong.

The telltale sign that there is a problem with (6) is exhibited in the premises and assumptions on which (5) depends. It is quite acceptable that the target formula should depend on *one* premise or assumption containing the name 'a', namely, the assumed instance, since we expect to use the instance we assume in deriving the target formula. But it is not acceptable that the target formula at (5) should depend on *other* premises or assumptions containing 'a', since this indicates that we deliberately chose an instance of the existential formula with 'a' in it in order to get interaction with those other premises or assumptions, interaction which would not have occurred if a different instance had been chosen, one with 'b', say, rather than 'a'. Thus (6) is incorrect because (5) depends not just on (3) but also on (2), and (2) contains 'a'. So, in general, reasoning from an assumed instance of an existential sentence is not generalizable if the individual constant *t* we use in forming the instance also occurs in a premise or other assumption that is involved in our derivation of the target formula from the instance.

What other indications that the reasoning is not generalizable might we be able to discern in proofs set out in our format? If we reflect on the form of ∃E as illustrated in the two previous proofs, there seem to be essentially two other situations in which reasoning from an instance *φt* of an existential sentence ⌜(∃ν)φν⌝ would not be generalizable, that is, two other situations in which a proof of *ψ* from the instance *φt* of ⌜(∃ν)φν⌝ would not guarantee the ability to derive *ψ* in the same way from every other instance *φt′*.[6]

The first of these is that *t* may occur in the existential sentence itself; thus if we were to use 'Fa & Ga' as an instance of '(∃x)(Fx & Gx)' and derive *ψ* from this instance by some sequence of steps, we would not in general expect to be able to derive *ψ* by the same steps from other instances of '(∃x)(Fa & Gx)', since the fact that the same name occurs in the two conjuncts of the instance we chose could be crucial to the steps we actually employ.

The second problem is that *t* may occur in the target formula *ψ* itself. If we use 'Fa' as an instance of '(∃x)Fx' when the target formula is, say, 'Ga', the same steps as we actually use could not in general be expected to deliver a proof of the target formula from other instances of '(∃x)Fx', since it may be crucial to the details that the name in the instance is the same as that in the target formula. For example, we may have used →E with 'Fa → Ga', as in the following unacceptable 'proof' of an invalid sequent:

---

[6]   We can prove rigorously that the restrictions on ∃E that we are inferring here are sufficient to guarantee this ability. See Forbes for a proof of this result. In that paper I refer to reasoning which justifies ∃E as reasoning which is *neutral* (as opposed to generalizable) with respect to assumed disjunct.

Example 3: Show (∃x)Fx, (∀x)(Fx → Gx) ⊢_{NK} Ga (?)

1	(1)	(∃x)Fx	Premise
2	(2)	(∀x)(Fx → Gx)	Premise
3	(3)	Fa	Assumption
2	(4)	Fa → Ga	2 ∀E
2,3	(5)	Ga	3,4 →E
1,2	(6)	Ga	1,3,5 ∃E (no!) ◆

In order to obtain 'Ga' we have to use 'a' for 'x' when we apply ∀E to (2), and consequently (3) is the *only* instance of (1) which allows us to proceed to the conclusion *via* →E. If instead we tried to use 'b', say, then although we could certainly proceed with ∀E using 'b' as well, and then apply →E, this would produce 'Gb' at (5), not 'Ga'. And if we assume 'Fb' at (3) and leave (4) as 'Fa → Ga', then of course we cannot apply →E, since we have no match for the antecedent of the conditional. Therefore the steps /Assumption/2,∀E/3,4→E/ do not constitute generalizable reasoning.

These examples illustrate three, and in fact all, of the ways in which reasoning for ∃E might fail to be generalizable with respect to assumed instance. With this motivation, we give a precise statement of the rule of ∃E in the following way:

Rule of ∃-Elimination: If an existential sentence ⌜(∃v)φv⌝ occurs at a line i and depends on lines $a_1,...,a_n$, and an instance of it φt is assumed at a line j, in which t replaces every occurrence of v, and if ψ is inferred at line k depending on lines $b_1,...,b_u$, then at line m we may infer ψ, labeling the line 'i,j,k ∃E' and writing on its left the line numbers X = $\{a_1,...,a_n\} \cup \{b_1,...,b_u\}/j$, *provided* (i) t does not occur in ⌜(∃v)φv⌝, (ii) t does not occur in ψ and (iii) t does not occur in any of the formulae on lines $b_1,...,b_u$ other than j. Schematically,

$$a_1,...,a_n \quad \text{(i)} \quad (\exists v)\phi v$$
$$\vdots$$
$$j \quad \text{(j)} \quad \phi t \qquad \text{Assumption}$$
$$\vdots$$
$$b_1,...,b_u \quad \text{(k)} \quad \psi$$
$$\vdots$$
$$X \quad \text{(m)} \quad \psi \qquad \text{i,j,k } \exists E$$

where t is not in (i) ⌜(∃v)φv⌝, (ii) ψ or (iii) any of the formulae at lines $b_1,...,b_u$ other than j. (As usual, we allow the case where φt is not used in the derivation of ψ, that is, the case where j ∉ $\{b_1,...,b_u\}$. In that case, X = $\{a_1,...,a_n\} \cup \{b_1,...,b_u\}$.)

Though complicated to state, the rule is easy enough to use, as for example in this proof:

Example 4: Show $(\forall x)(Fx \rightarrow Gx)$, $(\exists x)\sim Gx \vdash_{NK} (\exists x)\sim Fx$.

1	(1)	$(\forall x)(Fx \rightarrow Gx)$	Premise
2	(2)	$(\exists x)\sim Gx$	Premise
3	(3)	$\sim Ga$	Assumption
1	(4)	$Fa \rightarrow Ga$	1 $\forall$E
1,3	(5)	$\sim Fa$	3,4 SI (MT)
1,3	(6)	$(\exists x)\sim Fx$	5 $\exists$I
1,2	(7)	$(\exists x)\sim Fx$	2,3,6 $\exists$E ◆

(3) is an instance of (2), and we obtain the target formula '$(\exists x)\sim Fx$' at (6), using lines 1 and 3. This means that the target formula follows from lines 1 and 2, *provided* 'a' does not occur in the target formula, in the existential formula at (2) of which (3) is an instance, or in any premise or assumption used to derive the target formula at (6) other than (3). These provisos are all satisfied: 'a' occurs in line 5, to be sure, but what matters is whether it occurs *in the lines (5) depends on*, other than (3). So our application of $\exists$E at (7) is correct. Notice that it is important that $\exists$I be applied ahead of $\exists$E. It may not seem to make much difference if we apply $\exists$E at (6), claiming to derive '$\sim Fa$' from lines 1 and 2, and then apply $\exists$I to get '$(\exists x)\sim Fx$', but in fact, '$\sim Fa$' is not a semantic consequence of (1) and (2), so this way around would yield a 'proof' of an invalid sequent (proviso (ii) in the statement of the rule of $\exists$E would be violated).

With all our quantifier rules assembled, we can finish the proofs of the four QS sequents. We have the two sequents $(\exists x)\sim Fx \vdash_{NK} \sim(\forall x)Fx$ (the converse of Example 3.6 on page 192) and $(\forall x)\sim Fx \vdash_{NK} \sim(\exists x)Fx$ still to prove. The second is an exercise; here is a proof of the first:

Example 5: Show $(\exists x)\sim Fx \vdash_{NK} \sim(\forall x)Fx$.

The example is quite straightforward. Since the conclusion is a negation, we obtain it by $\sim$I, so we begin by assuming '$(\forall x)Fx$'. We then have the subgoal of proving '$\wedge$' from this assumption and the premise '$(\exists x)\sim Fx$', so we derive '$\wedge$' from the assumption and an instance of '$(\exists x)\sim Fx$', namely, '$\sim Fa$', assumed at line 3. Obtaining '$\wedge$' from (3) and (2) is easy, and after checking that the restrictions on $\exists$E are satisfied, we conclude that '$\wedge$' follows from (1) and (2) as well, as we assert at line 6.

1	(1)	$(\exists x)\sim Fx$	Premise
2	(2)	$(\forall x)Fx$	Assumption
3	(3)	$\sim Fa$	Assumption
2	(4)	Fa	2 $\forall$E
2,3	(5)	$\wedge$	3,4 $\sim$E
1,2	(6)	$\wedge$	1,3,5 $\exists$E
1	(7)	$\sim(\forall x)Fx$	2,6 $\sim$I ◆

Since LMPL includes LSL, it has sentence-letters in its lexicon. Here is an example of a proof of a sequent which contains sentence-letters:

Example 6: Show (∃x)(Fx → A) ⊢ₙₖ (∀x)Fx → A.

1	(1)	(∃x)(Fx → A)	Premise
2	(2)	(∀x)Fx	Assumption
3	(3)	Fa → A	Assumption
2	(4)	Fa	2 ∀E
2,3	(5)	A	3,4 →E
1,2	(6)	A	1,3,5 ∃E
1	(7)	(∀x)Fx → A	2,6 →I ◆

We have yet to see an example of a proof which involves more than one application of ∃E, as is required if a premise or assumption is an existential formula with another existential formula as its main subformula. Such proofs need not be technically difficult but require care when the successive applications of ∃E are made.

Example 7: Show (∃x)(∃y)(Fx & Gy) ⊢ₙₖ (∃y)(∃x)(Gy & Fx).

Since the premise is an existential sentence we would expect to use it for ∃E, so we should assume an instance, as at line 2 below. The instance itself is another existential sentence, so we would also expect to use *it* for ∃E, so we should next assume an instance of it, as at line 3. But we should not use the same name as was introduced at line 2, or else we will likely end up violating a restriction on ∃E. This leads to the following proof:

1	(1)	(∃x)(∃y)(Fx & Gy)	Premise
2	(2)	(∃y)(Fa & Gy)	Assumption
3	(3)	Fa & Gb	Assumption
3	(4)	Fa	3 &E
3	(5)	Gb	3 &E
3	(6)	Gb & Fa	5,4 &I
3	(7)	(∃x)(Gb & Fx)	6 ∃I
3	(8)	(∃y)(∃x)(Gy & Fx)	7 ∃I
2	(9)	(∃y)(∃x)(Gy & Fx)	2,3,8 ∃E
1	(10)	(∃y)(∃x)(Gy & Fx)	1,2,9 ∃E ◆

At line 8 we have obtained the target formula from the instance (3) of (2), so we can use ∃E to discharge (3) and make the target formula depend on (2) instead. But (2) is itself an instance of (1), so we apply ∃E again to discharge (2) and make the target formula depend on (1). Note two points about this proof: (i) the order of lines 7 and 8 is compulsory, since ∃I *prefixes* an existential quantifier—we cannot first introduce '(∃y)' at (7) and then at (8) insert '(∃x)' on its right; (ii) it would not have been acceptable to make the first application of ∃E after line 7, since 'b' is the term *t* to which the ∃E-restrictions apply for that application ('b' is the new term introduced in forming the instance (3) of (2)) and 'b' occurs in (7).

Some of the trickiest proofs in monadic predicate logic involve sequents where ∃E is required to reason from the premises and ∀I is needed to arrive at the conclusion, since both rules have special restrictions and it is not always straightforward to satisfy all the restrictions simultaneously. For instance, the following sequent is deceptively simple-looking:

Example 8: Show (∀x)(∃y)(Fx & Gy) ⊢$_{NK}$ (∃y)(∀x)(Fx & Gy).

In Part III of this book we shall see that in full predicate logic, reversing a two-quantifier prefix '(∀x)(∃y)' generally produces an invalid sequent. But monadic predicate logic is a special case, and Example 8 is valid, as the reader who tries to show it to be invalid will rapidly come to suspect. To prove the sequent, one naturally thinks of obtaining the conclusion by successive steps of ∀I and ∃I, ∀I being applied to a conjunction obtained from the premises. So a first attempt might look like this:

1	(1)	(∀x)(∃y)(Fx & Gy)	Premise
1	(2)	(∃y)(Fa & Gy)	1 ∀E
3	(3)	Fa & Gb	Assumption
3	(4)	(∀x)(Fx & Gb)	3 ∀I (**NO!**)
3	(5)	(∃y)(∀x)(Fx & Gy)	4 ∃I
1	(6)	(∃y)(∀x)(Fx & Gy)	2,3,5 ∃E ◆

But line 4 is incorrect, since the name 'a' which is being replaced in this application of ∀I occurs in an assumption on which (4) depends. This violates the restriction on ∀I. Nor is it possible to repair the proof by applying ∃I and ∃E first, leaving ∀I to last, since ∀I *prefixes* a quantifier—it does not allow us to insert universal quantifiers inside a formula.

If the problem with (4) is that the name being replaced occurs in (3), the solution is to derive some other formula ⌜Ft⌝ where t is neither 'a' nor 'b', conjoin it with 'Gb', and only *then* apply ∀I. We can obtain any formula ⌜Ft⌝ whatsoever from premise (1) using ∀E, although we have to work around the existential quantifier within whose scope ⌜Ft⌝ will lie. This will involve an inner application of ∃E as well as the main one with which the proof will end. So we arrive at the following:

1	(1)	(∀x)(∃y)(Fx & Gy)	Premise
1	(2)	(∃y)(Fa & Gy)	1 ∀E
3	(3)	Fa & Gb	Assumption
1	(4)	(∃y)(Fc & Gy)	1 ∀E
5	(5)	Fc & Gd	Assumption
5	(6)	Fc	5 &E
1	(7)	Fc	4,5,6 ∃E
3	(8)	Gb	3 &E
1,3	(9)	Fc & Gb	7,8 &I
1,3	(10)	(∀x)(Fx & Gb)	9 ∀I
1,3	(11)	(∃y)(∀x)(Fx & Gy)	10 ∃I
1	(12)	(∃y)(∀x)(Fx & Gy)	2,3,11 ∃E ◆

Why does this proof work where the previous one did not? The crucial difference is that at line 10, ∀I is being applied to the name 'c' in (9), and 'c' does not occur in either (1) or (3), on which (9) depends, so *this* application of ∀I is acceptable. Note also that the application of ∃E at line 7 is not in violation of any restriction on ∃E, for the term *t* to which restrictions apply is the one introduced in the instance at line 5, which is 'd', not 'c'; and (6) does not depend on any premise or assumption other than (5) containing 'd'. Finally, the key to this proof is the double application of ∀E to line 1. It is intrinsic to the meaning of the universal quantifier that we can infer as many instances of a universal sentence as we wish, replacing the bound variable with any name we please. This possibility is easy to overlook, but is often the way to solve harder problems.

❑ Exercises

I   Show the following:

     (1) (∃x)Fx, (∀x)(Fx → Gx) ⊢$_{NK}$ (∃x)Gx
   *(2) (∃x)Fx ∨ (∃x)Gx ⊣⊢$_{NK}$ (∃x)(Fx ∨ Gx) (only left-to-right solution given)
    (3) (∃x)(Fx & ~Gx), (∀x)(Hx → Gx) ⊢$_{NK}$ (∃x)(Fx & ~Hx)
    (4) (∀x)~Fx ⊢$_{NK}$ ~(∃x)Fx
    (5) (∀x)(Fx → (∀y)~Fy) ⊢$_{NK}$ ~(∃x)Fx
    (6) (∃x)(Fx & ~Gx) ⊢$_{NK}$ ~(∀x)(Fx → Gx)
    (7) (∃x)(Fx & Gx), (∀x)[(∃y)Fy → Rx], (∀x)[(∃y)Gy → Sx] ⊢$_{NK}$ (∀x)(Rx & Sx)
   *(8) (∃x)(Fx ∨ (Gx & Hx)), (∀x)(~Gx ∨ ~Hx) ⊢$_{NK}$ (∃x)Fx
    (9) (∃x)(Fx & (Gx ∨ Hx)) ⊢$_{NK}$ (∃x)(Fx & Gx) ∨ (∃x)(Fx & Hx)
   (10) (∃x)(Fx ↔ Gx), (∀x)(Gx → (Hx → Jx)) ⊢$_{NK}$
          (∃x)Jx ∨ ((∀x)Fx → (∃x)(Gx & ~Hx))
   (11) (∀x)(Fx & (∃y)Gy) ⊢$_{NK}$ (∃x)(Fx & Gx)
   (12) (∃x)(Fx & ~Fx) ⊣⊢$_{NK}$ (∀x)(Gx & ~Gx)
  *(13) (∃x)Gx ⊢$_{NK}$ (∀x)(∃y)(Fx → Gy)
   (14) (∀x)(Fx → (∃y)Gy), (∀x)(~Fx → (∃y)Gy) ⊢$_{NK}$ (∃z)Gz
   (15) ⊢$_{NK}$ (∀x)((Fx → Gx) ∨ (Gx → Fx))
   (16) (∃x)Fx → (∃x)Gx ⊢$_{NK}$ (∃x)(Fx → Gx)
  *(17) ⊢$_{NK}$ (∃x)(∀y)(Fx → Fy)
   (18) (∀x)(∃y)(Fx → Gy), (∀x)(∃y)(~Fx → Gy) ⊢$_{NK}$ (∃z)Gz
   (19) (∀x)(∀y)(Gy → Fx) ⊣⊢$_{NK}$ (∀x)((∃y)Gy → Fx)
   (20) (∃x)(∀y)(Fx → Gy) ⊣⊢$_{NK}$ (∃x)(Fx → (∀y)Gy)
   (21) (∃x)(∀y)(Fy → Gx) ⊢$_{NK}$ (∀x)(∃y)(Fx → Gy)
   (22) (∃x)(∀y)(Fx → Gy) ⊣⊢$_{NK}$ (∀x)Fx → (∀x)Gx

II   Show the following:

    (1) (∃x)(A & Fx) ⊣⊢$_{NK}$ A & (∃x)Fx
   *(2) (∃x)(A ∨ Fx) ⊣⊢$_{NK}$ A ∨ (∃x)Fx (only left-to-right solution given)
    (3) (∀x)(A ∨ Fx) ⊣⊢$_{NK}$ A ∨ (∀x)Fx
    (4) (∀x)Fx → A ⊢$_{NK}$ (∃x)(Fx → A)

III Show the following:

*(1) ⊢_{NK} (∃x)[(∃y)Fy → Fx]
(2) ⊢_{NK} (∃x)(Fx → (∀y)Fy)

(Hint: attempting to construct a counterexample is suggestive of a proof when one sees how the attempt breaks down.)

IV For each of the three provisos on ∃E, give your own example of an incorrect 'proof' of an invalid sequent which works by violating the proviso. In each case demonstrate that your sequent *is* invalid and say which restriction your 'proof' violates.

5 Extensions of sequent introduction

As we indicated at the end of §3, there are certain ways of extending Sequent Introduction which are convenient enough to warrant *ad hoc* additions to the strategy. The most important of these involves the interaction between negation and the quantifiers. We have already established the four *quantifier shift* sequents (two were exercises):

QS: (∀x)~Fx ⊣⊢_{NK} ~(∃x)Fx
(∃x)~Fx ⊣⊢_{NK} ~(∀x)Fx

and there is good reason to allow them to be used in SI. It would be painful if to prove the sequent ~(∀x)(Fx → Gx) ⊢_{NK} (∃x)~(Fx → Gx) we had to reiterate the steps of the proof of Example 3.6—and this is all it would take, since the '→' rules would not be needed—putting '(Fx → Gx)' in the new proof wherever we have 'Fx' in the proof of 3.6. But to obtain the sequent ~(∀x)(Fx → Gx) ⊢_{NK} (∃x)~(Fx → Gx) from the sequent ~(∀x)Fx ⊢_{NK} (∃x)~Fx we would have to substitute the *open* sentence 'Fx → Gx' for 'Fx', and we are only allowing substitution for *closed* sentences. So in order to have our cake and eat it, we just decree that, in addition to the applications of SI licensed by the rule stated in §8 of Chapter 4, another kind of application is licensed, using the quantifier shift sequents. More formally, we have the following:

> **Extension (1) to the Rule of Sequent Introduction:** If the formula at a line j in a proof has any of the forms ⌜~(∀v)φv⌝, ⌜(∃v)~φv⌝, ⌜~(∃v)φv⌝ or ⌜(∀v)~φv⌝, then at line k we may infer the provably equivalent formula of the form ⌜(∃v)~φv⌝, ⌜~(∀v)φv⌝, ⌜(∀v)~φv⌝ or ⌜~(∃v)φv⌝ respectively, labeling the line 'j SI (QS)' and writing on the left the same numbers as occur on the left of line j.

Note that the only change in a formula which this extension of SI licenses is that of shifting the quantifier or negation which is the *main* connective across the adjacent negation or quantifier and changing the quantifier from '∀' to '∃' or vice versa: we cannot get '(∀x)(∀y)~(Fx & Gy)' from '(∀x)~(∃y)(Fx & Gy)' in one step, since the quantifier over which we shift the negation here, '(∃y)', is not the main connective. However, it is simple enough to use ∀E on '(∀x)~(∃y)(Fx & Gy)' to obtain, say, '~(∃y)(Fa & Gy)', then apply SI (QS) to this formula to produce '(∀y)~(Fa & Gy)', and finally to apply ∀I to arrive at '(∀x)(∀y)~(Fx & Gy)'. This four-line proof is a considerable saving over any proof that does without SI (QS), and is typical in this respect, so the restriction of QS to the main negation or main quantifier in a formula is not a serious restriction.

The extended rule of SI allows us to give concise proofs such as the following, avoiding application of ∨E to the first premise:

Example 1: Show ~(∃x)(Fx & Gx) ∨ (∃x)~Gx, (∀y)Gy ⊢_NK (∀z)(Fz → ~Gz).

The second premise '(∀y)Gy' conflicts with the second disjunct of the first premise, and we can use QS to make it the *explicit* contradictory '~(∃x)~Gx'; then from the first disjunct, we derive the conclusion using QS and other applications of SI.

1	(1)	~(∃x)(Fx & Gx) ∨ (∃x)~Gx	Premise
2	(2)	(∀y)Gy	Premise
2	(3)	Ga	2 ∀E
2	(4)	~~Ga	3 SI (DN⁺)
2	(5)	(∀x)~~Gx	4 ∀I
2	(6)	~(∃x)~Gx	5 SI (QS)
1,2	(7)	~(∃x)(Fx & Gx)	1,6 SI (DS)
1,2	(8)	(∀x)~(Fx & Gx)	7 SI (QS)
1,2	(9)	~(Fa & Ga)	8 ∀E
1,2	(10)	~Fa ∨ ~Ga	9 SI (DeM)
1,2	(11)	Fa → ~Ga	10 SI (Imp)
1,2	(12)	(∀z)(Fz → ~Gz)	11 ∀I ◆

In general, SI (QS) is most useful in transforming the prefixes ⌜~(∀v)⌝ and ⌜~(∃v)⌝ into ⌜(∃v)~⌝ and ⌜(∀v)~⌝, which are typically easier to work with. This is particularly true when one cannot see how to derive a quantified formula directly and so one assumes its negation to use ~I and DN: usually the best thing to do with the assumed negative formula is to apply QS to it. For example:

Example 2: Show (∀x)Fx → A ⊢_NK (∃x)(Fx → A).

A first thought about strategy is that we should aim to derive '(∃x)(Fx → A)' by ∃I, so as a subgoal we want to derive, say, 'Fb → A'. In turn, then, we should assume 'Fb', derive 'A', and then use →I. 'A' is the consequent of the premise, so we can derive it if we can derive the antecedent '(∀x)Fx'. But here the strategy breaks down: we cannot obtain '(∀x)Fx' from our assumption 'Fb' by ∀I, since 'Fb' *is* an assumption and 'b' therefore occurs in a line on which 'Fb' depends (itself). So instead we try to obtain '(∃x)(Fx → A)' from '~~(∃x)(Fx → A)' and the latter by ~I. Hence our proof should use the assumption '~(∃x)(Fx → A)', and since this formula begins '~(∃x)', we can apply SI using QS to it:

1	(1)	(∀x)Fx → A	Premise
2	(2)	~(∃x)(Fx → A)	Assumption
2	(3)	(∀x)~(Fx → A)	2 SI (QS)
2	(4)	~(Fa → A)	3 ∀E
2	(5)	Fa & ~A	4 SI (Neg-Imp)
2	(6)	Fa	5 &E
2	(7)	~A	5 &E
2	(8)	(∀x)Fx	6 ∀I
1,2	(9)	A	1,8 →E
1,2	(10)	⋏	7,9 ~E
1	(11)	~~(∃x)(Fx → A)	2, 10 ~I
1	(12)	(∃x)(Fx → A)	11 DN ◆

Note that the application of ∀I at (8) is perfectly legal, since 'a' does not occur in (2). The rest of the proof is straightforward, provided, of course, that we remember such sequents as Neg-Imp. One moral of this example is that those who do not memorize such sequents are likely at best to produce enormously long proofs or at worst to flounder helplessly and make no progress.

The other *ad hoc* extension to SI which is worth making is motivated by consideration of the sequent (∀x)Fx, (∀y)Fy → (∀y)Gy ⊢ (∀z)Gz. This sequent is close to one which could be proved in one line by →E, but it is not of quite the right form, since the first premise is not the very same sentence as the antecedent of the second, nor is the consequent of the second premise the very same sentence as the conclusion. Therefore, before we can apply →E to the second premise, we have to obtain '(∀y)Fy' from the first, which can be done by consecutive steps of ∀E and ∀I; →E then delivers '(∀y)Gy', to which we must again apply ∀E and ∀I to obtain '(∀z)Gz'. In other words, in order to go from a formula to a *single-variable alphabetic variant* of that formula, a detour through quantifier elimination and introduction is needed, along with application of other rules if the quantifier for the variable is not the main connective (recall that φ and φ' are single-variable alphabetic variants if and only if there are variables v and v' such that the only difference between φ and φ' is that φ has v in exactly the positions in which φ' has v'). We can avoid such tedious detours if we extend SI by allowing ourselves to use any one-premise sequent whose conclusion is a single-variable alphabetic variant of the premise, such as '(∀y)Gy ⊢_NK (∀z)Gz' and '(∀x)(∃y)(Fx → Gy) ⊢_NK (∀x)(∃z)(Fx → Gz)'. Note that we are not restricting ourselves to sequents where the variables v and v' occur with quantifiers which are the main connectives of the relevant formulae.

Extension (2) to the Rule of Sequent Introduction: For any closed sentence φv, if φv has been inferred at a line j in a proof and φv' is a single-variable alphabetic variant of φv, then at line k we may write φv', labeling the line 'j SI (AV)' and carrying down on its left the same numbers as are on the left of line j.

We can now give the following proof:

Example 3: Show (∀x)Fx, (∀y)Fy → (∀y)Gy ⊢_NK (∀z)Gz.

1	(1)	(∀x)Fx	Premise
2	(2)	(∀y)Fy → (∀y)Gy	Premise
1	(3)	(∀y)Fy	1 SI (AV)
1,2	(4)	(∀y)Gy	2,3 →E
1,2	(5)	(∀z)Gz	4 SI (AV) ◆

AV has a rather different character from other uses of SI. In other cases there is a basic sequent which we have proved, and subsequent applications of SI are to substitution-instances of that basic sequent. But there is no single basic sequent for AV. Yet this is not to say that before we make a new use of AV, we must prove the very sequent that is being used. Instead, we can describe an algorithm for constructing proofs of AV sequents which will deliver any AV sequent we want. We do not actually *use* this algorithm but merely rely on its existence to justify AV. The description of the algorithm follows the same pattern as that required for the solution of Exercise 4.10.6.

One final comment: AV licenses only replacing a *variable* with another *variable*. AV cannot be used on individual constants. Otherwise we would be able to make such absurd derivations as 'Socrates is a city' from 'Paris is a city'.

❏ Exercises

Show the following, using QS and AV wherever they are helpful.

(1) ~(∀x)(Fx → Gx) ⊢_NK (∃x)(Fx & ~Gx)
(2) (∀x)(Fx → ~Gx) ⊢_NK ~(∃x)(Fx & Gx)
(3) (∃x)(Fx → (∃y)~Fy) ⊢_NK ~(∀x)Fx
(4) (∃x)Fx → ~(∀y)Gy, (∀x)(Kx → (∃y)Jy), (∃y)~Gy → (∃x)Kx ⊢_NK ~(∃x)Fx ∨ (∃y)Jy
(5) (∃x)Fx → (∀y)Gy ⊢_NK (∃x)(Fx → (∀y)Gy)
(6) (∀x)Fx ⊣⊢_NK ~(∃x)~Fx
(7) (∃x)Fx ⊣⊢_NK ~(∀x)~Fx
*(8) ⊢_NK (∃x)(∀y)(Fx → Fy)
(9) ⊢_NK (∃x)(∀y)(∀z)((Fy → Gz) → (Fx → Gx))
(10) (∀x)(∃y)(Fx → Gy) ⊢_NK (∃x)(∀y)(Fy → Gx)

6 Decision procedures

It can hardly have escaped the reader's attention that there is a significant difference between the use we made of the definition of validity in LSL and its use in LMPL. In LSL we had a procedure which, given an argument-form, could be applied to that argument-form to tell us *whether or not* it is valid. But in LMPL we have only developed a method for demonstrating the invalidity of argument-forms *that are in fact invalid:* we have no test which, when applied to an arbitrary argument-form in LMPL, tells us *whether or not* the argument-form is valid. Why this difference?

The difference is rooted in the finiteness of the exhaustive search procedure which was our basic test for validity in sentential logic. Given any argument-form (with finitely many premises) there are only finitely many interpretations to check to see if any demonstrates the argument's invalidity; more precisely, if the argument-form contains occurrences of n sentence-letters, then there are 2^n interpretations to examine. However, our procedure for demonstrating the invalidity of an invalid argument-form in monadic predicate logic gives no hint of any finite bound on the number of interpretations we have to check before we can pronounce that there is *no* counterexample to a given LMPL argument-form and that the argument-form is therefore valid. In particular, there is no hint of a finite bound on the size of the domain of discourse which has to be considered: if we have failed to find a counterexample to a specific argument-form using interpretations with domains containing one, two or three objects, it may be that we simply have to continue increasing the size of the domain. It is even conceivable that there are arguments which are invalid but whose invalidity can only be demonstrated by an interpretation with *infinitely many* objects in its domain.

The issues that we are raising here belong to a branch of logic and computer science called *decidability theory*. Decidability theory investigates the existence and efficiency of *algorithms* for the solutions of problems. Typically, the form of a problem is whether an object x of a certain kind $\mathcal{K}$ has, or does not have, a certain property P. For example, x might be an argument-form in a formal language such as LSL or LMPL and P the property of validity. The problem is said to be *solvable*, and the property P is said to *decidable for* objects of kind $\mathcal{K}$, if and only if there is an algorithm or *effective procedure* which, given an x of kind $\mathcal{K}$, resolves whether or not x has P. An algorithm or effective procedure is any technique that an idealized computer can be programmed to carry out; 'idealized' does not mean that the computer can perform miracles, but just that there are no fixed finite bounds on the time it has to run the program, the amount of memory it may use, and so on. An effective procedure which decides whether or not x has P is called a *decision procedure* for P, and if no decision procedure for P exists then P is said to be *undecidable* (for objects of kind $\mathcal{K}$) and the problem of whether an object of kind $\mathcal{K}$ has P or not is said to be *unsolvable*.[7]

The exhaustive search test for validity in LSL shows that the property of validity for LSL argument-forms is a decidable property, since it is simple to program a computer to carry out a truth-table test on an argument-form in LSL and to return the answer 'yes' if the form is valid and 'no' if the form is invalid.

---

[7] Decision problems arise in many branches of mathematics. A *graph* is a collection of points joined by edges. Given any graph G with n points, we can ask whether there is a path along its edges which passes through all points exactly once. The property of graphs of admitting such a path is one for which there is a decision procedure that has important affinities with the decision procedure for validity in sentential logic.

The truth-table test for validity is therefore a decision procedure for validity in LSL. The problem we are currently facing is whether or not the property of validity for LMPL is also decidable, and it seems *prima facie* that it is not. For if we use exhaustive search, we would program the machine, for each n, to try each possible interpretation of an input argument-form on a domain of n objects and if no counterexample is found, to move on to a domain of $n + 1$ objects. But if the input argument-form is *valid*, this means the machine will never output the answer 'yes', since it will never finish checking all the interpretations with finite domains, much less get around to considering any with infinite domains.

However, this reasoning is not a *proof* that validity in LMPL is undecidable, since there are two possibilities we have not considered. The first possibility is that there is some 'smarter' algorithm than exhaustive search which does eventually return the answer 'yes' if the input is a valid argument-form. The second possibility is that the exhaustive search algorithm can be modified to output the answer 'yes' if (and only if) the input is a valid argument-form, because for each form we can calculate an n such that if no counterexample appears by the time all interpretations with domains of $\leqslant n$ objects have been checked, it follows that the argument-form is valid; the machine would be programmed to calculate the value of n for its input and if no counterexample is found after all interpretations of size $\leqslant n$ have been checked, it would output 'yes' rather than move on to $n + 1$.

The second of these possibilities is actually the case. There is a well-known result about monadic predicate logic, proved by Löwenheim in 1915, that if an LMPL argument-form is invalid, then it has a counterexample whose domain is at most of size 2^k, where k is the number of distinct predicate-letters which occur in the form (see Boolos and Jeffrey Ch. 25). Consequently, if the argument-form has any counterexample at all, it has a counterexample whose domain is of exactly size 2^k (a smaller counterexample can be enlarged by the addition of 'irrelevant' objects without affecting the truth-values of any premise or the conclusion—see Exercise 2.II.2 on page 185). Therefore all we have to do is program our computer to count the number of predicate-letters in an input LMPL argument-form and then, for whatever value of k it arrives at, to check all possible interpretations of the argument-form with domain of size 2^k. If one of these is a counterexample the computer issues the answer 'no' (the form is invalid), otherwise it issues the answer 'yes'.

The decidability of validity for LSL and LMPL are 'in principle' results. The picture is in one way less rosy if we are interested not merely in the existence of decision procedures but also in their efficiency. There are essentially two measures of efficiency of an algorithm, corresponding to time taken to deliver the output and memory space required to store intermediate results in arriving at the output (see Harel Ch. 7). Here we will concentrate just on the time measure. Suppose we have some way of assigning a numerical value representing a quantity we can call the 'size' of the input—in the case of argument-forms in LSL, for example, the size of the input would be the number of sentence-letters in the argument-form. Then an algorithm determines a function which relates sizes to times, and the efficiency of the algorithm is reflected in how fast the function grows: if the function grows very fast this means that for each small increase in the size of the input there is a very large increase in the time it takes to compute the output.

We can make the idea of 'very fast' more precise, as 'faster than exponential growth', in the following way. If f and g are two functions of numbers, say that f is *bounded by* g if and only if there is some j such that $f(k) \leqslant g(k)$ for every $k \geqslant j$. For example, the function $f(x) = x + 50$ is bounded by the function $g(x) = 3x$, since for every $k \geqslant 25$, $f(k) \leqslant g(k)$. Every function of n bounded by some function of the form n^k for fixed k is said to be *polynomial*, and if the function which relates the size of an algorithm's input to the time the algorithm takes to compute the output is polynomial, then the problem the algorithm solves is said to be *solvable in polynomial time*. If the algorithm does not run in polynomial time, it is said to be *superpolynomial* or *nonpolynomial;* this will include all exponential functions, functions of n of the form k^n for fixed k. Problems which have polynomial-time solutions are said to be *tractable*, and problems whose only solutions run at best in exponential time are said to be *intractable*.

The truth-table test for validity runs in nonpolynomial time. Of course, in any particular case, an invalid argument-form may be input and the first interpretation that is checked may demonstrate its invalidity. This is the *best* case. But in classifying algorithms by their efficiency, we use worst-case performance, not best-case. In the worst case, the input is valid, or else invalid but only shown to be so by the last interpretation to be checked. If the input is of size n (has n sentence-letters), then $k = 2$ and there are k^n interpretations to check. So the number of interpretations to check grows exponentially with the size of the input (though each individual interpretation can be checked in polynomial time). Similarly, the decision procedure for validity in LMPL runs in nonpolynomial time, for if there are n predicate-letters, by Löwenheim's result we have to check every possible interpretation with a domain of size 2^n. This leads to a large number of interpretations, since (i) each subset of the domain is a possible extension for a predicate-letter, (ii) if there are m objects in the domain there are 2^m subsets, and (iii) if there are i predicate-letters and j possible extensions for them, there are j^i different ways of assigning the extensions to the predicate-letters. The inefficiency of the decision procedure for LMPL is the real reason why we only consider problems about demonstrating the invalidity of invalid arguments in LMPL, not problems of determining validity or invalidity.

If these exhaustive search procedures are essentially the best possible, then the decision problems for validity in sentential and monadic predicate logic are intractable: for inputs of relatively small sizes, machines a thousand times faster than today's fastest would require geological time to compute answers. But is exhaustive search the best possible algorithm? Modern work in a branch of artificial intelligence known as automatic theorem proving has uncovered algorithms which perform significantly better than exhaustive search for a substantial proportion of the possible inputs. But even these algorithms have the same worst-case performance as exhaustive search. So can worst-case performance be improved? The answer to this question is that no one knows. If we change our conception of what a computer is, then decision procedures can be made to run in polynomial time. In the sentential case, when establishing validity by constructing interpretations, we often arrive at branch points, where there is more than one case to be considered (e.g., there are two ways of making a biconditional false). Suppose we have a machine which always makes the best choice at such a decision-point, that is, if there are paths which can be followed to a demonstration of invalidity, then one of them is chosen. Such a machine is said to be *nondeterministic*, and validity for LSL is decidable on a nondeterministic machine in polynomial time. The problem of validity is therefore said to be *nondeterministic polynomial* ('NP'), and the central open question of decidability theory is whether problems that admit of a non-deterministic polynomial solution also admit of a deterministic polynomial solution. But the question is *open* (it is commonly called the 'P = NP?' problem); that is why no one knows if the worst-case behavior of the exhaustive search algorithms we have described can be improved upon.

❏ Exercise

Suppose a computer is programmed with the exhaustive search algorithm for validity in monadic predicate logic and is given as input a valid argument-form of LMPL with three predicate-letters. Find the number of interpretations it checks before pronouncing the argument-form valid. (Refer to (i), (ii) and (iii) on page 210.)

7 Tableaux for monadic predicate logic

In this section, we extend the semantic tableau test for validity in sentential logic (see §5 of Chapter Three) to monadic predicate logic. In view of the decidability of monadic predicate calculus, as discussed in the previous section, the reader might expect the extended tableau method to constitute a decision procedure. However, it is rather complicated to give such a procedure for full monadic predicate calculus, so we will content ourselves with an algorithm that tests the validity of those sequents *which do not contain nested quantifiers*, that is, quantifiers that occur within the scope of other quantifiers. With this restriction in mind, we add to the sentential rules four new rules, $\top\forall$, $\mathbb{F}\forall$, $\top\exists$ and $\mathbb{F}\exists$. These rules extend a path containing a quantified formula $\ulcorner(Qv)\phi v\urcorner$, signed $\top$ or $\mathbb{F}$, by adding a new node, or new nodes, that contain formulae signed in the same way as $\ulcorner(Qv)\phi v\urcorner$. Unfortunately, there are some complications.

Suppose '$\top$: $(\forall x)Fx$' occurs at a node. '$(\forall x)Fx$' requires for its truth on an interpretation $\mathcal{I}$ that every instance $\ulcorner Ft\urcorner$ be true, where t names an element of the domain of $\mathcal{I}$. But in constructing a tree, a domain is not given at the outset, so how do we know what instances to use? The answer is that we can use any instance we please, and any number of instances we please, since each time we write down an instance we can think of ourselves as adding a new object to the domain. On the other hand, it would be pointless just to add instances for the sake of doing so. We can certainly add instances that use names which already occur in the formulae on the paths in question, but we will want to be as parsimonious as possible in adding instances with new names. The same remarks apply to the rule 𝔽∃, since a false existential has universal import; for instance, '𝔽: (∃x)Fx' requires for its truth on an interpretation 𝘐 that *every*thing in 𝘐's domain fail to satisfy 'F', that is, that every formula ⌜~Ft⌝ be true, where t names an element of the domain of 𝘐. For the purposes of stating the tableau quantifier rules succinctly, we shall use the term *witness* to mean a formula with the signature required by its associated signed quantified formula. For example, for any t, the signed formula ⌜𝔽: Ft⌝ is a witness for '𝔽: (∃x)Fx', and the signed formula ⌜𝕋: Ft⌝ is a witness for '𝕋: (∀x)Fx'.

The rules 𝔽∀ and 𝕋∃ require special restrictions. A tableaux with, say, only '𝕋: (∃x)Fx', '𝕋: (∃x)~Fx' and '𝔽: ⋏' at its root node should not close, since (∃x)Fx, (∃x)~Fx ⊭ ⋏. But we could close the tableau by using the *same* name in witnesses for '𝕋: (∃x)Fx' and '𝕋: (∃x)~Fx', since we could get '𝕋: Fa' and '𝕋: ~Fa', hence '𝔽: Fa', in three steps. Evidently, the fallacy is in using 'a' twice. To be true on 𝘐, '(∃x)~Fx' requires the truth on 𝘐 of *some* instance, but there is no justification for choosing one that mentions the *same* object as figured in '(∃x)Fx's witness. To avoid such fallacious steps, we stipulate that when we apply 𝕋∃ to ⌜(∃ν)ϕν⌝ and extend a path 𝛱 by adding a node to 𝛱 with a witness ϕt, we never replace ν with a name t that already occurs in a formula on 𝛱. The same applies to the rule 𝔽∀, since a false universal has existential import; for instance, '𝔽: (∀x)Fx' requires for its truth on an interpretation 𝘐 that *some*thing in 𝘐's domain fail to satisfy 'F', that is, that some formula ⌜~Ft⌝ be true, where t names some element of the domain of 𝘐.

The tableaux rules for monadic predicate calculus are these:

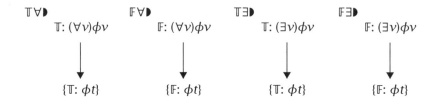

The rules say that if the tail formula is on a node n in a tree, then we may extend each path 𝛱 on which n lies by adding a new node to the bottom of 𝛱 which contains as many witnesses as we please (different t's). The braces indicate that we may label the new node with a *set* of formulae of the relevant form, not just a single formula. And we may use different sets for different nodes. However, we must observe the restriction that when applying 𝔽∀ and 𝕋∃, *we use only names which do not occur in formulae at earlier nodes of 𝛱*.

If we simply add these rules to the sentential system, there would be nothing to prevent us from failing to close a tableau which can be closed, since we could pointlessly continue applying the rules without ever making the *right* application. To avoid this, we adopt the following rules of order (see Jeffrey):

(1) Apply all possible sentential rules, checking off a formula whenever a rule is applied to it.
(2) If there are still open paths after (1), apply F∀ and T∃ once to each appropriate formula, using only one witness for each application and checking off a formula whenever a rule is applied to it.
(3) If there are still open paths after (2), apply T∀ and F∃ as often as possible. Either use only names in formulae already on the path, but do not *repeat* any formula already on the path; or, if there are no names on the path, write down one witness using any name. Do not check off the formulae.
(4) If there are still open paths after (3), go back to (1) and repeat the process until a pass through (1), (2) and (3) causes no changes in the tableau.

Example 1: Determine whether (∃x)(Fx & Gx) ⊨ (∃x)Fx.

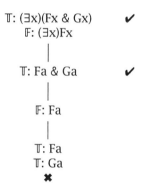

Note the absence of nested quantifiers. The tableau closes, hence (∃x)(Fx & Gx) ⊨ (∃x)Fx. We follow our rules of order, though it does not matter here whether we apply T& before or after F∃. Also, if it is indifferent which of two rules is applied first and one of the rules would cause branching, it is sometimes better to apply it second. Following the rules of order guarantees that if a tableau can be closed, it will be, and that if it cannot be closed, this will be established in finitely many steps. That is why the rules of order constitute a decision procedure for validity when applied to sequents without nested quantification.

Our next example is of a tableau which determines whether (∀x)Fx → (∀x)Gx ⊨ Fa → (∀x)Gx. The tableau is displayed on page 214. The left path in this tableau remains open, and therefore we have shown (∀x)Fx → (∀x)Gx ⊭ Fa → (∀x)Gx. As in the sentential case, we can read off a counterexample to the sequent from the signed atomic formulae on an open branch once we are finished. However, we will not go into the details of this procedure, since our primary method for demonstrating invalidity is the method of §2 of this chapter. (The interested reader who wishes to explore tableaux further might try applying our method to an invalid sequent containing nested quantifiers such as (∀x)[(∀y)Gy → Fx] ⊨ (∀x)(∀y)(Gy → Fx), to see exactly what goes wrong.)

Example 2: Determine whether $(\forall x)Fx \rightarrow (\forall x)Gx \vDash Fa \rightarrow (\forall x)Gx$.

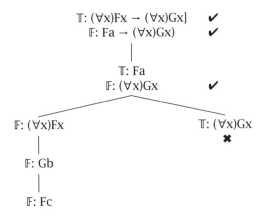

One final remark: since we added '$\wedge$' to the language, and '$\wedge$' is always false, we add to our definition of 'closed path' that a path closes if '$\mathbb{T}: \wedge$' is on it.

❑ Exercises

Repeat problems 1-19, 22 and 23 of Exercise I, §2, demonstrating invalidity by constructing open tableaux. Then for each of these sequents in which there is only one premise formula, test the converse semantic sequent for validity. For example, for problem 19, determine whether *$(\exists x)(Fx \leftrightarrow Gx) \vDash (\forall x)Fx \leftrightarrow (\forall x)Gx$.

8 Alternative formats

Like proofs in sentential NK, proofs in monadic NK may be presented in tree format or sequent-to-sequent format instead of Lemmon format. Here is a statement of the quantifier rules for tree-format proofs, along with an example:

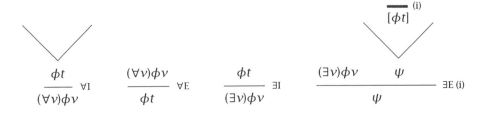

provided that no undischarged leaf formula above ϕt contains t.

provided that t is not in ψ, $(\exists v)\phi v$, or any undischarged leaf formula above the upper ψ except ϕt.

To illustrate, we repeat Example 4.6, $(\exists x)(Fx \rightarrow A) \vdash_{NK} (\forall x)Fx \rightarrow A$, as a tree.

It is also possible to regard the quantifier rules as specifying how to infer sequents from sequents. For this purpose, the rules would be formulated in the following way:

Rule of ∃I: From the sequent $\Gamma \vdash \phi t$ we may infer the sequent $\Gamma \vdash (\exists v)\phi v$, where v is not in ϕ.

Rule of ∃E: From the sequents $\Gamma \vdash (\exists v)\phi v$ and $\Sigma, \phi t \vdash p$, we may infer $\Gamma, \Sigma \vdash p$, provided t does not occur in p, $(\exists v)\phi v$ or $\Sigma/\phi t$.

Rule of ∀E: From the sequent $\Gamma \vdash (\forall v)\phi v$ we may infer the sequent $\Gamma \vdash \phi t$.

Rule of ∀I: From the sequent $\Gamma \vdash \phi t$ we may infer the sequent $\Gamma \vdash (\forall v)\phi v$, provided t does not occur in any formula in Γ.

A sequent-to-sequent proof of Example 4.6 looks like this:

(1)	$(\exists x)(Fx \rightarrow A) \vdash_{NK} (\exists x)(Fx \rightarrow A)$	Premise
(2)	$(\forall x)Fx \vdash_{NK} (\forall x)Fx$	Assumption
(3)	$Fa \rightarrow A \vdash_{NK} Fa \rightarrow A$	Assumption
(4)	$(\forall x)Fx \vdash_{NK} Fa$	2 ∀E
(5)	$(\forall x)Fx, Fa \rightarrow A \vdash_{NK} A$	3,4 →E
(6)	$(\exists x)(Fx \rightarrow A), (\forall x)Fx \vdash_{NK} A$	1,5 ∃E
(7)	$(\exists x)(Fx \rightarrow A) \vdash_{NK} (\forall x)Fx \rightarrow A$	6 →I ◆

❏ Exercises

(1) Arrange the solutions to Exercise 4.I in tree format (*(2)).
(2) Arrange the solutions to Exercise 4.I in sequent-to-sequent format.

9 Summary

- The definition of validity is the same for LMPL as it is for LSL: an LMPL argument-form is valid if and only if no interpretation makes its premises true and its conclusion false.
- What is new is the definition of interpretation. In addition to specifying truth-values for sentence-letters, an interpretation includes a nonempty domain of discourse, an assignment of subsets of the domain as extensions to predicate-letters and an assignment of elements of the domain to individual constants.
- We extend our system of natural deduction NK to LMPL by adding an introduction and an elimination rule for each quantifier.
- The new rules of ∀E and ∃I are straightforward, but the rules of ∀I and ∃E have special restrictions: ∀I may be applied to ϕt only if t does not occur in any premise or assumption on which ϕt depends, and ∃E may be applied to ψ inferred from (among other formulae) an instance ϕt of $\ulcorner(\exists v)\phi v\urcorner$ only if t does not occur in ψ, $\ulcorner(\exists v)\phi v\urcorner$ or any premise or assumption other than ϕt on which ψ depends.
- The point of the restriction on ∀I is to ensure that ∀I is applied to ϕt only when the reasoning which leads to ϕt would equally serve to establish $\phi t'$, for any other individual constant t'.
- The point of the restrictions on ∃E is to ensure that, if a target formula ψ has been derived from premises and assumptions including an instance ϕt of $\ulcorner(\exists v)\phi v\urcorner$, then ∃E is applied only when the steps of that derivation would equally serve to derive ψ from any other instance $\phi t'$ of $\ulcorner(\exists v)\phi v\urcorner$ in conjunction with the other premises and assumptions.

PART III

FIRST-ORDER LOGIC WITH IDENTITY

Advanced Symbolizations

1 N-place predicates

Though monadic predicate logic is considerably more powerful than sentential logic, like sentential logic it is subject to severe limitations in the range of arguments which can be adequately treated within it. For example, the best we can do in symbolizing the intuitively valid argument

> A: (1) Brutus killed Caesar
> (2) ∴ Someone killed Caesar

is to use 'K_' for '_ killed Caesar', so that we would get the symbolization

> B: Kb
> ∴ (∃x)Kx.

But of course the absorption of the proper name 'Caesar' into the predicate is something we cannot get away with, as the following valid argument illustrates:

> C: (1) Brutus killed Caesar
> (2) ∴ Brutus killed someone and someone killed Caesar.

A correct symbolization of the premise of C will distinguish *both* names, and for the verb 'killed' we will have a separate symbol which will recur twice in the conclusion.

'Killed' is our first example of a *two-place*, or *binary* or *dyadic*, predicate. The monadic predicates of LMPL apply to objects one at a time: it is single objects which satisfy predicates such as '_ is happy' or '_ is a tiger'. On the other hand, dyadic or two-place predicates are so-called because they apply to objects two at a time: it is *pairs* of objects which satisfy '_ killed _'. More exactly, it is *ordered* pairs: Brutus killed Caesar, not vice versa. We use pointed brackets to indicate an ordered sequence of objects, as in '⟨Brutus, Caesar⟩' and '⟨Caesar, Brutus⟩'. The first of these, and not the second, satisfies '_ killed _'. English has many other two-place predicates. For example, all comparatives are two-place: '_ is taller than _', '_ is wiser than _' and so on. To symbolize sentences which contain two-place predicates we use an uppercase letter for the predicate together with whatever other names, quantifiers and connectives are required. To start with the easiest case,

      (1) Brutus killed Caesar

is symbolized by

      (1.s) Kbc

in which the order of the individual constants reflects the order of the names in (1); 'Kcb' would mean 'Caesar killed Brutus'.

English also contains three-place predicates, predicates which apply to objects three at a time and are satisfied by *ordered triples* of objects. For example, '_ is between _ and _' is a three-place predicate, as in the sentence

      (2) 3 is between 2 and 4

which receives the symbolization

      (2.s) Babc

in which 'a' stands for 3, 'b' stands for 2 and 'c' stands for 4. In other words, the ordered triple $\langle 3,2,4 \rangle$ satisfies '_ is between _ and _'. Other examples of three-place predicates in English are '_ borrows _ from _' and '_ sells _ to _'. Notice that

      (3) 4 is larger than 3 and 2

does *not* contain a three-place predicate, since it is simply a conjunction, '4 is larger than 3 and 4 is larger than 2', which would receive the symbolization

      (3.s) Lca & Lcb.

A similar analysis cannot be made of (2): '3 is between 2 and 3 is between 4' is senseless.

As the number of places increases, it becomes harder to find natural examples of sample English predicates. There are apparent four-place predicates in English, such as '_ is closer to _ than _ is to _', as in the sentence

      (4) London is closer to Paris than New York is to Chicago

though one could always insist on analyzing (4) as involving just the two-place predicate 'is less than':

      (5) The distance between London and Paris is less than the distance between New York and Chicago.

So in place of

(4.s) Clpnc

we could have a sentence of the form

(5.s) Ltt'

where t is an expression for the distance between London and Paris in kilometers and t' is an expression for the distance between New York and Chicago in kilometers. However, if we are to adhere to the surface syntax of (4), (4.s) would be the correct symbolization. Natural examples of five-or-more-place predicates are even harder to come by, though they can always be gerrymandered. But in our formal language, there will be no finite limit on the number of places a predicate may have.

Quantifiers interact with two-place predicates in more complex ways than they do with monadic predicates. We can illustrate this with further considerations about the assassination of Julius Caesar. If Brutus killed Caesar then

(6) Someone killed Caesar

which we symbolize by first paraphrasing it in Loglish, expanding 'someone' into 'there is at least one _ such that' and relativizing it to the domain of people. This yields

(6.a) There is at least one x such that x killed Caesar.

'x killed Caesar' is like 'Brutus killed Caesar' except that it contains the free variable 'x' instead of the proper name 'Brutus'. So it is symbolized as 'Kxc' and (6.a) becomes

(6.s) ($\exists$x)Kxc.

But in English, 'someone' can also occupy the position which 'Caesar' occupies in (1):

(7) Brutus killed someone

while the existential quantifier cannot occupy the corresponding position in the formula. We paraphrase (7) *by moving the quantifier to the front of the phrase*, which in English results in 'there is someone whom Brutus killed' and in Loglish, relativizing 'there is at least one' to the domain of people, comes out as

(7.a) There is at least one x such that Brutus killed x.

This is easily transcribed into symbols as

(7.s) (∃x)Kbx.

A quantifier can also occupy both name positions, as in

(8) Someone killed someone.

In symbolizing sentences which contain more than one quantifier, it is highly advisable to approach the problem one quantifier at a time: we decide which quantifier is within the scope of which and take the one with wider scope first. In the case of (8), we start with the first 'someone' and obtain

(8.a) There is at least one x such that x killed someone

which we transcribe into symbols as

(8.b) (∃x) x killed someone.

Notice that *all* we do in the first step is paraphrase the first quantifier: the rest of the English remains the same, except that we insert the quantifier's variable where appropriate. This leaves us with the phrase 'x killed someone' to symbolize. 'x killed someone' is like 'Brutus killed someone' except that it contains 'x' instead of 'Brutus'. As with 'Brutus killed someone', we move the quantifier to the front of the phrase:

(8.c) x killed someone: there is at least one y such that x killed y: (∃y)Kxy

Notice that we have to use a new variable 'y' in (8.c), since if we were to use 'x' again, the 'x' of 'x killed someone' would be captured by the new quantifier, whereas we want it to be bound by the quantifier symbolized in (8.b). (In fact, '(∃x)(∃x)Kxx' will not be allowed as a wff by the syntactic rules we will give for the formal language we are using in these examples.) The final symbolization of (8) is obtained by substituting '(∃y)Kxy' for 'x killed someone' in (8.b):

(8.s) (∃x)(∃y)Kxy.

To repeat: the more complex the sentences we symbolize, the more important it is that we approach the problem in a sequence of steps, tackling one quantifier at a time.

In the formal language into which we are translating English, the distinction between active and passive voice disappears, in the sense that we can continue to use 'K_,_' for '_ killed _' even when the English uses 'was killed by'. (7) means the same as

(9) Someone was killed by Brutus

and the first step in symbolizing (9) using 'K_,_' would be to paraphrase (9) as (7), and then proceed as before. But correctly symbolizing different combinations of quantifiers with active or passive verbs requires care. Some of the differences which can arise are illustrated by the following two *vignettes* of everyday life in America:

Situation A

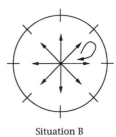

Situation B

In Situation A a group of people sit in a circle, each pointing a gun at the back of the head of the next person in the ring, and at a prearranged signal, they all fire simultaneously. In Situation B there is a person in the middle of a circle of people who is armed with a semiautomatic rifle, which he (it is always a he) has converted into a machine gun. This person shoots and kills everyone in the circle, then turns his weapon on himself. We wish to compare the truth-values of the following four sentences in Situations A and B:

(10) Everyone killed someone.
(11) Someone killed everyone.
(12) Everyone was killed by someone.
(13) Someone was killed by everyone.

The table below displays which sentences are true and which are false in the two situations. (10) is true in A because each person shoots the next person in the circle but false in B because the people on the perimeter do not kill anyone. (11) is false in situation A since there is no one who kills *everyone:* each person kills only one out of a possible eight people. But (11) is true in B since the man in the middle kills everyone. Note that (11) would be false if he did not turn his gun on himself—all that would then be true is that someone killed everyone *else.* (12) is true in both A and B, since in each situation, each person is killed by someone—in A, by the person behind, and in B, by the person in the middle. Finally, (13) is false in both A and B since there is no one person whom everyone killed ((13) would be true in a variant of B in which all the outward-pointing arrows are reversed). These results are tabulated as follows:

	A	B
10	T	⊥
11	⊥	T
12	T	T
13	⊥	⊥

Since no two of (10)–(13) have the same truth-values in both situations, they must all mean something different and hence must all have different symbolizations. We begin with (10), in which the first quantifier is 'everyone'. Relativizing 'every' to the domain of people, our first step is therefore

> (10.a) For every x, x killed someone

or

> (10.b) (∀x) x killed someone

This leaves us to fill in the symbolization of 'x killed someone':

> (10.c) x killed someone: there is at least one y such that x killed y:
> (∃y)Kxy

and substituting this formula into (10.b) yields

> (10.s) (∀x)(∃y)Kxy.

In (11) the initial quantifier is 'someone', so our first step is

> (11.a) There is at least one x such that x killed everyone

or

> (11.b) (∃x) x killed everyone.

This leaves us with 'x killed everyone' to symbolize. As usual, we move 'everyone' to the front:

> (11.c) x killed everyone: for every y, x killed y: (∀y)Kxy

which when substituted into (11.b) yields

> (11.s) (∃x)(∀y)Kxy.

(10.s) and (11.s) should be carefully contrasted: the crucial difference between them is in the order of the quantifiers, reflecting the difference in their order in the original English.

To symbolize (12) we have to convert the passive verb to active. The easiest stage at which to make the conversion is after we have dealt with the initial quantifier:

> (12.a) For every x, x was killed by someone

or

(12.b) ($\forall$x) x was killed by someone.

At the next step we convert before symbolizing:

(12.c) x was killed by someone: someone killed x: there is at least one y
such that y killed x: ($\exists$y)Kyx.

Note the order of the variables here: it is y, not x, who is the killer. Substituting into (12.b) produces

(12.s) ($\forall$x)($\exists$y)Kyx

which contrasts with (10.s) in the order of the variables attached to the predicate letter.

(13) is handled in a similar fashion. Beginning with the first quantifier we get

(13.a) There is at least one x such that x was killed by everyone

or

(13.b) ($\exists$x) x was killed by everyone

We then move the quantifier in 'x was killed by everyone' to the front of this phrase and convert to active:

(13.c) x was killed by everyone: everyone killed x: for every y, y killed x:
($\forall$y)Kyx

which yields the final symbolization

(13.s) ($\exists$x)($\forall$y)Kyx.

Again, (13.s) should be contrasted with (11.s): it is the difference in the order of the variables after the predicate letter which captures the active/passive distinction.

A two-place predicate may also be used to analyze predicates that are *prima facie* one-place. For example, a *killer* is someone who has *killed someone*. Thus, still restricting ourselves to the two-place predicate 'K_,_', we can symbolize sentences which are rather more involved than (10)–(13) because they contain nouns which unpack into verbs and quantifiers. For instance,

(14) Everyone killed a killer

which is true in Situation A but false in Situation B, can be paraphrased by

(15) Everyone killed someone who killed someone

for which we have the Loglish

(15.a) (∀x) x killed someone who killed someone

and in turn

(15.b) x killed someone who killed someone: (∃y)(x killed y and y killed
someone): (∃y)(Kxy & (∃z)Kyz)

so that for (14) and (15) we end up with

(15.s) (∀x)(∃y)(Kxy & (∃z)Kyz).

These examples involve only names, quantifiers and a single two-place
predicate, but the principles are the same in the symbolization of more complex sentences. For example, we can symbolize

(16) Caesar's assassins were friends of Brutus

using 'A_,_' for '_ is an assassin of _' and 'F_,_' for '_ is a friend of _'. First we
have to decide whether (16) is a universal sentence, an existential sentence or
a sentence whose main connective is truth-functional. Clearly, (16) is universal;
it does not mean merely that at least one of Caesar's assassins was a friend of
Brutus. Again, the relevant domain of discourse is people. So we begin with the
universal quantifier:

(16.a) For every x, if x is an assassin of Caesar, then x is a friend of Brutus

or

(16.b) (∀x)(x is an assassin of Caesar → x is a friend of Brutus).

This leaves us with the antecedent and consequent of the main subformula to
symbolize:

(16.c) x is an assassin of Caesar: Axc
(16.d) x is a friend of Brutus: Fxb.

Substituting these formulae in (16.b) gives the final symbolization

(16.s) (∀x)(Axc → Fxb).

Notice that the main parentheses in (16.b) are required; omitting them would
lead to ambiguity, since '(∀x)Axc → Fxb' could be read as the open sentence
'[(∀x)Axc] → Fxb', which means 'if everyone assassinated Caesar then x is a
friend of Brutus'.
    Next, let us try

(17) Caesar's assassins used knives supplied by Brutus

with 'K_' for '_ is a knife', 'U_,_' for '_ used _' and 'S_,_' for '_ supplied _'. Since the subject matter of the sentence involves not just Caesar, his assassins and Brutus, but also knives, the domain to which (17)'s quantifiers are relativized cannot be taken to be just the domain of people. In this situation we use the most general domain, that of things. As before, the sentence is universal, so we begin

(17.a) For every x, if x is an assassin of Caesar, then x used a knife supplied by Brutus

or

(17.b) ($\forall$x)(x is an assassin of Caesar → x used a knife supplied by Brutus).

The antecedent is the same as in (16.b), but the consequent is a little more complicated, since it contains a quantifier: 'a knife' means 'at least one knife'. Sometimes the indefinite article is a quantifier, as in this case, and sometimes not, as in (16.d). One test for when it occurs as a quantifier is that in such occurrences, replacing it with 'every' or 'at least one' results in natural English, and when it is not a quantifier, the results of such substitution will sound strained, as in 'x is at least one friend of Brutus'. Typically, quantificational occurrences of 'a' are in combination with a predicate and occupy one of the places in an n-place predicate. For example, in (17.b), 'a knife' occupies the second place of the two-place predicate 'used'.

Proceeding with the symbolization of (17), we move the quantifier 'a' of 'x used a knife supplied by Brutus' to the front of this phrase. A knife supplied by Brutus is a thing which is a knife *and* which Brutus supplied, *not* a thing which is a knife *if* Brutus supplied it: there is nothing conditional about being a knife supplied by Brutus. So we form a conjunction and convert the passive 'supplied by' to active as well:

(17.c) x used a knife supplied by Brutus: there is at least one y such that x used y and y is a knife and Brutus supplied y: ($\exists$y)((Uxy & Ky) & Sby).

Substituting formulae for the antecedent and consequent in (17.b) yields the final symbolization:

(17.s) ($\forall$x)(Axc → ($\exists$y)((Uxy & Ky) & Sby)).

Notice that we substitute the symbolization of the antecedent and consequent in (17.b) in exactly the same positions as the phrases which they symbolize occupy in (17.b). In particular, the existential quantifier remains as the main connective of the consequent; it does not wander into some other position earlier in the formula.

As a final example, we symbolize a three-quantifier sentence:

(18) Some of Caesar's advisers who knew all of Caesar's assassins suspected none of them.

We will use 'V_,_' for '_ advises _', 'S_,_' for '_ suspected _' and 'N_,_' for '_ knew
_'. The domain we relativize quantifiers to is again the domain of people, since
no other kinds of thing are mentioned in (18). (18) is an existential sentence,
since the initial 'some' is its main connective, and in Loglish reads:

(18.1) There is at least one x such that x is an adviser of Caesar who knew
all of Caesar's assassins, and x suspected none of Caesar's assassins

or

(18.b) ($\exists$x)(x is an adviser of Caesar who knew all of Caesar's assassins &
x suspected none of Caesar's assassins).

This leaves us to symbolize the two conjuncts within the parentheses, beginning with 'x is an adviser of Caesar who knew all of Caesar's assassins'. An
adviser of Caesar is a person who advises Caesar and an adviser *who* knew all
of the assassins is someone who is an adviser *and* knew all of them. Thus:

(18.c) x is an adviser of Caesar who knew all of Caesar's assassins: (Vxc &
x knew all of Caesar's assassins).

'x knew all of Caesar's assassins' contains another quantifier, which is moved
to the front of this phrase:

(18.d) x knew all of Caesar's assassins: for every y, if y is an assassin of
Caesar then x knew y: ($\forall$y)(Ayc $\rightarrow$ Nxy).

Observe that '($\forall$y)(Nxy $\rightarrow$ Ayc)' would be incorrect, since this formula says that
everyone x knew was an assassin of Caesar, not that x knew every assassin. Substituting (18.d) into (18.c) yields the correct symbolization of the first conjunct
of the conjunction in (18.b):

(18.e) (Vxc & ($\forall$y)(Ayc $\rightarrow$ Nxy)).

The second conjunct is 'x suspected none of Caesar's assassins', which contains the quantifier word 'none', which we can symbolize either as a universal
or as a negative existential. However, a drawback of the negative existential
symbolization will have become apparent to the reader who has worked
through §3–§5 of Chapter 6: if we are symbolizing sentences which are the premises and conclusion of an argument with a view to proving the argument, we
should avoid negative existential symbolizations since they are awkward to
work with, especially as premises. Typically, QS has to be applied to them, and
then after $\forall$E, SI has to be used to get the most straightforward formula, usu-

ally a conditional, to which sentential rules can be applied. By contrast, the universal quantifier analysis of 'none' gives a universal quantifier governing a conditional straight off, so that is how we shall treat this occurrence of 'none'. If x suspected none of the assassins, this means that for each person who is an assassin, x did not suspect that person. Thus:

(18.f)   x suspected none of Caesar's assassins: for every y, if y is an assassin of Caesar, then x did not suspect y: $(\forall y)(Ayc \rightarrow \sim Sxy)$.

We obtain the final symbolization of (18) by substituting (18.e) and (18.f) into their appropriate positions in (18.b):

(18.s) $(\exists x)[(Vxc \ \& \ (\forall y)(Ayc \rightarrow Nxy)) \ \& \ (\forall y)(Ayc \rightarrow \sim Sxy)]$.

There are three points to note about (18.s) and the process by which we arrived at it.

- The double occurrence of '$(\forall y)$' does not lead to double-binding, since the scope of the first '$(\forall y)$' is just '$(Ayc \rightarrow Nxy)$'. We could have used 'z' in place of 'y' in (18.f) but there is no need to.
- The use of two universal quantifiers in (18.s) is not essential, even though it reflects the English. The formula

  $(\exists x)[Vxc \ \& \ (\forall y)(Ayc \rightarrow (Nxy \ \& \ \sim Sxy))]$

  is logically equivalent to (18.s) and somewhat simpler. However, with complex formulae, logical equivalences are rarely discernible to the naked eye. The formulae we arrive at by our step-by-step procedure are likely to be correct and there is no need to attempt simplifications of them which may turn a correct answer into an incorrect one.
- For similar reasons, we do not allow quantifiers to wander out of the positions they are assigned in our step-by-step approach. In any reasonably complex formula, there are likely to be other positions the quantifiers could occupy without changing the meaning. For instance, it would not matter if the first universal quantifier in (18.s) were moved between the first bracket and the following parenthesis, giving a prefix '$(\exists x)[(\forall y)(Vxc \ \& \ ...$'. But with complex formulae it is often hard to tell if a repositioning of a quantifier changes the meaning, whereas the positions which the quantifiers are assigned in the step-by-step analysis are certain to be correct if the analysis has been carried out properly.

❑ Exercises

I   Relativizing '∃' and '∀' to the domain of people, symbolize each of the following sentences using 'L_,_' for '_ loves _' and no other predicate-letters. Show at least two intermediate steps in Loglish for each example.

    (1) Everyone loves someone.
    (2) No one loves everyone.
 *(3) There is no one who is unloved.
    (4) Everyone loves a lover.
    (5) Some love is unrequited.
 *(6) Only lovers love themselves.
    (7) Only the loved are lovers.
    (8) Everyone who is unloved loves a lover.

II   Relativizing '∃' and '∀' to the domain of people, symbolize each of the following sentences using 'K_,_' for '_ kills _' and no other predicate letters. Show at least two intermediate steps in Loglish for each example. Then classify with respect to Situations A and B on page 223, explaining your reasons.

    (1) Someone killed every killer.
    (2) Some killer killed a killer.
    (3) No killers killed themselves.
    (4) Everyone who killed a killer killed himself.
    (5) Only killers killed killers.

III   Symbolize each of the following sentences. Use the suggested symbols where given; where a complete dictionary is not given, state what (other) symbols you are using. Display at least two intermediate steps in Loglish and indicate the domain to which your quantifiers are relativized.

    (1) Every student reads at least one book. ('S_', 'R_,_', 'B_')
    (2) There is at least one book which every student reads.
    (3) There are students who read books.
 *(4) There are students who do not read books.
    (5) There are no students who read books.
    (6) Every student reads every book.
    (7) For every place, there's a road that leads to it. (See below)
    (8) No roads lead nowhere. ('R_', 'P_': _ is a place; 'L_,_': _ leads to _)
    (9) There's a place to which all roads that lead somewhere lead.
  (10) If anyone loves John, Mary will be jealous. ('a': J; 'b': M; 'J_': _is jealous)
  (11) If anyone loves John, Mary will be jealous of her. ('J_,_': _ is jealous of _)
 *(12) All plays attributed to Shakespeare were written by Marlowe.
  (13) If any of the plays attributed to Shakespeare was written by Marlowe, they all were.
  (14) Some politicians who took bribes from supporters were named by the *New York Times*. ('T_,_,_': _ took _ from _; 'S_,_': _ supports _; 'e': The *New York Times*)
  (15) Anyone who owns a house is in debt to someone. ('P_': _ is a person).
  (16) There is no largest prime number. ('Q_,_': _is at least as large as_; 'P_', 'N_')

*(17) There is a number which is prime just in case all numbers greater than or equal to it are composite. (Same dictionary as for (16))

(18) There are no movie stars who are not photographed. ('P_,_': _ photographs _)

(19) No student who fails all his courses will be readmitted. ('F_,_': _ fails _; 'C_,_': _ is a course of _)

(20) A person who uses a computer has an advantage over all those who don't. ('P_': _ is a person).

(21) No one who doesn't use a computer has an advantage over anyone who does.

*(22) No one will trust a politician who makes promises he or she can't keep. ('P_', 'T_,_', 'R_', 'M_,_', 'K_,_')

(23) People who don't return books they borrow are the moral equivalent of people who steal them. ('E_,_': _ is morally equivalent to _)

(24) Only valid arguments instantiate valid argument-forms. ('V_': _ is valid; 'A_': _is an argument; 'F_': _is an argument-form; 'I_,_': _ instantiates _)

(25) Not all acquaintance is mutual. (Use only 'A_,_': _ is acquainted with _)

(26) Some sequents have only infinite counterexamples. ('C_,_': _ is a counterexample to_)

*(27) There is a composer who is liked by anyone who likes any composer at all.

(28) People who give away everything they earn are either rich or foolish. ('G_,_,_': _ gives away _ to _)

(29) All the famous people whom she knows are photographed in books by Annie Liebowitz.

*(30) Only the lonely pity those who pity themselves. ('L_', 'P_,_')

IV Using the same symbols as in (26), translate the following formulae into *colloquial* English.

(1) $\sim(\exists x)(Sx \,\&\, (\exists y)Cyx)$

*(2) $(\exists x)(Sx \,\&\, (\forall y)(Cyx \to \sim Iy))$

(3) $(\forall x)[(\exists y)(Sy \,\&\, Cxy) \to (\forall z)(Sz \to Cxz)]$

(4) $(\forall x)[(Sx \,\&\, (\exists y)(Cyx \,\&\, Iy)) \to (\exists z)(Czx \,\&\, \sim Iz)]$

2 Identity, number and descriptions

In this section we introduce our final logical constant, identity, and show how it can be used to express sentences involving numbers and sentences involving name-like phrases called *singular definite descriptions*.

Identity

We noted in our discussion of the mass killer (Situation B, page 223) that unless he turns his gun on himself, it is not true that he kills everyone but only that he kills everyone *else* or everyone *but himself*. How can we get the effect of 'else' or 'but himself'? To express these notions we need to be able to express identity and nonidentity. To kill everyone else or everyone but oneself is to kill everyone who is *not identical to* oneself. Similar locutions can be dealt with similarly; for example, to love *no one but* oneself or *no one except* oneself or to love *only* oneself is to love a person if and only if that person is *identical to oneself*.[1]

'is identical to' appears to be a two-place predicate, so a first thought in approaching, say,

(1) Narcissus loves himself and no one else

is to use 'L_,_' for '_ loves _' and 'I_,_' for '_ is identical to _'. But treating '_ is identical to _' as just another two-place predicate misses an important affinity between it and the logical constants with which we are already familiar. For it will turn out that '_ is identical to _' has characteristic behavior in arguments which we can capture both semantically and in two rules of inference, an introduction and an elimination rule. In the next chapter we will develop the logic of identity. For the moment we just need a symbol to express identity, a symbol whose meaning will be constant in every interpretation. We use the symbol '=' for 'is identical to', and for convenience we abbreviate formulae of the form $\ulcorner \sim t = t' \urcorner$ by $\ulcorner t \neq t' \urcorner$; that is, we use the symbol '≠' for 'is not identical to'.

Returning to (1), we expand it into full sentential form as 'Narcissus loves himself and Narcissus loves no one else', which we commence symbolizing with the Loglish:

(1.a) Laa & a loves no one else.

The second conjunct contains the quantifier 'no one', which we analyze universally: if you love no one else, this means that for any person, if that person is not you, you do not love her or him; or expressing the same idea without negation, if you love that person then he or she must be yourself. In Loglish, relativizing 'every' to the domain of people, this is

(1.b) a loves no one else: for every x, if a loves x then x = a:
$(\forall x)(Lax \rightarrow x = a)$.

Substituting in (1.a) gives us

(1.s) Laa & $(\forall x)(Lax \rightarrow x = a)$.

Correspondingly,

(2) Narcissus loves someone else

means that Narcissus loves someone not identical to Narcissus. In the first step we move the existential quantifier to the front and obtain

(2.a) There is at least one x such that Narcissus loves x and x is not identical to Narcissus

---

[1] Most people seem to understand (i) 'John loves only himself' as a conjunction, (ii) 'John loves himself and no one else'. On another reading, the meaning of (i) is that of only the second conjunct of (ii). On this view, 'John loves only himself' does not literally entail that he loves himself. Either reading is acceptable.

which can be directly transcribed into symbols as

(2.s) $(\exists x)(Lax \ \& \ x \neq a)$.

The same approach works for more complex sentences, such as

(3) People who love only themselves are disliked by others.

Using 'K_,_' for '_ likes _', we start (3)'s symbolization by putting its first implicit 'all' into Loglish, relativizing to the domain of people:

(3.a) For every x, if x loves only x then x is disliked by others

or

(3.b) $(\forall x)[x$ loves only $x \to x$ is disliked by others]

We now symbolize the antecedent 'x loves only x' and the consequent 'x is disliked by others'. 'x loves only x' is a universal statement to the effect that anyone whom x loves *is* x, so, again relative to the domain of people, we have

(3.c) x loves only x: for every y, if x loves y then y is identical to x: $(\forall y)(Lxy \to y = x)$.

The consequent 'x is disliked by others' is ambiguous between 'all others' and 'some others'. The former seems more likely, so we have:

(3.d) x is disliked by others: for every y, if x is not identical to y then y does not like x: $(\forall y)(x \neq y \to \sim Kyx)$.

The symbolization of (3) is obtained by substituting (3.c) and (3.d) into (3.b) in the appropriate positions:

(3.s) $(\forall x)[(\forall y)(Lxy \to y = x) \to (\forall y)(x \neq y \to \sim Kyx)]$.

Notice that the universal quantifier '$(\forall y)$' is the main connective of the *antecedent* of the conditional which is the scope of '$(\forall x)$', so that the correct way of reading this conditional is '*if* everyone whom x loves is identical to x then...'. In other words, '$(\forall y)$' is *within* the scope of the main '$\to$' in (3.s). To move it in front of the bracket would put the main '$\to$' within the scope of '$(\forall y)$' and change the meaning of the conditional to something rather unnatural: 'everyone who is identical to x if loved by x, is disliked by others' (changing the 'y' in the consequent to avoid double-binding).

Number

Now that we see how to express distinctness, we can express some kinds of numerical sentences, such as 'exactly 14 nations are in the EEC', since plurality is just repeated distinctness. We treat the claim that exactly *n* things are thus-and-so as a combination of the claim that *at least n* are and *at most n* are; and for each value of *n*, it is straightforward to symbolize both claims. For example,

(4) At least one candidate qualified

is simply

(4.s) $(\exists x)(Cx \,\&\, Qx)$

using '$C\_$' for '$\_$ is a candidate', '$Q\_$' for '$\_$ qualified' and '$\exists$' relativized to the domain of people. To symbolize

(5) At most one candidate qualified

it helps to paraphrase the meaning of 'at most one' in terms of *choices with replacement*. Suppose we have a bag of colored balls. If there is at most one red ball in the bag and we choose balls from the bag one at a time, replacing each ball after it is drawn, then if on any two choices we draw a red ball, it must be the *same* ball we drew both times. This is a universal claim, so absent a special symbol for 'at most one', (5) is best approached as a universal sentence:

(5.a) For every x, for every y, if x is a candidate who qualified and y is a candidate who qualified, then x = y.

Here we can think of the two universal quantifiers as corresponding to two choices with replacement: if both choices produce candidates who qualified, then the candidate chosen first is identical to the one chosen second. The formalization as a universal sentence helps to emphasize that (5) is not existentially committing: it is true even if there are no candidates who qualified. (5.a) is easily transcribed into symbols:

(5.s) $(\forall x)(\forall y)\{[(Cx \,\&\, Qx) \,\&\, (Cy \,\&\, Qy)] \to x = y\}$.

Consequently, we can symbolize

(6) Exactly one candidate qualified

simply by conjoining (4.s) and (5.s). However, there is a better way. If there is exactly one candidate who qualified, then there is at least one, and *any* candidate who qualified is identical to *that* one. The first step in symbolizing (6) this way, relativizing quantifiers to the domain of people, is

(6.a) There is at least one x such that x is a candidate who qualified, and every candidate who qualified is identical to x

or

(6.b) $(\exists x)[(Cx \& Qx) \&$ every candidate who qualified is identical to x].

The conjunct 'every candidate who qualified is identical to x' is a universal open sentence with just one quantifier, and in Loglish is

(6.c) for every y, if y is a candidate who qualified then y is identical to x

or

(6.d) $(\forall y)((Cy \& Qy) \to y = x)$.

We get the complete symbolization of (6) by substituting (6.d) in (6.b):

(6.s) $(\exists x)[(Cx \& Qx) \& (\forall y)((Cy \& Qy) \to y = x)]$.

The same strategies suffice for 'there are exactly two...', 'there are exactly three...' and so on, though the formulae become increasingly unwieldy. First,

(7) There are at least two candidates who qualified

can be expressed using '$\neq$' and two existential quantifiers:

(7.s) $(\exists x)(\exists y)[(Cx \& Qx) \& (Cy \& Qy) \& x \neq y]$.

The third conjunct is necessary, for otherwise we would have an existential sentence which could be made true in the following way. Let Sue be the only candidate who qualified. Then 'Sue is a candidate who qualified and Sue is a candidate who qualified' is '$(Ca \& Qa) \& (Ca \& Qa)$', and so verifies '$(\exists y)[(Ca \& Qa) \& (Cy \& Qy)]$', which in turn verifies '$(\exists x)(\exists y)[(Cx \& Qx) \& (Cy \& Qy)]$'. So this formula, unlike (7.s), could be true if there is only one candidate who qualified.

Next, we express

(8) There are at most two candidates who qualified

again in terms of choices with replacement: if there are at most two candidates who qualified, this means that if any *three* choices with replacement produce candidates who qualified, then one candidate was chosen *at least twice*. So the Loglish version of (8) will begin with three universal quantifiers:

(8.a) For any x, for any y, for any z, if x is a candidate who qualified and y is a candidate who qualified and z is a candidate who qualified then x = y or y = z or x = z.

This is transcribed as

(8.s) $(\forall x)(\forall y)(\forall z)\{[(Cx \& Qx) \& (Cy \& Qy) \& (Cz \& Qz)] \to (x = y \lor x = z \lor y = z)\}$.

All three disjuncts in the consequent of the conditional subformula of (8.s) are required. For instance, if 'y = z' were omitted, the resulting formula could be false while (8) is true; this would be in a situation where the same candidate was chosen on second and third choices. Strictly, we should group the first two or the second two of the conjuncts in the antecedent of the conditional in (8.s) so that the antecedent is a two-conjunct conjunction, *mutatis mutandis* in the consequent, but our formulae are now achieving a length where the payoff from putting in all the parentheses is counteracted by the effort required to penetrate the clutter to detect the semantically significant matching pairs. So we will exploit another invisibility convention: whenever there are three or more consecutive conjuncts in a formula ϕ there are always invisible parentheses present which associate conjuncts so that ϕ can be produced by standard formation rules; similarly with disjuncts. For definiteness, we shall assume that invisible parentheses are grouped to the left: $\ulcorner(p \,\&\, q) \,\&\, r\urcorner$, not $\ulcorner p \,\&\, (q \,\&\, r)\urcorner$. But when symbolizing English with a view to giving a proof, it is a good idea to leave as many semantically redundant parentheses invisible as possible, and then to go back in the course of constructing the proof and make them visible in the way that is most convenient for applying the rules.

To express

    (9) There are exactly two candidates who qualified

we again have the option of just conjoining (7.s) and (8.s), but there is a more economical symbolization on the model of (6.s): if there are exactly two, this means that there are at least two, and any candidate who qualified is identical to one or other of those two. So in the style of (6.s) we have:

    (9.a) There is at least one x such that there is at least one y such that x is a candidate who qualified and y is a candidate who qualified and for any z, if z is a candidate who qualified, then z is identical to x or to y.

In symbols, this is

    (9.s) $(\exists x)(\exists y)\{[(Cx \,\&\, Qx) \,\&\, (Cy \,\&\, Qy) \,\&\, x \neq y] \,\&\, (\forall z)[(Cz \,\&\, Qz) \to (z = x \lor z = y)]\}$.

It should be clear how to proceed from here with 'at least...', 'at most...', and 'exactly three candidates who qualified'. For instance, to say

    (10) There are at least three candidates who qualified

is to say that there are x, y and z who are candidates who qualified and all three are distinct from each other. In symbols, this is

    (10.s) $(\exists x)(\exists y)(\exists z)\{[(Cx \,\&\, Qx) \,\&\, (Cy \,\&\, Qy) \,\&\, (Cz \,\&\, Qz)] \,\&\, (x \neq y \,\&\, x \neq z \,\&\, y \neq z)\}$.

Definite Descriptions

The other use of identity which we will cover here is in symbolizing a kind of noun phrase called a *singular definite description*. A singular definite description in English is a noun phrase beginning with the determiner 'the', and which picks out a single individual by means of describing its properties. For example, 'the first person on the moon' picks out Neil Armstrong, 'the first computer programmer' picks out Ada Lovelace, 'the fastest human' picks out Ben Johnson (unofficially) and so on. A singular definite description may be *improper* in two ways: it may fail to pick out any individual, as with 'the present king of France', or it may fail to identify an individual *uniquely;* for example, on many occasions in the recent past there has been more than one heavyweight boxing champion at a single time—a WBA champion, a WBF champion and so on—and at such times 'the heavyweight champion of the world' would be an improper singular definite description (though the plural definite description 'the heavyweight champions of the world' would be proper—however, we will not discuss plural definite descriptions).

It may occur to the reader that since singular definite descriptions—henceforth 'descriptions' since it is only the singular definite ones we will be concerned with—are like proper names in that they pick out single individuals, we could just translate them by individual constants. But that would be logically inadequate. Whether or not the definite description 'the discoverer of America' is proper, the sentence 'the discoverer of America was a European' logically entails that there are discoverers and there is such a thing as America. But if we collapse the definite description 'the discoverer of America' into an individual constant 'a', so that the sentence 'the discoverer of America was a European' is symbolized 'Ea', there is no way of drawing the conclusion that there are discoverers or that America exists. A logically adequate treatment of the description must therefore contain constituents corresponding to 'discovers' and 'America'.

'Discovers' is a two-place predicate and 'America' is a proper name, so these words belong to syntactic categories we have already met. However, the definite article 'the' is something new. 'The' occupies the same position in sentences as the indefinite article 'a', and since we use one or another of our quantifier symbols for the latter (depending on the meaning of 'a' in the context), perhaps we should introduce a new quantifier symbol for 'the'. There are good arguments for this approach, but the traditional procedure is to get the effect of 'the' using apparatus already at our disposal, as we did with 'no one' and 'none'. Comparing the following two sentences,

(11) Alice knows a candidate who qualified

and

(12) Alice knows the candidate who qualified

we see that the central difference between them is the implication of uniqueness in (12). If we paraphrase (11) in Loglish as

(11.a) There is at least one x such that x is a candidate who qualified and Alice knows x

then we can express the implication of uniqueness in (12) simply by changing 'at least one' to 'exactly one':

(12.a) There is exactly one candidate who qualified and Alice knows her

or in Loglish

(12.b) There is exactly one x such that x is a candidate who qualified and Alice knows x.

So the symbolization of (12) is the same as that of (6) but for the extra conjunct 'Alice knows x', 'Kax':

(12.s) $(\exists x)[(Cx \ \& \ Qx) \ \& \ (\forall y)((Cy \ \& \ Qy) \rightarrow y = x) \ \& \ Kax]$.

Sometimes definite descriptions occur in English without an explicit 'the', most commonly as possessives. For example, in symbolizing

(13) John's sister is famous

we treat 'John's sister' as a syntactic variant of 'the sister of John', so that (13) is paraphrased as

(13.a) The sister of John is famous

and then using our 'exactly one' approach to 'the' we rephrase (13.a) as

(13.b) There is exactly one sister of John and she is famous.

Using 'S_,_' for '_ is a sister of _', (13.b) becomes

(13.c) There is exactly one x such that Sxa and Fx

and so the final symbolization is

(13.s) $(\exists x)(Sxa \ \& \ (\forall y)(Sya \rightarrow y = x) \ \& \ Fx)$.

These examples illustrate the general form of a translation of an English sentence containing a definite description into our formal language: '...the ϕ...' becomes 'there is exactly one thing which is ϕ and ...it...', and we express 'there is exactly one thing which is ϕ' in the style of (6.s). This approach is originally due to Bertrand Russell and is known as 'Russell's Theory of Descriptions'; however, it is far from uncontroversial (see Russell, Lecture 6; Strawson, Paper 1). Note that the theory is only meant to apply to phrases of the form 'the ϕ' in which the object picked out is determined by the predicates in ϕ. Sometimes a description 'grows capital letters', and picks out an object which does not have the properties mentioned in the description. Russell gives the example 'the Holy Roman Empire', which, he points out, was neither holy, nor Roman, nor an empire. Such 'descriptions' should be treated as proper names.

In symbolizing sentences with descriptions in Russell's style, there is a contrast to look out for, according to how the ellipsis in 'the ϕ is...' is filled. One might think that

(14) The sheriff is corrupt

and

(15) The sheriff is the local tax inspector

both predicate properties of the sheriff, but though predication occurs in (14), the 'is' of (15) is *not* the 'is' of predication. The 'is' of (15) is often called the 'is' of identity, since it means 'is identical to', which the 'is' of (14) does not. Though it is controversial how deep this distinction goes (see Hintikka), we can draw it for our purposes by using the criterion that the 'is' of identity is always replaceable by the phrase 'is the same person/thing as', which the 'is' of (14) is clearly not. An alternative test is that the 'is' of identity is always followed by a proper name, a singular definite description or some other expression which purports to pick out a unique thing.

Applying Russell's strategy, we paraphrase (14) as

(14.a) There is exactly one sheriff and he/she is corrupt

which in Loglish, relativizing quantifiers to the domain of people, becomes

(14.b) There is exactly one x such that x is a sheriff and x is corrupt

and then in symbols, using 'S_' for '_ is a sheriff' and 'C_' for '_ is corrupt',

(14.s) $(\exists x)[Sx \,\&\, (\forall y)(Sy \to y = x) \,\&\, Cx]$.

On the other hand, (15) gets the Russellian paraphrase

(15.a) There is exactly one sheriff and he/she is identical to the local tax inspector

in which the phrase 'is identical to the local tax inspector' contains another definite description. We transcribe (15.a) into Loglish one description at a time:

(15.b) There is exactly one x such that x is a sheriff and x is identical to the local tax inspector: $(\exists x)(Sx \,\&\, (\forall y)(Sy \to y = x) \,\&\, x$ is identical to the local tax inspector).

This leaves 'x is identical to the local tax inspector', the first step in symbolizing which is to move 'the local tax inspector' to the front, giving

(15.c) the local tax inspector is such that x is identical to him/her.

Then by Russell's analysis, relativizing to the domain of people, we obtain

(15.d) there is exactly one local tax inspector and x is identical to him/her

or

(15.e) there is exactly one z such that z is a local tax inspector and z is identical to x: $(\exists z)(Tz \ \& \ (\forall y)(Ty \to y = z) \ \& \ z = x)$.

We get the symbolization of (15) by substituting (15.e) into (15.b):

(15.s) $(\exists x)[Sx \ \& \ (\forall y)(Sy \to y = x) \ \& \ (\exists z)(Tz \ \& \ (\forall y)(Ty \to y = z) \ \& \ z = x)]$.

The case of definite descriptions is one of many where we can exploit the flexibility of our formal language to express locutions for which the language was not explicitly designed. But as we shall see later, there are definite limits on how far we can take this.

❑ Exercises

I   Symbolize each of the following sentences relativizing quantifiers to the domain of people. Use the suggested symbols where given; if a complete dictionary is not given, state what (other) symbols you are using. Display at least two intermediate steps in Loglish (more complex problems require three or four).

   (1)   Only John likes John.
   (2)   Only John likes himself.
   (3)   John likes only himself.
   (4)   Everyone likes someone else.
   (5)   There is someone who is disliked by everyone else.
  *(6)   No one is wiser than anyone else.
   (7)   No one is wiser than everyone else.
   (8)   There are exactly two candidates.
   (9)   There are at most three questions.
  (10)   At least one candidate will answer at least two questions.
  (11)   At least two candidates will answer at least one question.
  (12)   At least two candidates will answer no questions. ('C_', 'Q_', 'A_,_')
 *(13)   At most two candidates will answer every question.
  (14)   At least two candidates will answer the same questions.
  (15)   Exactly one candidate will answer exactly one question
  (16)   The author of *Hamlet* was Marlowe. ('A_,_': _ authored _)
  (17)   John's sister loves Susan's brother.
 *(18)   The composer liked by anyone who likes any composer at all is Mozart.

(19) The only woman appointed to the Supreme Court is an ally of conservatives on some issues and liberals on others.

(20) Only the lonely pity those who pity no one but themselves.

3 The syntax of LFOL

The formal language into which we have been translating English in the previous two sections is an extension of the language of monadic predicate calculus obtained by adding (i) *n*-place predicate-letters for all values of *n* greater than 1 and (ii) the new logical constant '='. As a terminological simplification in giving a complete account of the lexicon of the extended language, we shall refer to sentence-letters including 'λ' as *zero-place* predicate-letters, since they contain no places into which an individual constant or variable can be inserted. We call the language 'LFOL' for 'the language of first-order logic'.

The lexicon of LFOL:

> For every $n \geqslant 0$, an unlimited supply of *n*-place predicate-letters λ, A_n, B_n, C_n,..., A_n', B_n', C_n',...; an unlimited supply of individual variables and individual constants; the sentential connectives '~', '&', '$\lor$', '$\rightarrow$' and '$\leftrightarrow$', the quantifiers '$\forall$' and '$\exists$'; the identity sign '='; and punctuation marks '(' and ')'.

As before, we do not display subscripts explicitly in formulae, since it is always clear by inspection whether a predicate-letter is 0-place or 1-place or whatever. We continue to use 'term' to mean either an individual constant or an individual variable. The formation rules are then almost exactly the same as those of LMPL, except that we wish to allow the formation of atomic formulae consisting in an *n*-place predicate-letter followed by *n* terms, $n > 1$. We retain LMPL's proscription on double-binding and redundant quantification. And the definitions of all the standard syntactic concepts—subformula, scope of a connective, main connective and so on—remain the same, so we do not repeat them here (see page 168 for the details).

The formation rules of LFOL:

> (*f-at*): Any sentence-letter is an atomic wff. For any *n*-place predicate-letter λ and *n* terms $t_1,...,t_n$ (all from the lexicon of LMPL), $\lambda t_1,...,t_n$ is an (atomic) wff.
>
> (*f-=*): If t_1 and t_2 are terms, $\ulcorner t_1 = t_2 \urcorner$ is a wff.
>
> (*f-con*): If ϕ and ψ are wffs (atomic or otherwise) so are $\ulcorner \sim\phi \urcorner$, $\ulcorner (\phi \And \psi) \urcorner$, $\ulcorner (\phi \lor \psi) \urcorner$, $\ulcorner (\phi \rightarrow \psi) \urcorner$ and $\ulcorner (\phi \leftrightarrow \psi) \urcorner$.
>
> (*f-q*): If ϕv is an open wff with all occurrences of the individual variable v free, then $\ulcorner (\exists v)\phi v \urcorner$ and $\ulcorner (\forall v)\phi v \urcorner$ are wffs.
>
> (*f!*): Nothing is a wff unless it is certified as such by the previous rules.

(*f-con*) is for any of the sentential connectives. (*f-=*) is a new rule and (*f-at*) is slightly different from its counterpart in LMPL. In exhibiting the construction of a wff in a parse tree, the bottom nodes will occur in pairs, with the left node of each pair containing an *n*-place predicate-letter and the right node an ordered sequence of *n* terms, while the node immediately above such a pair will contain the atomic formula which results from writing the predicate-letter on the left followed by the terms on the right, in the given order. To illustrate, here is the parse tree for the formula '($\forall$x)(Rxx → ($\exists$y)(Haxy & (Rxy → y = a)))'.

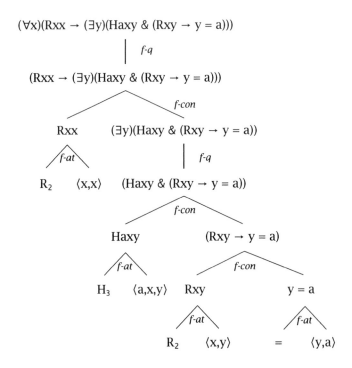

One aspect of our treatment of the identity symbol bears comment. Though semantically the symbol is treated as a logical constant, it combines syntactically like a two-place predicate-letter, as the leaf nodes of this tree reveal. But when an atomic formula is formed with it, the symbol occurs between the terms instead of as a prefix to them. Some writers put *all* two-place predicate-letters between their accompanying terms in atomic formulae. This style is called *infix* notation, as opposed to the *prefix* notation we have employed until now. In the remainder of this book we will continue to use prefix notation with the exception that if a symbol is a familiar one which is standardly infixed, we will follow the standard practice: just as we would not write '=(y,a)' instead of 'y = a', we will not write '<(x,y)' but rather 'x < y' and so on. We also assume that standard mathematical notation belongs to LFOL.

❏ Exercises

I   Identify which of the following formulae are non-wffs and in each case explain why the formula could not be constructed by the formation rules.

     (1) (∃x)[(∀y)Fxy → (∃z)(∃x)Hzx]
     (2) (∀x)[(∃y)(Rxy & Ryz) → Fx]
   *(3) (∀x)((∀y)(Rxy → (∃z)Hzx)
     (4) (∀x)((∀y)Rxy) → (∃z)Hzx)

II   Draw the parse tree for

$$(∃x)(Px \text{ \& } (∀y)((Cy → (∃z)(Pz \text{ \& } Dyz)) → (A ∨ Dyx)))$$

What is the scope of the universal quantifier in this formula?
Is the first occurrence of '→' within the scope of the second or vice versa?

4 Ambiguity

One of the problems we face in translating English into LFOL is that while every wff of LFOL has exactly one meaning (given an interpretation), it is often difficult, and sometimes impossible, to determine a unique meaning for some English sentences: English can be *ambiguous*. One familiar kind of ambiguity is *lexical* ambiguity, when the same word has two or more meanings and we try to use context or our general knowledge of the world to determine which meaning the speaker intends. For instance, though the word 'bank' can mean 'edge of a river' or 'financial institution', conversational context can be expected to fix which is intended by the statement 'John is at the bank'. A famous example in which it is general knowledge of the world which settles the meaning is

     (1) Time flies like an arrow, fruit flies like a banana.

     Another sort of lexical ambiguity has already been encountered, the ambiguity of the indefinite article 'a' which may express either existential or universal quantification. For instance, in the symbolization problem I.4 of §1, 'Everyone loves a lover', it is just as reasonable to take the indefinite article 'a' as a universal quantifier (as in 'Everyone likes a nice surprise') as it is to take it as existential, and the only way of determining which is meant in a conversational context may be to ask the speaker.
     The other main kind of ambiguity in English, and the one which concerns us most here, is *structural* ambiguity. Structural ambiguity arises when an English sentence can be understood in more than one way by attributing more than one syntactic structure to it. Usually there are ways of paraphrasing the English by other English sentences which bring out the ambiguity. For instance, a very common kind of structural ambiguity involves the relative scope of con-

nectives, and is called a *scope* ambiguity. We have already encountered some examples of this, such as the symbolization problem 3.I.7 of Chapter 5,

> (2)   All that glitters is not gold.

If we symbolize this sentence (relative to the domain of things) by following the English word order, we would get the Loglish

> (2.a)   (∀x)(if x glitters then x is not gold).

But this means that *nothing* which glitters is gold, and although that is a possible reading of (2), it is not the normal understanding of the sentence. The normal understanding is 'not everything which glitters is gold', in other words

> (2.b)   ~(∀x)(if x glitters then x is gold).

Symbolizing (2.a) and (2.b) gives us respectively

> (2.c)   (∀x)(Gx → ~Ox)

and

> (2.d)   ~(∀x)(Gx → Ox)

In (2.c) the scope of '~' is only the atomic formula 'Ox', while in (2.d) its scope is the main subformula '(∀x)(Gx → Ox)'. Hence the terminology 'scope-ambiguity': in (2), the scope of '~' is ambiguous.

The sentence

> (3)   Mary learned something from each of her teachers

exhibits a similar ambiguity. If taken at face value, the universal quantifier 'each' is *within* the scope of the existential quantifier 'something'. But this is an unnatural reading of (3), since it interprets (3) as saying

> (4)   There is something Mary learned from all her teachers.

In other words, there is some piece of knowledge, or quality of character, which Mary absorbed over and over again from all her teachers. But it is more natural to read (3) as allowing that Mary learned different things from different teachers without there necessarily being any one thing she learned from all of them. To raise the salience of this reading we can paraphrase (3) as

> (5)   From each of her teachers, Mary learned something

which is *consistent* with there being one thing she learned from them all but, unlike (4), does not *require* it. So the natural understanding of (3) involves interpreting the existential quantifier as being within the scope of the universal, as (5) makes explicit, rather than vice versa. This illustrates again how English word order is not always a reliable guide to the order in which we should take quantifiers in symbolizing English sentences.

We symbolize the two readings of (3) as follows, using 'T_,_' for '_ teaches _', 'L_,_,_' for '_ learns _ from _' and 'a' for 'Mary'. The first, less natural, reading is the one articulated by (4), in symbolizing which we begin with the existential quantifier:

(4.a) There is at least one x such that Mary learned x from all of her teachers

or

(4.b) ($\exists$x) Mary learned x from all of her teachers.

To symbolize 'Mary learned x from all of her teachers' we move the 'all' to the front:

(4.c) Mary learned x from all of her teachers: $(\forall y)(Tya \rightarrow Laxy)$.

We then substitute (4.c) in (4.b) to obtain the symbolization of (4):

(4.s) $(\exists x)(\forall y)(Tya \rightarrow Laxy)$.

On the other hand, the more natural interpretation of (3), spelled out in (5), begins with the universal quantifier:

(5.a) For every x, if x is a teacher of Mary then Mary learned something from x

or

(5.b) $(\forall x)(Txa \rightarrow$ Mary learned something from x).

To symbolize 'Mary learned something from x' we move the quantifier to the front of this phrase:

(5.c) Mary learned something from x: there is something Mary learned from x: there is at least one y such that Mary learned y from x: $(\exists y)Layx$.

Substitution in (5.b) yields the preferred reading of (3):

(5.s) $(\forall x)(Txa \rightarrow (\exists y)Layx)$.

(4.s) and (5.s) should be carefully contrasted. In terms of our understanding of quantifiers as choices with replacement, (4.s) says that we can pick something such that, choosing teachers of Mary at random, we find that Mary learned that something from all of them. In other words, the choice of thing learned is made first, and then it is held fixed while teachers are chosen. But in (5.s) we choose a teacher first, and for *each* such choice, (5.s) says we can·find something Mary learned from the teacher; this allows the thing learned to vary with choice of teacher.

The main difference between the two readings of (3) concerns the scope of the quantifier '($\exists$y)', but because the scope difference produces a difference in the order of the quantifiers, the structural ambiguity which (3) exhibits is sometimes called the $\forall\exists/\exists\forall$ ambiguity. The most famous example of this ambiguity is in a remark attributed to Abraham Lincoln:

    (6) You can fool some of the people all of the time.

The $\forall\exists$ reading of this remark is given by the paraphrase

    (7) At each time, there are people you can fool then

but perhaps different people at different times. The $\exists\forall$ reading is given by

    (8) There are people such that you can fool them at all times

that is, people who are perennially dim-witted. We symbolize the two readings using 'T_' for '_ is a time' and 'F_,_,_' for '_ can fool _ at _'. Both (7) and (8) contain three quantifiers, since we take 'you' to mean 'anyone' ('you' is lexically ambiguous between 'you' in the sense of an arbitrary person and 'you' in the demonstrative sense, the person to whom (6) is addressed; we choose the former, 'any person' sense, since it is unlikely that Lincoln was addressing some specific person). Since (7) and (8) concern both people and times, we have to relativize their quantifiers to the inclusive domain of things and so we must use special predicates for 'person' and 'time'.

To begin (7) we process the initial quantifier:

    (7.a) For every x, if x is a time then there are people anyone can fool at x

or

    (7.b) ($\forall$x)(Tx $\rightarrow$ there are people anyone can fool at x).

The main quantifier in 'there are people anyone can fool at x' is the existential one, so for this subformula we have

    (7.c) there is at least one y such that y is a person and anyone can fool y at x

or

(7.d) $(\exists y)(Py$ & anyone can fool y at x).

Note that if we do not include 'y is a person and' in (7.c), it will not say that there is some*one* whom anyone can fool at x, only that there is some*thing*, since the 'there is at least one' is relativized only to the domain of things. Finally, 'anyone can fool y at x' is a universal formula:

(7.e) for every z, if z is a person, z can fool y at x: $(\forall z)(Pz \rightarrow Fzyx)$.

Substituting (7.e) into (7.d) yields '$(\exists y)(Py$ & $(\forall z)(Pz \rightarrow Fzyx))$', and substituting this into (7.b) produces the $\forall\exists$ interpretation of (6):

(7.s) $(\forall x)(Tx \rightarrow (\exists y)(Py$ & $(\forall z)(Pz \rightarrow Fzyx)))$.

To say that at each time there are people you can fool at that time does not mean that there is someone who is fooled all the time, the latter being the $\exists\forall$ reading of Lincoln's remark. The $\exists\forall$ interpretation begins with the existential quantifier, so our first step is

(8.a) There is at least one x such that x is a person and anyone can fool x all of the time

or

(8.b) $(\exists x)(Px$ & anyone can fool x at all times).

In 'anyone can fool x at all times' there are two universal quantifiers governing an atomic formula, and we take them in the order in which they occur in the Loglish:

(8.c) $(\forall y)(Py \rightarrow$ y can fool x at all times)

and then for 'y can fool x at all times' we have

(8.d) $(\forall z)(Tz \rightarrow Fyxz)$.

Making the appropriate substitutions, the formalization of (6) on the $\exists\forall$ interpretation at which we arrive is

(8.s) $(\exists x)(Px$ & $(\forall y)(Py \rightarrow (\forall z)(Tz \rightarrow Fyxz)))$.

It is open to speculation which of (7.s) and (8.s) Lincoln intended. Usually, when English is ambiguous between a $\forall\exists$ and a $\exists\forall$ reading, the $\forall\exists$ reading is the likely one. But in this example, given the context of the rest of Lincoln's remark, the $\exists\forall$ reading has some salience (according to this writer's informal poll).

Two other frequent sources of structural ambiguity in English are the word 'only' and prepositional-phrase modifiers (e.g. 'at noon'). By combining them, we can construct multiply ambiguous sentences, such as:

> (9) John only reads books in libraries.

(9) admits of at least the following eight (!) interpretations, some of which can be distinguished from others by appropriate use of emphasis.

> (9.a) John *only* reads books in libraries.

That is to say, John's sole activity in life is that of reading books in libraries.

> (9.b) John only *reads* books in libraries.

That is to say, he does not burn them, deface them and so on.

> (9.c) John only reads *books* in libraries.

That is to say, he ignores periodicals and newspapers.

> (9.d) *John only* reads books in libraries.

That is to say, John is the only person who reads books in libraries.

There is also an ambiguity involving 'in libraries', which can modify 'books' or 'reads', as in 'John only reads library books' versus 'It's only in libraries that John reads books'. This ambiguity can be read back into each of (9.a)-(9.d). Thus (9.a) can be interpreted as saying either that his sole activity in life is being in libraries and reading books, or that his sole activity in life is reading (anywhere) books which are to be found in libraries. (9.b) can be interpreted as saying either that all he does to books when he is in a library is read them, or that all he does to library books is to read them. (9.c) can be interpreted as saying either that the only things he reads when he is in a library are books, or that his reading matter (at any location) always excludes periodicals and newspapers that are found in libraries. And (9.d) can be interpreted as saying either that John is the only person who reads library books or that he is the only person who, when in libraries, reads books. Symbolizing these different readings is left as an exercise.

❏ Exercises

I   Explain what ambiguities there are in the following sentences, and if the ambiguity is structural, symbolize two possible readings, explaining which, if either, seems the more likely to you.

(1) There's a solution to every problem. ('S_,_', 'P_')
(2) Every man owes something to a certain woman.
(3) A winner will always be envied. ('W_,_', 'E_,_', 'T_': _ is a time)
*(4) John would not sell a picture to anyone.

(5)   People like others if they like themselves.
(6)   There are games which only two people can play.
(7)   Delilah had a ring on every finger and a finger in every pie. ('d'; 'R_'; 'O_,_': _ is on _; 'F_,_': _ is a finger of _; 'P_'; 'I_,_': _ is in _; [this example is from Gustason and Ulrich])
(8)   You can fool all of the people some of the time.

Is there any sentence here where the main ambiguity is lexical?

II   Explain the three different readings of each of the following sentences and symbolize each reading:

(1)   Some students like only absentminded professors.
*(2)   Only private universities are expensive.

III   Symbolize the various possible interpretations of (9) on page 248. [You will want to quantify over John's activities and to describe them by type and location. Thus the library-book version of (9.a) should have the Loglish '(∀x)(if x is an activity of John's then (∃y)(y is a book & (∃z)((z is a library & y is in z) & x is a reading of y by z)))'.]

5 Summary

- The full system of first-order logic is obtained from monadic predicate logic by adding predicate-letters of two or more places and the new logical constant of identity.
- The addition of identity allows us to symbolize sentences about number and sentences containing singular definite descriptions. We use Russell's approach for the latter: '...the ϕ...' becomes 'there is exactly one ϕ and ...it...'.
- The syntax of the new formal language is essentially the same as that of LMPL, with modifications to allow for n-place predicates, $n \geqslant 2$, and for the identity symbol.
- We distinguish two kinds of ambiguity, lexical and structural. Some structural ambiguities are scope ambiguities, and the most common of these is the ∀∃/∃∀ ambiguity.

8 Validity and Provability in First-Order Logic with Identity

1 Interpretations in LFOL

If an argument of English (or some other natural language) has a valid form in LFOL (the language of first-order logic) the English argument is said to be *first-order* valid; a first-order valid English argument is valid absolutely. If the form of the argument[1] is *invalid,* then the English argument is first-order invalid. But it is conceivable that it is still not invalid absolutely, since there may be further, more powerful, systems of logic in which it would be represented as valid. This issue is discussed in §8 of this chapter.

In order to apply these definitions to natural-language arguments, we need to explain how to apply 'valid' and 'invalid' to LFOL forms, which in turn requires that we extend the apparatus of monadic predicate logic so that it can accommodate many-place predicates and identity. The following argument is intuitively invalid:

> A: (1) Someone killed someone.
> (2) ∴ Someone killed himself.

The essential structure of a situation in which the premise of A is true and its conclusion false is one in which there are two people, one of whom kills the other but not himself. The translation of A into LFOL is

> B: $(\exists x)(\exists y)Kxy$
> ∴ $(\exists x)Kxx$

To show that A is first-order invalid we have to show that B is invalid, which requires as usual that we describe an interpretation on which B has a true premise and a false conclusion. We will need two objects, so as before we can write $D = \{\alpha, \beta\}$. But how do we specify the extension of the predicate 'K'? Since 'K' applies to objects two at a time, it will not do to give the extension of 'K' as a set of individual objects. What we should do instead is give it as a set of *pairs*

---

[1] The *caveat* of note 3 on page 164 remains in force.

of objects. Moreover, each pair should be an *ordered* pair, for we want to distinguish the killer from the victim. There are four ordered pairs we can form from the domain $\{\alpha,\beta\}$, namely, $\langle\alpha,\alpha\rangle$, $\langle\alpha,\beta\rangle$, $\langle\beta,\alpha\rangle$ and $\langle\beta,\beta\rangle$ (angled brackets indicate order). To put the first or the fourth of these pairs into the extension of 'K' would make the conclusion true (it would make 'a killed a' or 'b killed b' true), so instead we put the second, or the third (or both). For definiteness, let us say that Ext(K) = $\{\langle\beta,\alpha\rangle\}$. We can then explain why this interpretation works using the same evaluation rules ($\forall\top$), ($\forall\bot$), ($\exists\top$) and ($\exists\bot$) as in monadic predicate logic (page 173). We set all this out as follows.

Example 1: Show $(\exists x)(\exists y)Kxy \not\vDash (\exists x)Kxx$.

Interpretation: D = $\{\alpha,\beta\}$, Ext(K) = $\{\langle\beta,\alpha\rangle\}$. *Explanation:* The premise is true because '$(\exists y)Kby$' is true, in turn because 'Kba' is true, since 'b' denotes β and 'a' denotes α (wherever possible we will omit mentioning this) and $\langle\beta,\alpha\rangle \in$ Ext(K). The conclusion is false because 'Kaa' and 'Kbb' are both false, since $\langle\alpha,\alpha\rangle \notin$ Ext(K) and $\langle\beta,\beta\rangle \notin$ Ext(K).

The main point of interest here is that in explaining the truth of a multiply quantified sentence such as the premise of B, we take *each quantifier in turn* and look at the instances. This method has the potential to become rather cumbersome and later we will employ shortcuts, but to begin with we will adhere to it, as the reader should.

With monadic predicate logic we could also represent interpretations by tables. This procedure is less useful now, since to handle two-place predicates we would either have to write out all pairs of objects in the left column or else make the table three-dimensional, a style of representation that is rather hard to display adequately on a two-dimensional surface. But so long as we are only concerned with two-place predicates, and only one or two of these, there is a useful way of picturing interpretations. The picture below represents the inter-

$$\alpha \longleftarrow \beta$$

pretation just given. In this picture the directed line connecting elements of the domain indicates a pair of objects in the extension of the two-place predicate 'K' and their order. Had the argument contained more than one two-place predicate we would have had to label these lines, but when there is only one such predicate, labeling is not necessary. Obviously, such pictures become hard to read when there are many two-place predicates to be assigned extensions and when the extensions contain many pairs of objects.

In addition to many-place predicates, we also have to accommodate identity in our semantics. Doing this requires that we give up the alphabetic convention which simplified the description of interpretations in monadic predicate logic, the convention that 'a' denotes α, that 'b' denotes β and so on. The complication is that we need to be able to construct interpretations which verify identity sentences, sentences in which '=' is flanked with names, such as the

English 'Marilyn Monroe = Norma Jean Baker' and the LFOL 'a = b'. If we want to make 'a = b' true, 'a' must be assigned the same reference as 'b', violating the alphabetic convention. A further complication is that when different names can stand for the same object, the quantifier evaluation rules ($\forall\top$), ($\forall\bot$), ($\exists\top$) and ($\exists\bot$) (page 173) are no longer quite so natural. Suppose 'a' and 'b' are coreferential. Our rule ($\forall\top$) says that for '($\forall$x)Fx' to be true on an interpretation, all its instances on that interpretation must be true. But this is now an uneconomic way to put it, since 'Fa' and 'Fb' are either both true or both untrue and it suffices to consider just one of them. So we introduce the notion of a *representative* instance. From all the object language names for an object x in a particular interpretation, we choose one as a representative, conventionally the first in alphabetic order, and consider only instances formed with that representative name, as opposed to others denoting the same object, when evaluating quantified sentences. We can regard our previous procedure as a special case of this one, the case when each domain element has just one object language name.

Here is an example of an invalid argument-form whose invalidity turns on the fact that the object-language individual constants 'a' and b' may refer to the same thing. Recall that we use 'Ref(_)' to abbreviate "the reference of '_'", the parentheses embodying quotation.

Example 2: Show Fa & Fb $\not\vDash$ ($\exists$x)($\exists$y)(Fx & Fy & x ≠ y).

Interpretation: D = {α}, Ext(F) = {α}, Ref(a) = α, Ref(b) = α. *Explanation:* 'Fa & Fb' is true since both conjuncts are true; 'Fa' is true since Ref(a) = α and $\alpha \in$ Ext(F), and 'Fb' is true since Ref(b) = α and $\alpha \in$ Ext(F). But '($\exists$x)($\exists$y)(Fx & Fy & x ≠ y)' ('at least two things are F') is false since its representative instance '($\exists$y)(Fa & Fy & a ≠ y)' is false, in turn because *its* representative instance 'Fa & Fa & a ≠ a' is false, since 'a ≠ a' is false. 'a ≠ a' is false because, obviously, Ref(a) = Ref(a).

Here are two points to note about this example.

- There is a semantic rule for identity which we use in evaluating identity sentences, and which we already employ in saying above that 'a ≠ a' is false. The rule is that for any individual constants t_1 and t_2 of LFOL, $\ulcorner t_1 = t_2 \urcorner$ is true if Ref(t_1) = Ref(t_2) and false if Ref(t_1) ≠ Ref(t_2). This is obvious enough; for instance, 'Marilyn Monroe = Norma Jean Baker' is true because the reference of 'Marilyn Monroe' is the same as the reference of 'Norma Jean Baker', while 'Mark Twain = Charles Dickens' is false since the reference of 'Mark Twain' is not identical to the reference of 'Charles Dickens'. In the special case when t_1 and t_2 are the *same* name, the truth-condition Ref(t_1) = Ref(t_2) is automatically satisfied. Thus 'a ≠ a' is false because 'a = a' *has* to be true.
- In evaluating the existential sentences '($\exists$x)($\exists$y)(Fx & Fy & x ≠ y)' and '($\exists$y)(Fa & Fy & a ≠ y)', we do not consider instances of them formed using 'b', since 'a' and 'b' denote the same object, and so the 'a'-instances are the representatives.

We now present a complete account of the semantics of first-order logic with identity.

> An *interpretation* I of an argument-form in LFOL consists in (i) a specification of a nonempty domain of discourse; (ii) an assignment of referents from the domain to the individual constants occurring in the form; and (iii) for each n-place predicate-letter in the form, an assignment of a set of ordered n-tuples of objects drawn from the domain, except that when $n = 1$ we assign subsets of the domain and when $n = 0$ we assign $\top$ or $\bot$. 'λ' is assigned $\bot$ by every interpretation I.

> An argument-form in LFOL is *valid* if and only if no interpretation makes its premises all true and its conclusion false. Similarly, if $p_1,...,p_n$ and q are sentences of LFOL, then $p_1,...,p_n \vDash q$ if and only if no interpretation makes $p_1,...,p_n$ all true and q false.

The rules for evaluating sentences in interpretations are these:

(1) An n-place closed atomic sentence $\lambda t_1...t_n$ is true on an interpretation I if and only if the n-tuple of objects $\langle \text{Ref}(t_1),...,\text{Ref}(t_n) \rangle$ belongs to the extension of λ in I (if $n = 1$, we write '$\text{Ref}(t_1)$' instead of '$\langle \text{Ref}(t_1) \rangle$'). If $n = 0$, λ is a sentence-letter and its truth-value on I is whatever truth-value it is assigned by I.

(2) A closed sentence of the form $\ulcorner t_1 = t_2 \urcorner$ is true on I if and only if $\text{Ref}(t_1) = \text{Ref}(t_2)$ in I.

(3) For all LFOL closed wffs ϕ and ψ, $\ulcorner{\sim}\phi\urcorner$ is true on I if and only if ϕ is false on I; $\ulcorner(\phi \mathbin{\&} \psi)\urcorner$ is true on I if and only if ϕ is true on I and ψ is true on I; $\ulcorner(\phi \vee \psi)\urcorner$ is true on I if and only if ϕ is true on I or ψ is true on I; $\ulcorner(\phi \rightarrow \psi)\urcorner$ is true on I if and only if ψ is true on I if ϕ is true on I; and $\ulcorner(\phi \leftrightarrow \psi)\urcorner$ is true on I if and only if ϕ and ψ are both true on I or both false on I.

(4) A closed existential sentence $\ulcorner(\exists v)\phi v\urcorner$ is true on I if and only if at least one representative instance ϕt is true on I.

(5) A closed universal sentence $\ulcorner(\forall v)\phi v\urcorner$ is true on I if and only if every representative instance ϕt is true on I.

It is no coincidence that the evaluation rules for LFOL closely correspond to the formation rules of LFOL's syntax. This is also the case in sentential logic, but there the correspondence is even closer. Say that a semantic evaluation rule *reflects* a syntactic formation rule if and only if (i) the semantic rule ascribes a semantic property (such as a condition for having each truth-value) to a formula ϕ whose formation is described by the syntactic rule, and (ii) the constituents of ϕ whose semantic properties determine the semantic property of ϕ according to the evaluation rule are the constituents on which the syntactic rule operates to form ϕ. Then a language is said to have a *fully compositional* semantics if and only if (i) for each syntactic formation rule there is a semantic evaluation rule which reflects it, and (ii) there are no other evaluation rules.

Inspection reveals that LSL's semantics are fully compositional, but the semantics of LMPL and LFOL are not. The problem is with the evaluation rules for quantified sentences. According to the syntactic rules, a quantified sentence is formed from an *open* sentence by prefixing a quantifier with the relevant variable. But the truth-value of a quantified sentence is not determined by any semantic property of the *open* sentence obtained from the quantified one by deleting the main quantifier; indeed, we have not ascribed semantic properties to open sentences at all. However, this departure from full compositionality is simply a matter of convenience: there is a way of giving the semantics of quantified sentences on which full compositionality is preserved, but it is more difficult than the method of instances we have been using (see Hodges, §14).

We can increase our familiarity with the evaluation rules by evaluating some sentences in randomly chosen interpretations.

Example 3: Interpretation: $D = \{\alpha, \beta, \gamma\}$, $\mathrm{Ext}(R) = \{\langle\alpha,\beta\rangle, \langle\alpha,\gamma\rangle, \langle\alpha,\alpha\rangle, \langle\beta,\beta\rangle\}$, $\mathrm{Ext}(F) = \{\beta, \gamma\}$, $\mathrm{Ref}(d) = \alpha$.

(i)   '$(\exists x)(Fx\ \&\ Rxx)$' is true because 'Fb & Rbb' is true, since $\beta \in \mathrm{Ext}(F)$ and $\langle\beta,\beta\rangle \in \mathrm{Ext}(R)$.

(ii) '$(\exists x){\sim}Rdx$' is false because $\mathrm{Ref}(d) = \alpha$ and '~Raa', '~Rab' and '~Rac' are all false.

(iii) '$(\exists x)({\sim}Fx\ \&\ {\sim}Rxx)$' is false since '~Fa & ~Raa' is false (because $\langle\alpha,\alpha\rangle \in \mathrm{Ext}(R)$), '~Fb & ~Rbb' is false (because $\langle\beta,\beta\rangle \in \mathrm{Ext}(R)$), and '~Fc & ~Rcc' is false (because $\gamma \in \mathrm{Ext}(F)$).

(iv) '$(\forall x)(Rxb \rightarrow Fx)$' is false since 'Rab → Fa' is false, since $\langle\alpha,\beta\rangle \in \mathrm{Ext}(R)$ and $\alpha \notin \mathrm{Ext}(F)$.

(v)   '$(\forall x)(Fx \rightarrow (\exists y)Rxy)$' is false because 'Fc → (∃y)Rcy' is false; 'Fc' is true, and '(∃y)Rcy' is false since 'Rca' is false, 'Rcb' is false and 'Rcc' is false.

(vi) '$(\forall x)[(\exists y)Rxy \rightarrow (\exists y)(Rxy\ \&\ x \neq y)]$' is false since '(∃y)Rby → (∃y)(Rby & b ≠ y)' is false. For '(∃y)Rby' is true since $\langle\beta,\beta\rangle \in \mathrm{Ext}(R)$, but '(∃y)(Rby & b ≠ y)' is false since 'Rba & b ≠ a' is false ('Rba' is false), 'Rbb & b ≠ b' is false ('b ≠ b' is false), and 'Rbc & b ≠ c' is false ('Rbc' is false).

These examples are a simple extension to LFOL of the kind of example which arose in monadic predicate logic. However, we also want to be able to consider interpretations whose domains are too large for the truth-values of quantified sentences to be explicable by enumerating instances. For example, we want to be able to consider mathematical interpretations whose domains are infinite. In such domains we cannot specify the extensions of predicate symbols by a complete list, so we have to be more general. In Example 4 below, for instance, where the domain is the set $\mathcal{N}$ of natural numbers, we specify the extension of 'B' to be $\{\langle x,y,z\rangle : x \text{ is between } y \text{ and } z\}$. This set-theoretic expression is read 'the set of all ordered triples $\langle x,y,z\rangle$ such that x is between y and z'. We also assume that the object language has a numeral for each number, the standard Hindu-Arabic numeral for that number (we do not give ref-clauses for these numerals). When explanation of truth-value by citing instances is not feasible, we substitute an explanation in words, trying to give a mathematical reason (usually an extremely simple one) for our verdict. Generally, explanation in terms of instances in large domains is not possible when it is the truth of a universal sentence or the falsity of an existential sentence that we are explaining. On the other hand, the method of instances still works if the problem is to explain why a universal sentence is false, since we can cite a falsifying instance. Similarly, if the problem is to explain why an existential sentence is true, we can cite a verifying instance. In the examples immediately following, which illustrate problems of this sort, we use infix notation because of its familiarity (recall the discussion on page 242).

Example 4: Interpretation: $D = \mathcal{N}$, the set of natural numbers $\{0,1,2,3,...\}$. Ext(B) $= \{\langle x,y,z \rangle : x$ is between y and $z\}$, Ext($<$) $= \{\langle x,y \rangle : x$ is less than $y\}$, Ext($\leqslant$) $= \{\langle x,y \rangle : x$ is less than y or $x = y\}$.

(i) '$(\forall x)(\exists y)x < y$' is true, since for any choice of x we can always find a bigger y in the domain, for example $x + 1$.

(ii) '$(\forall x)(\exists y)y < x$' is false, since one of its instances, '$(\exists y)y < 0$', is false.

(iii) '$(\forall x)(\exists y)(\exists z)Bxyz$' is false since it has a false instance, '$(\exists y)(\exists z)B0yz$'.

(iv) '$(\exists x)(\forall y)x < y$' is false since each of its instances $\ulcorner(\forall y)\mathbf{n} < y\urcorner$, for every n, is false.[2] That is, putting any numeral for 'x' in '$(\forall y)x < y$' produces a false sentence, because there is no number less than every number, since no number is less than itself. (Suppose someone suggested that, say, '$(\forall y)6 < y$' is a true instance of (iv); then '6 < 6' would have to be true.)

(v) '$(\exists x)(\forall y)y < x$' is false since there is no number which every number is less than, for any such number would have to be less than itself.

(vi) '$(\exists x)(\forall y)y \leqslant x$' is false since there is no number to which every number is less than or equal, for any such number would have to be the biggest or *last* natural number. But the natural numbers continue on without end. In other words, there is no number n such that $\ulcorner(\forall y)y \leqslant \mathbf{n}\urcorner$ is true, since for any number n, $y = n + 1$ is a counterexample.

It is important to observe that it would be incorrect in (v) to cite the unending nature of the domain $\mathcal{N}$ as the reason why '$(\exists x)(\forall y)y < x$' is false, for this sentence would be false even if the domain were some finite subset of $\mathcal{N}$. Of course, (vi), '$(\exists x)(\forall y)y \leqslant x$', would be true in such a domain; this is one significant difference between '$<$' and '$\leqslant$'.

---

[2] Boldface 'n' abbreviates 'the standard numeral for the number n'. Thus $\ulcorner(\forall y)\mathbf{n} < y\urcorner$ abbreviates "the result of writing '$(\forall y)$' followed by the standard numeral for n followed by '$< y$'".

❏ Exercises

I   Let the interpretation I be as follows:

D = {α,β,γ,δ}; Ext(F) = {α,β,γ}; Ext(G) = {β,γ,δ}; Ref(e) = β
Ext(R) = {⟨α,α⟩,⟨β,β⟩,⟨γ,γ⟩,⟨δ,δ⟩}; Ext(S) = {⟨α,β⟩,⟨β,γ⟩,⟨γ,δ⟩}

What are the truth-values of the following sentences on I? Adequately justify your answer in each case.

  (1) (∃x)Sxe
  (2) (∀x)(Rex ∨ Sxe)
  (3) (∀x)(Fx → Gx)
  (4) (∀x)(Fx ↔ Rxx)
 *(5) (∀x)(Rxx → (∃z)Sxz)
  (6) (∀x)[(∃y)Fy → (∃y)Syx]
  (7) (∀x)[(∃y)Sxy → (∃z)Rxz]
  (8) (∃x)[(Fx ∨ Gx) & (Rxx & ~(∃y)Sxy)]

II Let the interpretation I be as follows:

D = {α,β,γ,δ}; Ext(F) = {α,β,γ}
Ext(R) = {⟨α,β⟩,⟨β,γ⟩,⟨δ,α⟩,⟨α,γ⟩}; Ext(S) = {⟨β,α⟩,⟨γ,β⟩,⟨α,δ⟩}

What are the truth-values of the following sentences on I? Adequately justify your answer in each case.

  (1) (∀x)(Fx → (∃y)Syx)
  (2) (∃x)[(∃y)Syx & (∃y)Sxy]
  (3) (∀x)(∀y)(∀z)((Rxy & Ryz) → Rxz)
 *(4) (∀x)(∀y)(Rxy ↔ ~Syx)
  (5) (∀x)[~(∃y)Ryx → ~Fx].

III Let the interpretation I be as follows:

D = {1,2,3,4,5,6,7,8,9,10}, Ref(a) = 10
Ext(<) = {⟨x,y⟩: x is less than y}, x ∈ D, y ∈ D
Ext(S) = {⟨x,y,z⟩: z = x + y}, x, y, z ∈ D
Ext(B) = {⟨y,x,z⟩: x < y and y < z}, x, y, z ∈ D.
(Read 'S_,_,_' as '_ and _ sum to _' and 'B_,_,_' as '_ is between _ and _'.)

What are the truth-values of the following sentences on I? Adequately justify your answer in each case.

  (1) (∀x)(∀y)(∃z)Sxyz
  (2) (∃x)(∀y)~(∃z)Sxyz

(3) $(\forall x)(\forall y)(\forall z)(Sxyz \rightarrow (\exists w)Bwxy)$
*(4) $(\exists x)(\forall y)(y \neq a \rightarrow (\exists z)Sxyz)$
(5) $(\forall x)(\forall y)(\forall z)(Sxyz \leftrightarrow Syxz)$

IV Let D be the segment of the rational line from 0 to 1 inclusive (i.e., the closed interval [0,1]). Where '$B_{\_,\_,\_}$' means '_ is between _ and _' and Ref(a) = 1, evaluate the following with enough explanation to show your answer is not a guess:

(1) $(\exists x)(\forall y)x < y$
(2) $(\forall x)(\exists y)y < x$
*(3) $(\forall x)(x \neq a \rightarrow (\exists y)y > x)$
(4) $(\forall x)(\forall y)(x \neq y \rightarrow (\exists z)Bzxy)$
(5) $(\forall x)(\forall y)(\forall z)(Bzxy \rightarrow (\exists w)Bwxz)$

2 Demonstrating invalidity

The procedure for constructing counterexamples in first-order logic is much the same as the procedure we used for monadic predicate logic. We begin with a nonempty domain. If the conclusion of the argument-form is universal, we set up a falsifying instance, and if it is existential, we note what is required to avoid making it true. We then make all the premises true, either by direct entry in the extensions of predicates for existential premises or by noting, for universal premises, what is required to avoid making them false. Here are some examples.

Example 1: Show $(\forall x)(\exists y)Kxy \nvdash (\exists x)(\forall y)Kxy$.

This is the formal counterpart of our discussion in §1 of Chapter 7 of the difference between 'everyone killed someone' and 'someone killed everyone'. To make the conclusion false, we must avoid creating an object x such that for every object y in the domain (including x), $\langle x,y \rangle \in$ Ext(K). On the other hand, to make the premise true, we must ensure that for each x in D there is a y in D such that $\langle x,y \rangle \in$ Ext(K). The simplest solution is an interpretation with D = $\{\alpha,\beta\}$ and Ext(K) = $\{\langle \alpha,\beta \rangle,\langle \beta,\alpha \rangle\}$. We picture this below.

$$\alpha \; \rightleftarrows \; \beta$$

Explanation: '$(\forall x)(\exists y)Kxy$' is true since its two instances '$(\exists y)Kay$' and '$(\exists y)Kby$' are true. '$(\exists y)Kay$' is true because 'Kab' is true, since $\langle \alpha,\beta \rangle \in$ Ext(K); and '$(\exists y)Kby$' is true because 'Kba' is true, since $\langle \beta,\alpha \rangle \in$ Ext(K). But '$(\exists x)(\forall y)Kxy$' is false since (i) '$(\forall y)Kay$' is false (because 'Kaa' is false) and (ii) '$(\forall y)Kby$' is false (because 'Kbb' is false). This is a special case of Situation A in §1 of Chapter Seven (page 223), the case where there are only two people in the circle.

Example 2: Show (∃x)(Fx & (∀y)Gxy) ⊭ (∀x)(Fx → (∃y)Gxy).

To make the conclusion false, we require some object x in Ext(F) such that for no y in D is ⟨x,y⟩ in Ext(G). To make the premise true we require some object x in Ext(F) such that for every y in D, ⟨x,y⟩ is in Ext(G). Consequently, we need at least two objects in the domain, one for the false instance of the conclusion and the other for the true instance of the premise. An interpretation which works is D = {$α,β$}, Ext(F) = {$α,β$}, Ext(G) = {⟨$α,α$⟩, ⟨$α,β$⟩}. We can represent the extension of a one-place predicate in a picture simply by writing the predicate above the objects which belong to its extension. This gives the picture below, the

F   F

$α$ ⟶ $β$

arrows indicating the extension of 'G'. *Explanation:* '(∀x)(Fx → (∃y)Gxy)' is false since 'Fb → (∃y)Gby' is false, since 'Fb' is true and '(∃y)Gby' is false ('Gba' and 'Gbb' are both false). '(∃x)(Fx & (∀y)Gxy)' is true because 'Fa & (∀y)Gay' is true ('Gaa' is true and 'Gab' is true).

Example 3: Show (∀x)(∀y)(∀z)(Rxyz → (∃w)(Fxw & Fyw)), (∃x)(∃y)(∃z)Rxyz ⊭ (∃x)(∀y)Fxy.

Example 3 contains a three-place predicate 'R', whose extension will be a collection of *triples* of objects from the domain of discourse. To make premise 2 of (3) true, we must assign at least one triple of objects to the extension of 'R', and to make the first premise true we will then have to make appropriate entries in Ext(F). To make the conclusion false, however, we shall have to ensure that for each object x, there is some y such that ⟨x,y⟩ is *not* in Ext(F). We could try the simplest possible approach first, that is, set D = {$α$} and assign ⟨$α,α,α$⟩ to Ext(R); this means ⟨$α,α$⟩ would have to go into Ext(F), but if we leave things there, the conclusion is true, since '(∀y)Fay' is true. However, this can be circumvented by adding a new object $β$ to the domain and leaving Ext(R) and Ext(F) as they are. So we get the interpretation D = {$α,β$}, Ext(R) = {⟨$α,α,α$⟩}, Ext(F) = {⟨$α,α$⟩}. *Explanation:* '(∃x)(∃y)(∃z)Rxyz' is true because '(∃y)(∃z)Rayz' is true, in turn because '(∃z)Raaz' is true, since 'Raaa' is true, since (at last!) ⟨$α,α,α$⟩ ∈ Ext(R). '(∀x)(∀y)(∀z)(Rxyz → (∃w)(Fxw & Fyw))' is true because once the three leading quantifiers are evaluated we arrive at a conditional which either has a false antecedent (so the conditional is true) or else has 'Raaa' as antecedent. And 'Raaa → (∃w)(Faw & Faw)' is true because its consequent '(∃w)(Faw & Faw)' is true, since 'Faa & Faa' is true, because ⟨$α,α$⟩ ∈ Ext(F). Finally, the conclusion '(∃x)(∀y)Fxy' is false because '(∀y)Fay' is false and '(∀y)Fby' is false (recall that the falsity of an existential sentence requires the falsity of every instance). '(∀y)Fay' is false because 'Fab' is false, since ⟨$α,β$⟩ ∉ Ext(F), and '(∀y)Fby' is false since, for example, 'Fbb' is false, because ⟨$β,β$⟩ ∉ Ext(F).

Notice two features of this explanation. First, it would be unreasonable to consider every instance of premise 1 separately, taking each '∀' in turn, since with two objects and three places in 'R_,_,_' there would be eight instances; besides, it is only the instances where the antecedent is true which really interest us. In general, when quantifiers of the same sort are stacked, we take them all at once and consider only the relevant instances (this policy allows us to say that premise 2 is true because 'Raaa' is true). But secondly, when quantifiers of *different* sorts are stacked, it will still be necessary to take them one at a time, as in our explanation of the falsity of the conclusion of (3); the combinations '∃∀' and '∀∃' tend to be the ones which give most trouble in demonstrating invalidity. Note then that the instances of '(∃x)(∀y)Fxy' on *1* are '(∀y)Fay' and '(∀y)Fby'—we do not remove two quantifiers of different types in one step.

Example 4: Show (∀x)[(∀y)Rxy ↔ (∃z)(∀w)Rwz] ⊭ (∀x)(∃y)(∀z)(Rxy ↔ Rzy).

To make the conclusion false, we need a false instance, say '(∃y)(∀z)(Ray ↔ Rzy)'. If D = {α}, then this means '(∀z)(Raa ↔ Rza)' has to be false, so we need a false instance of it, and consequently we need another object β in the domain. The false instance can then be 'Raa ↔ Rba'. But adding β to the domain means that we require '(∀z)(Rab ↔ Rzb)' to be false as well (otherwise '(∃y)(∀z)(Ray ↔ Rzy)' would be true). To make '(∀z)(Rab ↔ Rzb)' false, our only option is to have 'Rab ↔ Rbb' false. So 'Raa ↔ Rba' and 'Rab ↔ Rbb' are both to be false. There are various ways of arranging this, and to help choose among them we consider what will be required to verify the premise: both '(∀y)Ray ↔ (∃z)(∀w)Rwz' and '(∀y)Rby ↔ (∃z)(∀w)Rwz' must be true. Since both biconditionals have the same right-hand side, '(∀y)Ray' and '(∀y)Rby' will have to have the same truth-value. If both were true then '(∃z)(∀w)Rwz' would have to be true, which means that for some z in D, we require ⟨w,z⟩ in Ext(R) for every w in D; but this is inconsistent with our condition for the falsity of the conclusion, that 'Raa ↔ Rba' and 'Rab ↔ Rbb' are both to be false. Thus we must make the premise true by making '(∀y)Ray' and '(∀y)Rby' both false. In sum: we want an interpretation where 'Raa ↔ Rba' and 'Rab ↔ Rbb' are both false and '(∀y)Ray' and '(∀y)Rby' are both false. When D = {α,β} there is more than one such interpretation, but one that works is D = {α,β}, Ext(R) = {⟨α,α⟩,⟨β,β⟩}, displayed below (notice that

α β

if we had entered ⟨α,β⟩ in place of ⟨β,β⟩, '(∀y)Ray' would be true). *Explanation:* '(∀x)(∃y)(∀z)(Rxy ↔ Rzy)' is false because '(∃y)(∀z)(Ray ↔ Rzy)' is false, in turn because '(∀z)(Raa ↔ Rza)' is false (since 'Raa ↔ Rba' is false) and '(∀z)(Rab ↔ Rzb)' is false ('Rab ↔ Rbb' is false). However, '(∀x)[(∀y)Rxy ↔ (∃z)(∀w)Rwz]' is true since '(∀y)Ray ↔ (∃z)(∀w)Rwz' and '(∀y)Rby ↔ (∃z)(∀w)Rwz' are both true. The right-hand side of both is false, since '(∀w)Rwa' is false ('Rba' is false) *and* '(∀w)Rwb' is false ('Rab' is false); and '(∀y)Ray' and '(∀y)Rby' are both false (consider 'Rab' and 'Rba' respectively).

Our next example, like Example 2 of the previous section, involves identity.

Example 5: Show (∀x)(Fx → (Gx & Hx)), Ga, ~Hb ⊭ a ≠ b.

By premise 1, anything in the extension of 'F' has to be in both Ext(G) and Ext(H). Consequently, if Ref(a) were in the extension of 'F', premises 2 and 3 would guarantee the truth of the conclusion. So the key to constructing a counterexample is to ensure Ref(a) ∉ Ext(F). An interpretation which works is D = {α}, Ref(a) = Ref(b) = α, Ext(F) = ∅, Ext(G) = {α}, Ext(H) = ∅. *Explanation:* Premise 1 is true since 'Fa → (Ga & Ha)' is true, because α ∉ Ext(F). Premises 2 and 3 are true since Ref(a) = Ref(b) = α, α ∈ Ext(G) and α ∉ Ext(H). The conclusion 'a ≠ b' is false since Ref(a) = Ref(b).

Our final example is of great theoretical significance, an instance of a phenomenon unknown in monadic predicate logic: an argument-form with finitely many premises, any counterexample to which has an infinite domain.

Example 6: Show (∀x)(∃y)Rxy, (∀x)(∀y)(∀z)((Rxy & Ryz) → Rxz) ⊭ (∃x)Rxx.

Let us first see why any finite domain which verifies the premises also verifies the conclusion. If we begin with D = {α}, then to make '(∀x)(∃y)Rxy' true we have to have ⟨α,α⟩ ∈ Ext(R). But of course this makes the conclusion true. So we consider D = {α,β}. Every instance of '(∀x)(∃y)Rxy' has to be true, so '(∃y)Ray' and '(∃y)Rby' have to be true. To keep the conclusion false, we have to let Ext(R) be {⟨α,β⟩, ⟨β,α⟩}. But now the second premise is false, since putting 'a' for 'x' and 'z' and 'b' for 'y' gives us the instance '(Rab & Rba) → Raa', which has a true antecedent and false consequent as Ext(R) stands. In other words, premise 2 compels the addition of ⟨α,α⟩ *to* Ext(R), making the conclusion true.

One way out of this difficulty is to drop ⟨β,α⟩ from Ext(R), making '(Rab & Rba) → Raa' true without having to put ⟨α,α⟩ into Ext(R), and to add a new object γ to D to provide the true instance of '(∃y)Rby'; that is, we put ⟨β,γ⟩ in Ext(R) in place of ⟨β,α⟩. But since the first premise makes a claim about every object, the same difficulty arises over γ as arose over β: we need to make '(∃y)Rcy' true, and any way of doing this when D = {α,β,γ} will make the conclusion true. For we have ⟨α,β⟩ and ⟨β,γ⟩ in Ext(R). If we add ⟨γ,γ⟩ to make '(∃y)Rcy' true, the conclusion is true immediately. If we add ⟨γ,β⟩, then since '(Rbc & Rcb) → Rbb' is an instance of the second premise with true antecedent, the second premise compels us to add ⟨β,β⟩ *to* Ext(R), making the conclusion true again. And if instead we add ⟨γ,α⟩ to Ext(R) to make '(∃y)Rcy' true, we will have to add ⟨α,α⟩ to Ext(R). For

       (i)   (Rab & Rbc) → Rac

and

       (ii) (Rac & Rca) → Raa

are both instances of the second premise. With $\langle\alpha,\beta\rangle$ and $\langle\beta,\gamma\rangle$ in Ext(R), we must add $\langle\alpha,\gamma\rangle$ to keep the second premise true, in view of (i), but then the antecedent of (ii) is true, so we must add $\langle\alpha,\alpha\rangle$ to keep the second premise true, and this makes the conclusion true. Therefore, in order to make '(∃y)Rcy' true, keeping the other premises true and the conclusion false, we have to add a fourth object δ to D and put $\langle\gamma,\delta\rangle$ in Ext(R). But with δ in D, the first premise requires that '(∃y)Rdy' be true, and a similar argument shows that the only way to effect this without making the conclusion true is to add a fifth object ϵ and put $\langle\delta,\epsilon\rangle$ in Ext(R). But then '(∃y)Rey' has to be true, and so on. The moral is that to make the conclusion false while keeping all the premises true, we need an infinite domain D = $\{\alpha_1, \alpha_2, \alpha_3,...\}$ and Ext(R) should be the set of all pairs $\langle\alpha_i,\alpha_j\rangle$ such that $i < j$. Note that it is not sufficient for the truth of the second premise that Ext(R) be just the pairs $\langle\alpha_i,\alpha_j\rangle$ such that $j = i + 1$. For then '(Ra$_1$a$_2$ & Ra$_2$a$_3$) → Ra$_1$a$_3$' would be a false instance of the second premise.

Obviously, we cannot explain the truth of the premises and the falsity of the conclusion on this interpretation by citing instances. The simplest procedure is to specify the interpretation as an *arithmetical* interpretation, that is, one where D = N (the natural numbers $\{0,1,2,3,...\}$) and Ext(R) is defined in terms of arithmetical properties. We can then explain the truth-values of the premises and conclusion in the same way as we did for the arithmetical interpretation in the previous section. This leads to the following counterexample, with D = $\{0,1,2,3,...\}$, Ext(R) = $\{\langle x,y\rangle: x < y\}$. Explanation: '(∀x)(∃y)Rxy' is true since for every number x there is a number y greater than x, for example $x + 1$. '(∀x)(∀y)(∀z)((Rxy & Ryz) → Rxz)' is true since if x is less than y and y less than z, then x is less than z. But '(∃x)Rxx' is false since no number is less than itself. In §7 we will consider the significance of this and similar examples further, in connection with the issue of decidability.

❏ Exercises

I   Show the following:

   (1) (∃x)Rax, (∃x)Rbx ⊭ (∃x)(Rax & Rbx)
   (2) (∀x)(∀y)(Rxy → Ryx) ⊭ (∀x)Rxx
   (3) (∃y)(∀x)Rxy ⊭ (∀y)Ryy
  *(4) (∀x)(Rxa → ~Rxb), (∀x)(∃y)Rxy ⊭ (∃x)~Rxb
   (5) (∀x)(∀y)(∀z)((Rxx & Rxz) → (Rxy ∨ Ryz)) ⊭ (∃y)(∀z)Ryz
   (6) (∃x)(∀y)(∀z)(Fx → ~Ryz) ⊭ (∀x)Fx ↔ ~(∃x)(∃y)Rxy
   (7) (∃x)(Hx & (∀y)(Fy → Gxy)), (∀x)(∀y)(Hx → (Gxy ↔ Ky)) ⊭ (∀x)(Kx ↔ Fx)
   (8) (∀x)(∀y)((∀z)Hyz → Hxy), (∀x)(∃y)Hxy ⊭ (∀x)(∀y)Hxy
   (9) (∀x)(∀y)((∀z)Hyz → Hxy) ⊭ (∃x)(∀y)Hxy → (∀x)(∀y)(Hyx → Hxy)
  (10) (∀x)((Fx → Gx) → (∃y)Hyy) ⊭ (∀x)(Fx & Gx) → (∀x)(∃y)Hxy
  (11) (∀x)(∃y)Rxy, (∀x)(∀y)(Rxy → Ryx) ⊭ (∃x)(∃y)(∃z)((Rxy & Ryx) & ~Rxz)
 *(12) (∀x)(Fx → (∀y)(Gy → Rxy)), (∀x)(Hx → Gx) ⊭ (∀x)(∀y)((Fx & Hx) → Rxy)

(13) $(\forall x)(\forall y)(\forall z)(Rxy \vee Rzz) \nvDash (\exists x)(\forall y)Rxy$
(14) $\nvDash (\exists x)(\forall y)((Rxy \,\&\, {\sim}Ryx) \rightarrow (Rxx \leftrightarrow Ryy))$
(15) $(\forall x)[(\exists y)Rxy \,\&\, (\exists y)Ryx], (\forall x)(\forall y)(\forall z)((Rxy \,\&\, Ryz) \rightarrow Rxz) \nvDash$
       $(\exists x)(\exists y)(Rxy \,\&\, Ryx)$

((14) is from Mendelson and can be solved with a domain of four objects. To
show that the formula is not valid, find an interpretation in which its negation
is true. The negation of the formula is equivalent to '$(\forall x)(\exists y)((Rxy \,\&\, {\sim}Ryx) \,\&\,$
${\sim}(Rxx \leftrightarrow Ryy))$' (why?). For (15), make sure you understand the discussion of
Example 6, though the solution is not quite the same.)

II   Show the following:

(1) $a \neq b, Fa \nvDash {\sim}Fb$
(2) $(\exists x)(\exists y)(x \neq y) \nvDash a \neq b$
*(3) $(\exists x)(Fx \,\&\, (\forall y)(Fy \rightarrow y = x) \,\&\, x = a), (\forall x)(Gx \rightarrow Fx) \nvDash Ga$
(4) $(\exists y)(Ray \,\&\, Gy) \rightarrow Gb \nvDash (\forall y)(Ray \rightarrow y = b)$
(5) $(\exists x)(Fx \,\&\, (\forall y)(Fy \rightarrow y = x) \,\&\, Gx) \nvDash (\exists x)(Gx \,\&\, (\forall y)(Gy \rightarrow y = x))$

III   In the branch of mathematics known as set theory, the Axiom of Extensionality says that it is necessary and sufficient for x and y to be the same set that
they have exactly the same members. In Loglish, relativizing quantifiers to the
domain of things, this becomes

(E.1) For any x, for any y, if x and y are sets, then x = y if and only if x
       and y have the same members.

Using '$\in$' for 'is a member of', 'x and y have the same members' is expressed by
the formula '$(\forall z)(z \in x \leftrightarrow z \in y)$', which says that a thing belongs to x if and
only if it belongs to y. So the entire axiom can be written as:

(E) $(\forall x)(\forall y)((Sx \,\&\, Sy) \rightarrow (x = y \leftrightarrow (\forall z)(z \in x \leftrightarrow z \in y)))$

Show the following:

(1) $E \nvDash (\forall x)(\forall y)((Sx \,\&\, Sy) \rightarrow x = y)$
*(2) $E \nvDash (\exists x)(Sx \,\&\, (\forall y)y \notin x)$
(3) $E \nvDash (\exists x)(\exists y)((Sx \,\&\, Sy) \,\&\, x \neq y)$

((1) says that the axiom does not entail that there is at most one set; (2) says
that it does not entail that there is a set with no members, that is, does not
entail that $\varnothing$ exists; (3) says that it does not entail that there are at least two
sets.)

3 Proofs in NK

When we add a new logical constant to the language, we have to extend not only our semantics but also our deductive system; so we now need an introduction rule and an elimination rule for '='. However, the reader will be pleased to hear that this is the only addition to NK which we require: the rules already in place for the quantifiers suffice to handle arguments with many-place predicates, provided they contain no occurrence of '='. So we begin with some examples of this latter sort, and we will deal with identity in the next section. First, we recall the four quantifier rules schematically:

∀-Elimination:

$$a_1,...,a_n \quad (j) \quad (\forall v)\phi v$$
$$\vdots$$
$$a_1,...,a_n \quad (k) \quad \phi t \qquad j \; \forall E$$

ϕt is obtained syntactically from $\ulcorner(\forall v)\phi v\urcorner$ by deleting the quantifier prefix $\ulcorner(\forall v)\urcorner$ and then replacing *every* occurrence of v in the open sentence ϕv with *one and the same* individual constant t.

∀-Introduction:

$$a_1,...,a_n \quad (j) \quad \phi t$$
$$\vdots$$
$$a_1,...,a_n \quad (k) \quad (\forall v)\phi v \qquad j \; \forall I$$

where t is not in any of the formulae on lines $a_1,...,a_n$. $\ulcorner(\forall v)\phi v\urcorner$ is obtained by replacing *every* occurrence of t in ϕt with a variable v not in ϕt and then prefixing $\ulcorner(\forall v)\urcorner$.

∃-Introduction:

$$a_1,...,a_n \quad (j) \quad \phi t$$
$$\vdots$$
$$a_1,...,a_n \quad (k) \quad (\exists v)\phi v \qquad j \; \exists I$$

$\ulcorner(\exists v)\phi v\urcorner$ is obtained syntactically from ϕt by replacing one or more occurrences of t in ϕt with an individual variable v not in ϕt, and then by prefixing $\ulcorner(\exists v)\urcorner$.

∃-Elimination:

$$a_1,...,a_n \quad (i) \quad (\exists v)\phi v$$
$$\vdots$$
$$j \quad (j) \quad \phi t \qquad \text{Assumption}$$
$$\vdots$$
$$b_1,...,b_u \quad (k) \quad \psi$$
$$\vdots$$
$$X \quad (m) \quad \psi \qquad i,j,k \; \exists E$$

where t is not in (i) $\ulcorner(\exists v)\phi v\urcorner$, (ii) ψ, or (iii) any of the formulae at lines $b_1,...,b_u$ other than j. $X = \{a_1,...,a_n\} \cup \{b_1,...,b_u\}/j$.

The reason we do not need to change these rules in any way is that there is nothing in their formulation which requires that the formula ϕv contain only monadic predicates. However, in our discussion of $\forall$I and $\exists$I, we remarked that one of the contrasts between them, that with $\exists$I one need not replace every occurrence of t in ϕt while with $\forall$I one must replace every occurrence, is best motivated with many-place predicates. The motivation is that if we have derived, say, 'Kaa' from assumptions none of which contain 'a', then the same steps would lead to proofs of 'Kbb', 'Kcc' and so on. So what we have shown is that an arbitrary person killed himself, that is, everyone committed suicide ('$(\forall x)Kxx$'), *not* that an arbitrary person killed everyone ('$(\forall x)Kax$') or that everyone killed an arbitrary person ('$(\forall x)Kxa$'). On the other hand, 'Kaa' *does* entail all of '$(\exists x)Kxa$', '$(\exists x)Kax$' and '$(\exists x)Kxx$'.

Our first example of a proof in LFOL is of the valid direction in which one can interchange '$\forall$' and '$\exists$':

Example 1: Show $(\exists x)(\forall y)Lxy \vdash_{NK} (\forall y)(\exists x)Lxy$.

We approach this problem in the same way as we approached problems in monadic predicate logic. The conclusion is universal and so will be obtained from an instance of (1) by $\forall$I applied to a formula of the form '$(\exists x)Lx\_$', where the blank is filled by some individual constant. So our subgoal is to derive such a formula, presumably using $\exists$I. This leads to the following proof:

1	(1)	$(\exists x)(\forall y)Lxy$	Premise
2	(2)	$(\forall y)Lay$	Assumption
2	(3)	Lab	2 $\forall$E
2	(4)	$(\exists x)Lxb$	3 $\exists$I
2	(5)	$(\forall y)(\exists x)Lxy$	4 $\forall$I
1	(6)	$(\forall y)(\exists x)Lxy$	1,2,5 $\exists$E $\quad\blacklozenge$

Note that the order of the steps $\forall$I and $\exists$E could be reversed without violating any restrictions on either rule.

A slightly more complex example is:

Example 2: Show $(\forall x)(\exists y)Rxy, (\forall x)(\forall y)(Rxy \rightarrow Ryx) \vdash_{NK} (\forall x)(\exists y)(Rxy \mathbin{\&} Ryx)$

whose proof illustrates a situation where the order in which we apply $\forall$I and $\exists$E *does* matter. To obtain the conclusion of Example 2 from its two premises, we have to apply $\forall$E to each premise. In the case of the first premise we obtain an existential sentence, say '$(\exists y)Ray$', so we try to derive the conclusion using an instance of this, say 'Rab', and then apply $\exists$E. To reach the conclusion we must presumably obtain a formula of the form '$(\exists y)(R\_y \mathbin{\&} Ry\_)$' and then use $\forall$I on it. But since our formula '$(\exists y)(R\_y \mathbin{\&} Ry\_)$' will likely have either 'a' or 'b' in the blanks, we cannot apply $\forall$I while it depends on 'Rab'. Consequently, we have to use ∃E first, to remove dependence on 'Rab'. This requires that the blank in '(∃y)(R_y & Ry_)' not be filled with 'b', since 'b' is the name we will use to form the instance 'Rab' of '(∃y)Ray'. So we should aim to derive '(∃y)(Ray & Rya)', which we can presumably obtain by &I and ∃I, then use ∃E and lastly ∀I. This leads to the following proof:

1	(1)	(∀x)(∃y)Rxy	Premise
2	(2)	(∀x)(∀y)(Rxy → Ryx)	Premise
1	(3)	(∃y)Ray	1 ∀E
4	(4)	Rab	Assumption
2	(5)	(∀y)(Ray → Rya)	2 ∀E
2	(6)	Rab → Rba	5 ∀E
2,4	(7)	Rba	6,4 →E
2,4	(8)	Rab & Rba	4,7 &I
2,4	(9)	(∃y)(Ray & Rya)	8 ∃I
1,2	(10)	(∃y)(Ray & Rya)	3,4,9 ∃E
1,2	(11)	(∀x)(∃y)(Rxy & Ryx)	10 ∀I ◆

Another fairly straightforward example:

Example 3: Show (∀x)(Fx → Gx), (∀x)(∀y)(Rxy → Syx), (∀x)(∀y)(Sxy → Syx) ⊢_NK (∀x)[(∃y)(Fx & Rxy) → (∃y)(Gx & Sxy)].

The conclusion is a universal sentence, so we will obtain it by ∀I. Inspecting the conclusion, we see that to apply ∀I we need a conditional of the form '(∃y)(F_ & R_y) → (∃y)(G_ & S_y)' in which all four blanks are filled by the same individual constant. So our subgoal is to derive such a conditional, which as usual requires that we assume the antecedent, derive the consequent and use →I. Since the main connective of the antecedent is '∃', we will immediately make another assumption, an instance of the antecedent, with a view to using ∃E. Given the shape of the instance, we then make appropriate applications of ∀E to the three premises to ensure that our reasoning is generalizable with respect to the assumption. The proof is:

1	(1)	(∀x)(Fx → Gx)	Premise
2	(2)	(∀x)(∀y)(Rxy → Syx)	Premise
3	(3)	(∀x)(∀y)(Sxy → Syx)	Premise
4	(4)	(∃y)(Fa & Ray)	Assumption
5	(5)	Fa & Rab	Assumption
5	(6)	Fa	5 &E
1	(7)	Fa → Ga	1 ∀E
1,5	(8)	Ga	6,7 →E
5	(9)	Rab	5 &E
2	(10)	(∀y)(Ray → Sya)	2 ∀E
2	(11)	Rab → Sba	10 ∀E
2,5	(12)	Sba	9,11 →E
3	(13)	(∀y)(Sby → Syb)	3 ∀E

3	(14)	Sba → Sab	13 ∀E
2,3,5	(15)	Sab	12, 14 →E
1,2,3,5	(16)	Ga & Sab	8,15 &I
1,2,3,5	(17)	(∃y)(Ga & Say)	16 ∃I
1,2,3,4	(18)	(∃y)(Ga & Say)	4,5,17 ∃E
1,2,3	(19)	(∃y)(Fa & Ray) → (∃y)(Ga & Say)	4,18 →I
1,2,3	(20)	(∀x)[(∃y)(Fx & Rxy) → (∃y)(Gx & Sxy)]	19 ∀I ◆

At line 12 we see that the individual constants in 'Sba' are in the wrong order: conjoining with 'Ga' will not yield what we want. But (3) allows us to change the order around, provided we make appropriate applications of ∀E.

One last example, slightly more complicated than the previous two, is the symbolization of the following English argument:

> A: No one who is taught is a teacher. All who are taught teach themselves. Therefore no one teaches anyone.

To symbolize A relative to the domain of people, we use just 'T_,_' for '_ teaches _'. The first premise says: 'for any x, if someone teaches x then x is not a teacher', where 'x is not a teacher' means 'there is no one whom x teaches'. The second premise says 'for any x, if someone teaches x, then x teaches x', for which a straightforward symbolization is '(∀x)[(∃y)Tyx → Txx]'. But this is equivalent to 'for any x and y, if y teaches x then x teaches x', which yields '(∀x)(∀y)(Tyx → Txx)', a formulation that is easier to work with in a proof (the other version requires deriving an existential antecedent for use in →E). So we obtain the following formal problem from A:

Example 4: Show (∀x)[(∃y)Tyx → (∀z)~Txz], (∀x)(∀y)(Tyx → Txx) ⊢_NK (∀x)(∀y)~Txy.

The conclusion appears to be derivable by two applications of ∀I from a formula of the form '~T_,_' in which the blanks are filled by different individual constants (why could we not have the same individual constant in both positions?). Therefore, the most obvious procedure would be to obtain '~T_,_' using ~I, deriving a contradiction from the assumption 'T_,_' and the two conditionals (lines 3 and 9 below) which can be inferred by ∀E from the premises. This proof can be carried through in fourteen lines, without SI. However, for the sake of variety, we give a proof which uses LEM and ∨E instead. The particular instance of LEM which we choose is motivated by the antecedent of the conditional we get from (1) by ∀E.

1	(1)	(∀x)[(∃y)Tyx → (∀z)~Txz]	Premise
2	(2)	(∀x)(∀y)(Tyx → Txx)	Premise
1	(3)	(∃y)Tya → (∀z)~Taz	1 ∀E
	(4)	(∃y)Tya ∨ ~(∃y)Tya	TI (LEM)
5	(5)	(∃y)Tya	Assumption
1,5	(6)	(∀z)~Taz	3,5 →E

1,5	(7)	~Taa	6 ∀E
2	(8)	(∀y)(Tya → Taa)	2 ∀E
2	(9)	Tba → Taa	8 ∀E
1,2,5	(10)	~Tba	7,9 SI (MT)
11	(11)	~(∃y)Tya	Assumption
11	(12)	(∀y)~Tya	11 SI (QS)
11	(13)	~Tba	12 ∀E
1,2	(14)	~Tba	4,5,10,11,13 ∨E
1,2	(15)	(∀y)~Tby	14 ∀I
1,2	(16)	(∀x)(∀y)~Txy	15 ∀I ◆

Two points about this proof bear emphasis. First, observe that line 3 is a conditional: *we cannot apply quantifier elimination rules to (3)*. Rather, since (3) is a conditional, we must use →E or some appropriate application of SI on it. Second, observe that we cannot obtain the conclusion '(∀x)(∀y)~Txy' directly by ∨E, since this would require applications of ∀I replacing the individual constant 'a' in formulae which depend on one or the other disjunct of our instance of ⌜p ∨ ~p⌝; but both disjuncts contain 'a', so ∀I would be illegal.

As these examples illustrate, we can derive a strategy for a proof by considering how the last line will be inferred from the second last, how it will be inferred in turn, and so on, just as in monadic predicate logic and sentential logic. Sometimes, however, the quantificational structure of the premises is so complex that it can be difficult to see precisely how to proceed toward a formula or formulae from which the conclusion can be obtained in a few steps. Consider the following intuitively valid English argument:

> B: There is someone who teaches everyone who is taught by anyone. There is someone who teaches everyone who teaches anyone. Everyone who does not teach anyone is taught by someone. Therefore there is someone who teaches everyone.

The symbolization and proof of this argument is an exercise. But to gain an idea of how the formal proof ought to go, it is useful to consider how we could establish the conclusion by informal reasoning, as we did in §1 of Chapter 4 with the argument about who helped open the safe. The premises of B are

> (i) Someone teaches all who are taught.
> (ii) Someone teaches all teachers.
> (iii) All nonteachers are taught.

The natural way to reason informally to the conclusion that someone teaches everyone is this. (ii) means (iv), all teachers are taught, and (iii) says that all nonteachers are taught. By the Law of Excluded Middle, a given person α is either a teacher or a nonteacher, so by (iii) and (iv), α is taught; the reasoning generalizes to any other person; hence *everyone* is taught. Thus by (i), someone teaches everyone. So the most direct formal proof of B will use an instance of LEM with '(∃z)Tcz', say, for p. The details are left to the reader.

❑ Exercises

I   Show the following:

(1) $(\forall x)(Fx \rightarrow (\forall y)Rxy)$, ~Rab $\vdash_{NK}$ ~Fa

(2) $(\forall x)(Fx \rightarrow (\forall y)(Gy \rightarrow Rxy))$, $(\forall x)(Hx \rightarrow Gx)$ $\vdash_{NK}$
     $(\forall x)(\forall y)((Fx \& Hy) \rightarrow Rxy)$

*(3) $(\exists x)Fx \rightarrow (\forall x)(Fx \rightarrow Gxa)$, $(\exists x)Hx \rightarrow (\forall x)(Hx \rightarrow Jxa)$ $\vdash_{NK}$
     $(\exists x)(Fx \& Hx) \rightarrow (\exists x)(\exists y)(Gxy \& Jxy)$

(4) $(\forall y)(\exists x)((Fyy \& Fyx) \vee Fxy)$ $\vdash_{NK}$ $(\forall y)(\exists x)Fxy$

(5) $(\forall x)(\forall y)(Rxy \leftrightarrow (Fy \rightarrow Gx))$, $(\forall z)Raz$ $\vdash_{NK}$ $(\exists x)Fx \rightarrow (\exists x)Gx$

(6) $(\forall x)[(\exists y)Rxy \rightarrow (\forall z)Rzx]$ $\vdash_{NK}$ $(\forall x)(\forall y)(Rxy \rightarrow Ryx)$

*(7) $(\forall x)(\forall y)(Exy \rightarrow Eyx)$, $(\forall x)(\forall y)(Exy \rightarrow Exx)$ $\vdash_{NK}$ $(\forall x)[(\exists y)Eyx \rightarrow Exx]$

(8) $(\forall x)(Fx \rightarrow Gx)$ $\vdash_{NK}$ $(\forall x)[(\forall y)(Gy \rightarrow Rxy) \rightarrow (\forall z)(Fz \rightarrow Rxz)]$

(9) $(\exists y)(\forall x)Rxy$, $(\forall x)(Fx \rightarrow (\exists y)Syx)$, $(\forall x)(\forall y)(Rxy \rightarrow \sim Sxy)$ $\vdash_{NK}$ $(\exists x)\sim Fx$

(10) $(\exists x)(\forall y)((\exists z)Ryz \rightarrow Ryx)$, $(\forall x)(\exists y)Rxy$ $\vdash_{NK}$ $(\exists x)(\forall y)Ryx$

(11) $(\exists x)(Fx \& (\forall y)(Gy \rightarrow Rxy))$, $(\forall x)(Fx \rightarrow (\forall y)(Ky \rightarrow \sim Rxy))$ $\vdash_{NK}$
     $(\forall x)(Gx \rightarrow \sim Kx)$

*(12) $(\exists x)(\forall y)((Fx \vee Gy) \rightarrow (\forall z)(Hxy \rightarrow Hyz))$, $(\exists z)(\forall x)\sim Hxz$ $\vdash_{NK}$
     $(\exists y)(\forall x)(Fy \rightarrow \sim Hyx)$

(13) $(\forall x)[(\exists y)Tyx \rightarrow (\forall z)\sim Txz]$, $(\forall x)[(\exists y)Tyx \rightarrow Txx]$ $\vdash_{NK}$ $(\forall x)(\forall y)\sim Txy$
     ((13) is Example 4 with the alternative symbolization; give a proof using ~I.)

(14) $(\exists x)[Sx \& (\forall y)((Py \& (\exists z)(Sz \& Dyz)) \rightarrow Dyx)]$, $(\forall x)(Px \rightarrow (\exists y)(Sy \& Dxy))$
     $\vdash_{NK}$ $(\exists x)(Sx \& (\forall y)(Py \rightarrow Dyx))$

(15) $(\forall x)(\forall y)[(\exists z)Hyz \rightarrow Hxy]$ $\vdash_{NK}$ $(\exists x)(\exists y)Hxy \rightarrow (\forall x)(\forall y)Hxy$

(16) $(\exists x)(Fx \& (\forall y)((Gy \& Hy) \rightarrow \sim Sxy))$, $(\forall x)(\forall y)(((Fx \& Gy) \& Jy) \rightarrow \sim Sxy)$,
     $(\forall x)(\forall y)(((Fx \& Gy) \& Rxy) \rightarrow Sxy)$, $(\exists x)(Gx \& (Jx \vee Hx))$ $\vdash_{NK}$ $(\exists x)(\exists y)((Fx$
     $\& Gy) \& \sim Rxy)$

(17) $(\forall x)(\forall y)(\forall z)(Rxy \rightarrow \sim Ryz)$ $\vdash_{NK}$ $(\exists y)(\forall x)\sim Rxy$

(18) $(\forall x)Fx \leftrightarrow \sim(\exists x)(\exists y)Rxy$ $\vdash_{NK}$ $(\exists x)(\forall y)(\forall z)(Fx \rightarrow \sim Ryz)$

(19) $(\exists x)(\forall y)[(\exists z)(Fzy \rightarrow (\exists w)Fyw) \rightarrow Fxy]$ $\vdash_{NK}$ $(\exists x)Fxx$

(20) $\vdash_{NK}$ $(\forall x)(\exists y)(\forall z)[(\exists u)Txyu \rightarrow (\exists v)Txzv]$

II   Symbolize the following arguments, then give proofs of their forms. State your dictionary. Note that in (3) and (9), quantifiers must be relativized to the domain of things, so your symbolization should employ the predicate 'P_' for '_ is a person'.

(1) All Caesar's assassins were senators. All senators were known to Calpurnia. Therefore all Caesar's assassins were known to Calpurnia.

(2) All horses are animals. Therefore whatever is the head of a horse is the head of an animal.

(3) Babies are illogical persons. No person is despised who can manage a crocodile. Illogical persons are despised. Therefore babies cannot manage crocodiles.[3]

---

[3]   This famous example is taken from Lewis Carroll's (C. L. Dodgson's) *Symbolic Logic* (1896).

*(4) No tall people applied. John is taller than Mary and Mary is tall. A person is tall if he or she is taller than a tall person. Therefore John did not apply.

(5) If there are any barbers then there is a barber who shaves all and only those who do not shave themselves. Therefore there are no barbers.

(6) There's a professor who likes John, and John likes all professors. Anyone who likes John likes everyone John likes. Therefore there's a professor who likes all professors.

*(7) Every student who cheated bribed at least one professor. Some students were accused and so was each professor they bribed. All accused students cheated. Therefore some professor either isn't a student or is a cheat.

(8) Everyone is honest with anyone who is honest with everyone. So if there is at least one person who is honest with everyone, everyone is honest with at least one person who is honest with them.

(9) There is a statement assented to by everyone who assents to any statement at all. Everyone assents to at least one statement. Therefore there is a statement to which everyone assents.

(10) There is someone who teaches everyone who is taught by anyone. There is someone who teaches everyone who teaches anyone. Everyone who does not teach anyone is taught by someone. Therefore there is someone who teaches everyone. (Use only 'T_,_': _ teaches _)

4 Rules for identity in NK

We turn now to the matter of making our final extension to NK by adding rules for the new logical constant '='. What inferential principles governing identity are fundamental to its meaning and could be embodied in an elimination rule and an introduction rule? So far as an elimination rule is concerned, the simplest way of using identity sentences in inferences is illustrated by:

A: (1) Superman is an extraterrestrial.
   (2) Superman is identical to Clark Kent.
   (3) ∴ Clark Kent is an extraterrestrial.

The conclusion is obtained by using the identity sentence (2) to replace the occurrence in (1) of the name on the left of (2) with the name on the right of (2). More generally, we formulate the rule as follows:

Rule of Identity Elimination: For any sentences ϕt_1 and $\ulcorner t_1 = t_2 \urcorner$ of LFOL, where t_1 and t_2 are individual constants, if $\ulcorner t_1 = t_2 \urcorner$ occurs at a line j in a proof and ϕt_1 occurs at a line k, then at a line m we may infer ϕt_2, writing on the left of m the numbers which appear on the left of j and k.[4] ϕt_2 results from ϕt_1 by replacing *at least one* occurrence of t_1 in ϕt_1 by t_2. Schematically:

---

[4] Notice that the rule allows substition of the term on the right of the identity sentence by the term on the left but not vice versa. This is a little artificial but is done to maintain consistency with *MacLogic*.

$$a_1,\ldots,a_n \quad (j) \qquad t_1 = t_2$$
$$\vdots$$
$$b_1,\ldots,b_u \quad (k) \qquad \phi t_1$$
$$\vdots$$
$$a_1,\ldots,a_n b_1,\ldots,b_u \quad (m) \qquad \phi t_2 \qquad\qquad j,k \; =\!E$$

Identity Elimination is sometimes called Leibniz's Law or the Indiscernibility of Identicals, though these labels are more properly applied to the intuitive principle which the rule of inference encapsulates. This is the principle that if x and y are the same object, then whatever is true of x is true of y and vice versa, or in other words, the properties of x are the same as the properties of y. Here is a simple example of a proof employing =E:

Example 1: Show $(\exists x)(Fx \; \& \; x = a) \vdash_{NK} Fa$.

1	(1)	$(\exists x)(Fx \; \& \; x = a)$	Premise
2	(2)	$Fb \; \& \; b = a$	Assumption
2	(3)	$b = a$	2 &E
2	(4)	Fb	2 &E
2	(5)	Fa	3,4 =E
1	(6)	Fa	1,2,5 $\exists$E ◆

Since the name used to form the instance of (1) at (2) is 'b', no restriction on $\exists$E is violated at line 6 (if we had used 'a' for 'x' at (2), which restrictions would have been violated?).

In applying the rule of =E, we will take care that the first number j in the label is the line number of the identity sentence whose left name is replaced in line k with its right name, and the second number k is the line number of the formula in which the replacement of one name for another is made. Without some such convention, confusion could arise, since the line in which the substitution is made may *itself* be an identity sentence, as happens in the following example:

Example 2: Show $a = b, b = c \vdash_{NK} a = c$.

1	(1)	$a = b$	Premise
2	(2)	$b = c$	Premise
1,2	(3)	$a = c$	2,1 =E ◆

In this proof we replace the occurrence of 'b' in (1) with 'c', using the identity sentence at line 2; so (1) is the line in which the replacement is made and (2) is the identity sentence used. Labeling line 3 '1,2 =E' would be incorrect, since this would mean that we have replaced an occurrence of 'a' in (2) using (1); but there is no occurrence of 'a' in (2).

The rule of Identity Introduction is even more straightforward. Like Theorem Introduction, it licenses the writing of a new line in a proof, a line which depends on no line, not even itself.

Rule of Identity Introduction: At any line j in a proof, we may write down any formula of the form ⌜t = t⌝, where t is an individual constant. No numbers are written on the left of j, and the line is labeled '=I'.

It is an immediate consequence of this rule that the use of an individual constant in LFOL is existentially committing, since by =I we can write, say, 'b = b', and then by ∃I immediately infer '(∃x)(x = b)', which is one way of expressing 'b exists'. Consequently, 'nondenoting names' and 'names of nonexistents' are not possible in LFOL, though apparently they are possible in ordinary language (e.g., the names of characters in myths, such as Pegasus, the winged horse of Greek mythology). The existentially committing nature of names in LFOL is exactly what one would expect in view of the semantics, where we stipulated that (i) only *nonempty* domains are allowed, (ii) in interpreting sentences of LFOL, every individual constant must be assigned a reference from the domain, and (iii) an existential sentence is true if and only if it has a true instance generated by an object in the domain. There are systems of logic which allow for names that are not existentially committing by changing some of these features of the semantics. For instance, one kind of 'free logic' splits the domain into two subdomains, one of existents and the other of nonexistents, and allows names to be assigned referents from either subdomain, but requires for an existential sentence to be true that it have a true instance generated by an object in the subdomain of *existents*. The rules ∃I and ∀E are correspondingly adjusted. However, we will not concern ourselves with these alternatives at this point (free logic is discussed in §6 of Chapter 9).

A simple example of a proof which uses =I is the following, the converse of our first example:

Example 3: Show Fa ⊢$_{NK}$ (∃x)(Fx & x = a).

1	(1)	Fa	Premise
	(2)	a = a	= I
1	(3)	Fa & a = a	1,2 &I
1	(4)	(∃x)(Fx & x = a)	3 ∃I ◆

Of course, not all proofs using the identity rules are as simple as these. Here is a rather more complex example:

Example 4: Show (∀x)(Fx → (∃y)(Gyx & ~Gxy)), (∀x)(∀y)((Fx & Fy) → x = y) ⊢$_{NK}$ (∀x)(Fx → (∃y)~Fy).

The conclusion will be inferred by ∀I applied to a formula of the form 'F_ → (∃y)~Fy', so our subgoal is to derive such a formula. We do this by assuming an appropriate antecedent, say 'Fa', then derive the consequent and apply →I. To derive the consequent '(∃y)~Fy' we must presumably derive a sentence of the form '~F_', say '~Fb', which in turn is likely to be obtained by assuming 'Fb' and using ~E and ~I. However, we delay this assumption until line 7, since to execute this part of the proof we need to use (5), '(∃y)(Gya & ~Gay)', which requires assuming an instance 'Gba & ~Gab' and obtaining the target formula '(∃y)~Fy' by ∃E, and the name used to form the instance is the one which should appear in the '~F_' to which we are going to apply ∃I.

1	(1)	(∀x)(Fx → (∃y)(Gyx & ~Gxy))	Premise
2	(2)	(∀x)(∀y)((Fx & Fy) → x = y)	Premise
3	(3)	Fa	Assumption
1	(4)	Fa → (∃y)(Gya & ~Gay)	1 ∀E
1,3	(5)	(∃y)(Gya & ~Gay)	3,4 →E
6	(6)	Gba & ~Gab	Assumption
7	(7)	Fb	Assumption
3,7	(8)	Fa & Fb	3,7 &I
2	(9)	(∀y)((Fa & Fy) → a = y)	2 ∀E
2	(10)	(Fa & Fb) → a = b	9 ∀E
2,3,7	(11)	a = b	8,10 →E
6	(12)	Gba	6 &E
2,3,6,7	(13)	Gbb	11,12 =E
6	(14)	~Gab	6 &E
2,3,6,7	(15)	~Gbb	11,14 =E
2,3,6,7	(16)	Λ	13,15 ~E
2,3,6	(17)	~Fb	7,16 ~I
2,3,6	(18)	(∃y)~Fy	17 ∃I
1,2,3	(19)	(∃y)~Fy	5,6,18 ∃E
1,2	(20)	Fa → (∃y)~Fy	3,19 →I
1,2	(21)	(∀x)(Fx → (∃y)~Fy)	20 ∀I ◆

Though the main idea of the proof is straightforward—to use 'a = b' to derive the two halves of a contradiction from the instance (6) of (5)—care must be taken at the end of the proof to discharge (7) and use ∃I before applying ∃E, otherwise a constraint on ∃E will be violated.

Our final example is based on our discussion of (2.6), where we established:

(2.6) (∀x)(∃y)Rxy, (∀x)(∀y)(∀z)((Rxy & Ryz) → Rxz) ⊭ (∃x)Rxx.

We noted that only an interpretation with an infinite domain can make both premises true and the conclusion false. Therefore, in any finite domain, if the premises are true, so is the conclusion. Consequently, if we were to add to (2.6) a premise implying that only finitely many objects exist, the result would be a valid argument-form. For each finite n, we know how to say that at most n objects exist. So for any finite n, adding 'at most n objects exist' to the premises of (2.6) will result in a provable sequent. We demonstrate this for the case $n = 2$.

Example 5: Show (∀x)(∃y)Rxy, (∀x)(∀y)(∀z)((Rxy & Ryz) → Rxz),
(∀x)(∀y)(∀z)(x = y ∨ (y = z ∨ x = z)) ⊢_{NK} (∃x)Rxx.

The reasoning of the proof follows our intuitive discussion of why there is no finite interpretation which establishes (2.6). We use premise 1 repeatedly, and premise 2, to obtain atomic sentences of the form 'R_,_', where the blanks are filled by different names, and we then use identity sentences obtained from (3) to make substitutions which give us atomic sentences of the form 'R_,_' where the same name fills the blanks; and from these we can infer the conclusion by ∃I. The proof is complicated because using premise 1 to obtain atomic sentences of the form 'R_,_', with different names in the blanks, involves going through existential sentences (instances of (1)), assuming corresponding instances which then need to be discharged with ∃E. Additionally, using premise 3 to make substitutions will require nested applications of ∨E. The proof follows below; in order to shorten it, we will allow ourselves to make multiple applications of ∀E in a single step when universal quantifiers are stacked in sequence in the line to which ∀E is being applied, as in (8) and (17) below:

1	(1)	$(\forall x)(\exists y)Rxy$	Premise
2	(2)	$(\forall x)(\forall y)(\forall z)((Rxy \,\&\, Ryz) \to Rxz)$	Premise
3	(3)	$(\forall x)(\forall y)(\forall z)(x = y \lor (y = z \lor x = z))$	Premise
1	(4)	$(\exists y)Ray$	1 ∀E
1	(5)	$(\exists y)Rby$	1 ∀E
6	(6)	Rab	Assumption
7	(7)	Rbc	Assumption
3	(8)	$a = b \lor (b = c \lor a = c)$	3 ∀E
9	(9)	$a = b$	Assumption
6,9	(10)	Rbb	9,6 =E
6,9	(11)	$(\exists x)Rxx$	10 ∃I
12	(12)	$b = c \lor a = c$	Assumption
13	(13)	$b = c$	Assumption
7,13	(14)	Rcc	13,7 =E
7,13	(15)	$(\exists x)Rxx$	14 ∃I
16	(16)	$a = c$	Assumption
2	(17)	$(Rab \,\&\, Rbc) \to Rac$	2 ∀E
6,7	(18)	Rab & Rbc	6,7 &I
2,6,7	(19)	Rac	17,18 →E
2,6,7,16	(20)	Rcc	16,19 =E
2,6,7,16	(21)	$(\exists x)Rxx$	20 ∃I
2,6,7,12	(22)	$(\exists x)Rxx$	12,13,15,16,21 ∨E
2,3,6,7	(23)	$(\exists x)Rxx$	8,9,11,12,22 ∨E
1,2,3,6	(24)	$(\exists x)Rxx$	5,7,23 ∃E
1,2,3	(25)	$(\exists x)Rxx$	4,6,24 ∃E ◆

(4) and (5) are existential sentences obtained from (1), and (6) and (7) are instances of them (a proof could be devised even if the individual constant introduced in (6) is different from the constant in (5)). (8) is the disjunction of identity sentences which we will use to infer 'R_,_' with the same name in the blanks, and then '(∃x)Rxx' by ∃I. We infer '(∃x)Rxx' from the first disjunct of (8) at line 11, and from the second disjunct (which is on line 12) at line 23. Since the second disjunct is itself a disjunction, we use a subsidiary ∨E on it to obtain '(∃x)Rxx' from it at line 22, having inferred '(∃x)Rxx' from its two disjuncts, at

lines 15 and 21 respectively. The '$(\exists x)Rxx$' at line 23 follows from (8) with the help of (6) and (7). It remains to use $\exists E$ to discharge (6) and (7), beginning with (7). The relevant individual constant in (7) is 'c', which does not occur in (5), or in any of (2), (3) and (6), or in (23); so the application of $\exists E$ at (24) is legal. Finally, the relevant individual constant in (6) is 'b', which does not occur in (4), or in any of (1), (2) and (3), or in (24). So the application of $\exists E$ at (25) is also legal.

❏ Exercises

I   Show the following:

(1) $\vdash_{NK} (\forall x)(\forall y)((Fx \ \& \sim Fy) \to x \neq y)$

(2) $(\forall x)(Fx \to (\exists y)(Gy \ \& \ x = y)) \dashv\vdash_{NK} (\forall x)(Fx \to Gx)$

(3) $\vdash_{NK} (\forall x)(\forall y)(x = y \to y = x)$

(4) $\vdash_{NK} (\forall x)(x = b \to x = c) \to b = c$

(5) $(\forall y)(Ray \to y = b) \vdash_{NK} (\exists y)(Ray \ \& \ Gy) \to Gb$

(6) $(\forall x)(Rxa \to x = c), (\forall x)(Rxb \to x = d), (\exists x)(Rxa \ \& \ Rxb) \vdash_{NK} c = d$

*(7) $(\exists x)(\exists y)Hxy, (\forall y)(\forall z)(Dyz \leftrightarrow Hzy), (\forall x)(\forall y)(\sim Hxy \lor x = y) \vdash_{NK}$
$(\exists x)(Hxx \ \& \ Dxx)$

(8) $(\forall x)(\forall y)((Rxy \ \& \ Ryx) \to x = y), (\forall x)(\forall y)(Rxy \to Ryx) \vdash_{NK}$
$(\forall x)[(\exists y)(Rxy \lor Ryx) \to Rxx]$

(9) $(\exists x)(\forall y)(x = y \leftrightarrow Fy), (\forall x)(Gx \to Fx) \vdash_{NK} (\forall x)(\forall y)((Gx \ \& \ Gy) \to x = y)$

(10) $(\exists x)Fx, \sim(\exists x)(\exists y)(x \neq y \ \& \ (Fx \ \& \ Fy)) \vdash_{NK} (\exists x)(\forall y)(x = y \leftrightarrow Fy)$

II   Symbolize the following arguments and give proofs of their forms. State your dictionary.

(1) There are at most two clever people and at least two geniuses. All geniuses are clever. Therefore there are exactly two clever people.

(2) Different women cannot give birth to the same person. Different women gave birth to John and James. So Bill cannot be both John and James.

(3) Everyone is happy if there is someone he loves and someone who loves him. No one who does not love himself loves anyone else. No one is happy if he loves no one. Therefore all and only those who love themselves are happy.

(4) Any candidate who answered at least one difficult question answered at most two questions. Some candidates answered every question. At least one question was difficult. Therefore there were at most two questions.

III   Show that the Axiom of Extensionality in set theory entails that distinct sets have different members and that there is at most one set with no members. In other words, where E is the formula discussed in Exercise 2.III on page 262, show

(1) $E \vdash_{NK} (\forall x)(\forall y)(((Sx \ \& \ Sy) \ \& \ x \neq y) \to$
$(\exists z)((z \in x \ \& \ z \notin y) \lor (z \in y \ \& \ z \notin x)))$

*(2) $E \vdash_{NK} (\forall x)(\forall y)[((Sx \ \& \ Sy) \ \& \ [\sim(\exists z)z \in x \ \& \ \sim(\exists z)z \in y]) \to x = y]$

5 Properties of binary relations

Many-place predicate symbols are sometimes called *relation* symbols, and the entities they stand for *relations*. Of particular interest are two-place, or *binary*, relations. Formally speaking, a binary relation is a collection of ordered pairs, in other words, the sort of entity which can be the extension of a two-place predicate in a domain D. A binary relation R can have a variety of structural properties, and the study of the logical relationships among these properties is the topic of this section. The relationships among properties of binary relations are of great importance in advanced logic and algebra; see, for instance, Chapter 10 on intuitionism.

The three basic properties of binary relations are *reflexivity*, *symmetry* and *transitivity*. Given that binary relations are collections of ordered pairs, the strictly correct way of defining these three properties is as follows:

- A binary relation R on a domain D is reflexive if and only if, for every object x in D, $\langle x,x \rangle$ belongs to R.
- A binary relation R on a domain D is symmetric if and only if, for all objects x and y in D, if $\langle x,y \rangle$ belongs to R then $\langle y,x \rangle$ belongs to R.
- A binary relation R on a domain D is transitive if and only if, for all objects x, y, and z in D, if $\langle x,y \rangle$ and $\langle y,z \rangle$ belong to R, then $\langle x,z \rangle$ belongs to R.

However, it will be convenient to mix metalanguage and object language in a way that allows us to use 'R' simultaneously as a metalanguage variable and as a two-place predicate symbol of the object language (in the latter, it is styled roman). We can then formulate the preceding three definitions in this way:

- R is reflexive if and only if: $(\forall x)Rxx$.
- R is symmetric if and only if: $(\forall x)(\forall y)(Rxy \to Ryx)$.
- R is transitive if and only if: $(\forall x)(\forall y)(\forall z)((Rxy\ \&\ Ryz) \to Rxz)$.

Here we are also suppressing mention of D, whose identity is usually clear from context. We have already encountered a number of binary relations, for instance the relation on natural numbers *less than or equal to*. This relation is reflexive, since every number is less than or equal to itself, and it is transitive, since if $x \leqslant y$ and $y \leqslant z$ then $x \leqslant z$, but it is not symmetric, since we can have $a \leqslant b$ while $b \not\leqslant a$. For comparison, the relation *less than* is transitive, but neither reflexive nor symmetric.

These two mathematical relations exemplify two different possibilities when a relation fails to have a certain structural property. A binary relation R might fail to be reflexive because there is *some* object x such that $\langle x,x \rangle$ does not belong to R, though there are also some x's such that $\langle x,x \rangle$ does belong to R. On the other hand, it might be that for *no* x does $\langle x,x \rangle$ belong to R. In the former case we say that R is *nonreflexive* and in the latter that it is *irreflexive*. In the style of our previous definitions,

- R is irreflexive if and only if: $(\forall x)\sim Rxx$.
- R is nonreflexive if and only if: $(\exists x)Rxx$ & $(\exists x)\sim Rxx$.

Thus *less than* is irreflexive rather than nonreflexive. We can draw an analogous distinction with respect to symmetry. If R is not symmetric, it might be that *whenever* $\langle x,y \rangle$ belongs to R then $\langle y,x \rangle$ does not belong to R, or it may just be that there are some pairs $\langle x,y \rangle$ where $\langle x,y \rangle$ and $\langle y,x \rangle$ both belong to R, and some pairs $\langle x,y \rangle$ which belong to R while $\langle y,x \rangle$ does not. The relevant terms here are *asymmetric* and *nonsymmetric*.

- R is asymmetric if and only if: $(\forall x)(\forall y)(Rxy \rightarrow \sim Ryx)$.
- R is nonsymmetric if and only if: $(\exists x)(\exists y)(Rxy$ & $Ryx)$ & $(\exists x)(\exists y)(Rxy$ & $\sim Ryx)$.

Hence *less than* ($<$) is asymmetric, while *less than or equal to* ($\leqslant$) is nonsymmetric, since, when $x = y$, both $\langle x,y \rangle$ and $\langle y,x \rangle$ belong to it.

A similar distinction arises with respect to transitivity: R may fail to be transitive because whenever $\langle x,y \rangle$ and $\langle y,z \rangle$ belong to it, $\langle x,z \rangle$ does not belong to it, or it may just be that for some x, y and z, $\langle x,y \rangle$, $\langle y,z \rangle$ and $\langle x,z \rangle$ all belong to it, and for some x, y and z, $\langle x,y \rangle$ and $\langle y,z \rangle$ belong to it while $\langle x,z \rangle$ does not. We say that R is *intransitive* and *nontransitive* respectively:

- R is intransitive if and only if: $(\forall x)(\forall y)(\forall z)((Rxy$ & $Ryz) \rightarrow \sim Rxz)$.
- R is nontransitive if and only if: $(\exists x)(\exists y)(\exists z)(Rxy$ & Ryz & $Rxz)$ & $(\exists x)(\exists y)(\exists z)(Rxy$ & Ryz & $\sim Rxz)$.

For example, *twice as large as* is an intransitive relation on the domain $D = \{1, 2, 3...\}$ of positive integers, since x would be four times as large as z. *Knows* is a nontransitive relation on the domain of people, since there are triples of people x, y and z where x knows y, y knows z and x knows z, but also triples where x knows y, y knows z but x does not know z.[5]

A structural property which does not fit into a pair is that of *antisymmetry*. We noted that *less than or equal to* is nonsymmetric because of the pairs $\langle x,y \rangle$ where $x = y$. But if we restrict ourselves to the pairs $\langle x,y \rangle$ where $x \neq y$, then $\langle x,y \rangle \in R$ implies $\langle y,x \rangle \notin R$. When the restriction of a relation to the pairs $\langle x,y \rangle$ where $x \neq y$ is asymmetric, the relation itself is said to be antisymmetric:

- R is antisymmetric if and only if: $(\forall x)(\forall y)(x \neq y \rightarrow (Rxy \rightarrow \sim Ryx))$.

An equivalent definition is:

- R is antisymmetric if and only if: $(\forall x)(\forall y)((Rxy$ & $Ryx) \rightarrow x = y)$.

We will introduce other properties of relations later, but let us start with these.

---

5 Some authors use 'nonreflexive', 'nonsymmetric' and 'nontransitive' as synonyms of 'not reflexive', 'not symmetric' and 'not transitive'. I follow Lemmon's usage.

We have three groups of structural properties:

(F): Reflexivity, irreflexivity and nonreflexivity.
(S): Symmetry, asymmetry, antisymmetry and nonsymmetry.
(T): Transitivity, intransitivity and nontransitivity.

Every relation has at least one property from each group, since each group is jointly exhaustive of the relevant possibilities. It is also possible for a relation to have more than one property from a given group. A relation R is *empty* in a domain D if there is no $\langle x,y \rangle$ with x and y in D such that $\langle x,y \rangle \in R$. If D is the domain of points on the Euclidean plane, then the relation of being an infinite distance apart is an example: although the Euclidean plane is of infinite extent, any two points on it are only a finite distance apart. An empty relation is symmetric, asymmetric, anti-symmetric, transitive and intransitive, for these properties have definitions which are universally quantified conditionals, and when the quantifiers are evaluated in a domain where the relation is empty, the resulting conditionals have false antecedents and so are all true. A *nonempty* relation may be asymmetric and anti-symmetric, or non-symmetric and anti-symmetric ($\leqslant$), or symmetric and anti-symmetric (identity). And so long as no instance of 'Rxy & Ryz' is true, it may be transitive and intransitive.

These remarks concern what combinations of properties from within the same group are logically possible. The rest of this section is devoted to the question of what combinations of properties from different groups are logically possible. If we pick a property at random from each group, are we guaranteed a relation which has the three chosen properties, whatever they are? We can answer such questions using proofs and counterexamples. For instance, it is not possible to have an asymmetric, reflexive relation, or an asymmetric, nonreflexive one, which we show by proving in NK that any asymmetric relation is irreflexive. We shall continue to take liberties with the object language/metalanguage distinction by specifying the sequent to be proved in the metalanguage, though the proof itself will be perfectly standard:

Example 1: Show R is asymmetric $\vdash_{NK} R$ is irreflexive [i.e. $(\forall x)(\forall y)(Rxy \rightarrow {\sim}Ryx)$ $\vdash_{NK} (\forall x){\sim}Rxx$].

1	(1)	$(\forall x)(\forall y)(Rxy \rightarrow {\sim}Ryx)$	Premise
2	(2)	Raa	Assumption
1	(3)	Raa $\rightarrow$ ~Raa	1 $\forall$E
1,2	(4)	~Raa	2,3 $\rightarrow$E
1,2	(5)	$\curlywedge$	2,4 ~E
1	(6)	~Raa	2,5 ~I
1	(7)	$(\forall x){\sim}Rxx$	6 $\forall$I ◆

Thus the sequent to be proved in NK is derived from the definitions of the property terms in the specification of the sequent.

The converse of the previous sequent is incorrect. To show this we use semantic methods.

Example 2: Show R is irreflexive $\nvdash R$ is asymmetric $[(\forall x)\sim Rxx \nvdash (\forall x)(\forall y)(Rxy \to \sim Ryx)]$.

Interpretation: $D = \{\alpha, \beta\}$, $Ext(R) = \{\langle\alpha,\beta\rangle, \langle\beta,\alpha\rangle\}$. *Explanation:* '$(\forall x)\sim Rxx$' is true, since $\langle\alpha,\alpha\rangle \notin Ext(R)$ and $\langle\beta,\beta\rangle \notin Ext(R)$, but '$(\forall x)(\forall y)(Rxy \to \sim Ryx)$' is false, since 'Rab $\to \sim$Rba' is false.

We can say more about the relationship between irreflexivity and asymmetry:

Example 3: Show R is transitive $\vdash_{NK} R$ is irreflexive if and only if R is asymmetric.

In other words, from '$(\forall x)(\forall y)(\forall z)((Rxy \,\&\, Ryz) \to Rxz)$' we have to derive the biconditional '$(\forall x)(\forall y)(Rxy \to \sim Ryx) \leftrightarrow (\forall x)\sim Rxx$', one direction of which is simply the previous proof, so we get it by SI; hence it will be the other direction which uses transitivity. To derive '$(\forall x)(\forall y)(Rxy \to \sim Ryx)$' from '$(\forall x)\sim Rxx$' we try to derive, say, 'Rab $\to \sim$Rba' and then use $\forall$I twice. To derive 'Rab $\to \sim$Rba' we assume 'Rab', then assume 'Rba' with a view to obtaining '$\sim$Rba' by $\sim$E and $\sim$I.

1	(1)	$(\forall x)(\forall y)(\forall z)((Rxy \,\&\, Ryz) \to Rxz)$	Premise
2	(2)	$(\forall x)(\forall y)(Rxy \to \sim Ryx)$	Assumption
2	(3)	$(\forall x)\sim Rxx$	2 SI (Example 1)
	(4)	$(\forall x)(\forall y)(Rxy \to \sim Ryx) \to (\forall x)\sim Rxx$	2,3 $\to$I
5	(5)	$(\forall x)\sim Rxx$	Assumption
6	(6)	Rab	Assumption
7	(7)	Rba	Assumption
6,7	(8)	Rab & Rba	6,7 &I
1	(9)	(Rab & Rba) $\to$ Raa	1 $\forall$E
1,6,7	(10)	Raa	8,9 $\to$E
5	(11)	$\sim$Raa	5 $\forall$E
1,5,6,7	(12)	$\curlywedge$	10,11 $\sim$E
1,5,6	(13)	$\sim$Rba	7,12 $\sim$I
1,5	(14)	Rab $\to \sim$Rba	6,13 $\to$I
1,5	(15)	$(\forall y)(Ray \to \sim Rya)$	14 $\forall$I
1,5	(16)	$(\forall x)(\forall y)(Rxy \to \sim Ryx)$	15 $\forall$I
1	(17)	$(\forall x)\sim Rxx \to (\forall x)(\forall y)(Rxy \to \sim Ryx)$	5,16 $\to$I
1	(18)	$\ulcorner 4 \,\&\, 17 \urcorner$	4,17 &I
1	(19)	$(\forall x)(\forall y)(Rxy \to \sim Ryx) \leftrightarrow (\forall x)\sim Rxx$	18 Df ◆

Notice that we do not allow ourselves to make more than one application of $\forall$I at a line, since for each application we must check that the special restriction on $\forall$I is not violated. However, as (18) illustrates, we occasionally allow ourselves to use line numbers as names of the formulae at those lines to avoid having to write out very long formulae constructed from them: '$\ulcorner 4 \,\&\, 17 \urcorner$' means 'the result of writing the formula at line 4 followed by '&' followed by the formula at line 17' (the invisible outer parentheses at (4) and (17) would become visible in this process).

In discussing the properties of binary relations, it is usual to count identity as a relation, since we can think of it as the collection of all ordered pairs $\langle x,x \rangle$, for every object x. It is obvious that identity is reflexive, symmetric and transitive. However, it is also antisymmetric. In fact, identity is the *only* reflexive, symmetric and antisymmetric relation, as we can prove.

Example 4: Show R is reflexive, R is symmetric, R is antisymmetric $\vdash_{NK}$ $(\forall x)(\forall y)(Rxy \leftrightarrow x = y)$.

1	(1)	$(\forall x)Rxx$	Premise
2	(2)	$(\forall x)(\forall y)(Rxy \rightarrow Ryx)$	Premise
3	(3)	$(\forall x)(\forall y)((Rxy \ \& \ Ryx) \rightarrow x = y)$	Premise
4	(4)	Rab	Assumption
2	(5)	$Rab \rightarrow Rba$	2 $\forall$E
2,4	(6)	Rba	4,5 $\rightarrow$E
2,4	(7)	$Rab \ \& \ Rba$	4,6 &I
3	(8)	$(Rab \ \& \ Rba) \rightarrow a = b$	3 $\forall$E
2,3,4	(9)	$a = b$	7,8 $\rightarrow$E
2,3	(10)	$Rab \rightarrow a = b$	4,9 $\rightarrow$I
11	(11)	$a = b$	Assumption
1	(12)	Raa	1 $\forall$E
1,11	(13)	Rab	11,12 =E
1	(14)	$a = b \rightarrow Rab$	11,13 $\rightarrow$I
1,2,3	(15)	$\ulcorner 10 \ \& \ 14 \urcorner$	10,14 &I
1,2,3	(16)	$Rab \leftrightarrow a = b$	15 Df
1,2,3	(17)	$(\forall y)(Ray \leftrightarrow a = y)$	16 $\forall$I
1,2,3	(18)	$(\forall x)(\forall y)(Rxy \leftrightarrow x = y)$	17 $\forall$I ◆

However, identity is not the only reflexive, symmetric and transitive relation, as we can establish semantically:

Example 5: Show R is reflexive, R is symmetric, R is transitive $\not\vdash$ $(\forall x)(\forall y)(Rxy \leftrightarrow x = y)$.

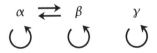

Interpretation: $D = \{\alpha,\beta,\gamma\}$, $Ext(R) = \{\langle \alpha,\alpha \rangle, \langle \beta,\beta \rangle, \langle \gamma,\gamma \rangle, \langle \alpha,\beta \rangle, \langle \beta,\alpha \rangle\}$. *Explanation:* '$(\forall x)Rxx$' is true since $\langle \alpha,\alpha \rangle$, $\langle \beta,\beta \rangle$ and $\langle \gamma,\gamma \rangle$ are all in $Ext(R)$. Symmetry of R holds since it is trivial when $x = y$ and though $\langle \alpha,\beta \rangle \in Ext(R)$, so is $\langle \beta,\alpha \rangle$. Transitivity holds as well, since all instances of '$(\forall x)(\forall y)(\forall z)((Rxy \ \& \ Ryz) \rightarrow Rxz)$' are true. In particular, '$(Rab \ \& \ Rba) \rightarrow Raa$' is true. But '$(\forall x)(\forall y)(Rxy \leftrightarrow x = y)$' is false since '$Rab \leftrightarrow a = b$' is false because $\langle \alpha,\beta \rangle \in Ext(R)$ but $\alpha \neq \beta$.

The combination of the properties of reflexivity, symmetry and transitivity has some mathematically important features, and there is a special term for any relation with this combination:

- A binary relation R is said to be an *equivalence* relation if and only if R is reflexive, symmetric and transitive.

An equivalence relation on a domain *partitions* the domain into mutually exclusive and jointly exhaustive *equivalence classes*. 'Mutually exclusive' means that no object belongs to more than one class, while 'jointly exhaustive' means that every object belongs to at least one class. The partitioning is effected by assigning objects x and y to the same class if and only if $\langle x,y \rangle \in R$. For example, in the previous interpretation, the equivalence classes induced by Ext(R) are $\{\alpha,\beta\}$ and $\{y\}$. In English, equivalence relations are usually expressed by phrases of the form '...is the same _ as...' or '...has the same _ as...', and this explains the word 'equivalence'. For example, 'is the same height as' expresses the equivalence relation *being the same height as*, which partitions the domain of people into equivalence classes, the class of all who are 6' high, the class of all who are 5' 10" high and so on, with a class for each specific height which any person has. Two members of the same class are *equivalent with respect to height*. But not all phrases which fit the schema '...is the same _ as...' express equivalence relations. For example, *being the same nationality as* is not an equivalence relation since a person can have dual nationality and thus belong to two different groups of people with the same nationality; there are also stateless individuals. It is an exercise to determine which of the three structural properties fails.[6]

There are alternative characterizations of 'equivalence relation', for instance, the following in terms of the property of being *Euclidean:*

- A binary relation R is said to be Euclidean if and only if:
  $(\forall x)(\forall y)(\forall z)((Rxy \ \& \ Rxz) \rightarrow Ryz)$.

R is an equivalence relation if and only if R is reflexive and Euclidean, thus:

Example 6: Show R is an equivalence relation $\dashv\vdash_{NK} R$ is reflexive and Euclidean.

That R is reflexive if R is an equivalence relation is true by definition. To show that R is Euclidean, we give a proof whose premises are two of the properties of equivalence relations, symmetry and transitivity:

1	(1)	$(\forall x)(\forall y)(Rxy \rightarrow Ryx)$	Premise
2	(2)	$(\forall x)(\forall y)(\forall z)((Rxy \ \& \ Ryz) \rightarrow Rxz)$	Premise
3	(3)	Rab & Rac	Assumption
3	(4)	Rab	3 &E
1	(5)	Rab $\rightarrow$ Rba	1 $\forall$E
1,3	(6)	Rba	4,5 $\rightarrow$E
3	(7)	Rac	3 &E

---

[6] Because *being the same nationality as* is not an equivalence relation, the pedantically correct thing to say is 'A and B have a nationality in common' rather than 'A and B are of the same nationality'.

1,3	(8)	Rba & Rac	6,7 &I
2	(9)	(Rba & Rac) → Rbc	2 ∀E
1,2,3	(10)	Rbc	8,9 →E
1,2	(11)	(Rab & Rac) → Rbc	3,10 →I
1,2	(12)	(∀z)((Rab & Raz) → Rbz)	11 ∀I
1,2	(13)	(∀y)(∀z)((Ray & Raz) → Ryz)	12 ∀I
1,2	(14)	(∀x)(∀y)(∀z)((Rxy & Rxz) → Ryz)	13 ∀I ◆

The converse sequent is left as an exercise.

Six other structural properties of relations are worth noting:

- R is *serial* if and only if: $(\forall x)(\exists y)Rxy$.
- R is *directed* if and only if: $(\forall x)(\forall y)(\forall z)((Rxy \,\&\, Rxz) \rightarrow (\exists w)(Ryw \,\&\, Rzw))$.
- R is *connected* if and only if: $(\forall x)(\forall y)(\forall z)((Rxy \,\&\, Rxz) \rightarrow (Ryz \vee Rzy))$.
- R is *totally connected* if and only if: $(\forall x)(\forall y)(Rxy \vee Ryx)$.
- R is *weakly linear* if and only if R is reflexive, transitive and totally connected.
- R is *linear* (or *strictly linear*) if and only if R is weakly linear and antisymmetric.

Some interconnections among these properties and the previous ones will be investigated in the exercises. Other interconnections we have already discovered. For example, in §3 of this chapter we discovered that every serial, transitive, finite relation is either reflexive or nonreflexive, or equivalently, that any serial, transitive, irreflexive relation is infinite.

❏ Exercises

I   With respect to the three groups of structural properties F, S and T distinguished on page 277, classify the following relations, explaining where necessary:

(1) *is a child of*
(2) *is a sister of*
(3) *is of the same nationality as*
*(4) *is similar in color to*
(5) *has the same parents as*
*(6) *semantically entails* (for sequents with one premise).

Why is (3) not an equivalence relation?

II   Show the following:

(1) R is antisymmetric ⊭ R is not symmetric
*(2) R is weakly linear ⊭ R is linear

    (3) R is serial, R is symmetric $\nvDash$ R is reflexive
    (4) R is serial, R is Euclidean $\nvDash$ R is transitive
    (5) R is transitive, R is antisymmetric, R is serial $\nvDash$ $(\exists y)(\forall x)Rxy$
  *(6) R is reflexive, R is transitive $\nvDash$ R is directed

III Show the following:

    (1) R is reflexive, R is Euclidean $\vdash_{NK}$ R is an equivalence relation
  *(2) R is transitive, R is intransitive $\vdash_{NK}$ R is asymmetric
    (3) R is Euclidean $\vdash_{NK}$ R is connected
    (4) R is serial, R is symmetric, R is Euclidean $\vdash_{NK}$ R is reflexive

*IV A relation R is *secondarily reflexive* if and only if $(\forall x)(\forall y)(Rxy \rightarrow Ryy)$. Show

    R is Euclidean $\vdash_{NK}$ R is secondarily reflexive

V A relation R is *cyclical* if and only if $(\forall x)(\forall y)(\forall z)((Rxy \; \& \; Ryz) \rightarrow Rzx)$. Show

    R is reflexive, R is cyclical $\vdash_{NK}$ R is symmetric

6 Alternative formats

No new issues of principle arise for converting proofs in Lemmon format to tree format or sequent-to-sequent format. In tree format, the identity rules may be stated as:

$$\frac{\rule{2cm}{1pt}}{t = t} \; {=}\mathrm{I} \qquad\qquad \frac{t_1 = t_2 \qquad \phi t_1}{\phi t_2} \; {=}\mathrm{E}$$

=I allows us to use any instance of $t = t$ as an automatically discharged assumption. To illustrate a simple proof in tree format, we convert Example 4.1:

Example 1: Show $(\exists x)(Fx \; \& \; x = a) \vdash_{NK} Fa$.

$$\frac{\begin{array}{cc} \dfrac{\overline{Fb \; \& \; b = a}^{\;(1)}}{b = a} \; {\&}\mathrm{E} & \dfrac{\overline{Fb \; \& \; b = a}^{\;(1)}}{Fb} \; {\&}\mathrm{E} \\[2ex] \end{array}}{}$$

For sequent-to-sequent proofs, the identity rules have the following form:

> **Rule of =I:** At any point in a proof we may include a line of the form $\vdash t = t$.
> **Rule of =E:** From the sequents $\Gamma \vdash t_1 = t_2$, $\Sigma \vdash \phi t_1$, we may infer the sequent $\Gamma, \Sigma \vdash \phi t_2$.

So Example 1 may be reformatted as a sequent-to-sequent proof as follows:

(1)	$(\exists x)(Fx \ \& \ x = a) \vdash (\exists x)(Fx \ \& \ x = a)$	Premise
(2)	$Fb \ \& \ b = a \vdash Fb \ \& \ b = a$	Assumption
(3)	$Fb \ \& \ b = a \vdash Fb$	2 &E
(4)	$Fb \ \& \ b = a \vdash b = a$	2 &E
(5)	$Fb \ \& \ b = a \vdash Fa$	3,4 =E
(6)	$(\exists x)(Fx \ \& \ x = a) \vdash Fa$	1,2,5 $\exists$E ◆

❑ Exercises

(1) Arrange the remaining examples in §4 in tree format.
(2) Arrange the remaining examples in §4 in sequent-to-sequent format.

7 Semantic consequence, deductive consequence and decidability

The same questions arise about the strength of our rules of inference for first-order logic as arose about the rules for the truth-functional connectives. We repeat the definitions here:

- A system S of rules of inference is said to be *sound* if and only if every S-provable argument-form is valid, that is, if and only if whenever $p_1,...,p_n \vdash_S q$, then $p_1,...,p_n \vDash q$.
- A system S of rules of inference is said to be *complete* if and only if every valid argument-form is S-provable, that is, if and only if whenever $p_1,...,p_n \vDash q$, then $p_1,...,p_n \vdash_S q$.

Our system of rules NK for first-order logic is both sound and complete. In other words, for any closed sentences $p_1,...,p_n$ and q of LFOL, the following statements are true:

> (Snd): If $p_1,...,p_n \vdash_{NK} q$ then $p_1,...,p_n \vDash q$.
> (Comp): If $p_1,...,p_n \vDash q$ then $p_1,...,p_n \vdash_{NK} q$.

This includes the case where $n = 0$; that is, by (Snd), every theorem q is logically valid (true in all interpretations), and by (Comp), every logically valid sentence *q* is a theorem. (Snd) and (Comp) are sometimes known as *finitary* soundness and *finitary* completeness respectively, because they are restricted to the cases where there are only finitely many premises. In first-order logic, cases where there are infinitely many premises are of considerable interest, so that soundness and completeness are often given their *strong* formulations: a system *S* is said to be *strongly sound* if and only if, for any set of LFOL-sentences Γ, if $\Gamma \vdash_S q$ then $\Gamma \vDash q$, and *strongly complete* if and only if, for any set of LFOL-sentences Γ, if $\Gamma \vDash q$ then $\Gamma \vdash_S q$. In these latter definitions, Γ may be either a finite or an infinite set of sentences. NK is strongly sound and strongly complete, though this is not implied merely by its finitary soundness and finitary completeness. These topics are pursued in more advanced texts.

In discussing sentential NK, we remarked that the soundness of the system of rules is obvious, while their completeness is not obvious, since although one can tell by inspecting the list of sentential rules that none are fallacious, one cannot tell by inspection that every rule needed in the proof of a valid sequent of LSL has been included on the list. In first-order logic, completeness is certainly at least as unobvious as in sentential logic, but soundness (finitary or strong) is no longer obvious, because of the need for restrictions on ∀I and ∃E. We motivated each restriction we imposed by indicating a problem that would arise without it, but this is no guarantee that we have thought of *all* the restrictions we need; perhaps some crucial restriction has been omitted, and as a result some invalid sequent has an NK-proof. In fact this is not so, but it requires a nontrivial proof to demonstrate that the rules are sound.

These remarks about the full system of first-order logic hold good of monadic predicate logic as well. The soundness of the rules for the full system entails the soundness of the rules for the monadic system, since the latter's rules are a subset of the former's. Consequently, a proof of an invalid sequent in LMPL would *ipso facto* be a proof of an invalid sequent in LFOL, and as there are no provable invalidities in first-order logic, it follows that there are none in monadic predicate logic. However, the completeness of the rules for the full system does not immediately entail the completeness of the rules for the monadic system, since it is conceivable that there is some valid sequent whose premises and conclusion contain only sentence-letters or one-place predicates, yet any NK-proof of it requires the introduction of some formula containing a many-place predicate. But there are results in a branch of logic known as *proof theory* which rule out this possibility (see Sundholm).

There remains the question of decidability. Recall from §6 of Chapter 6 that the validity problem for a kind of logic is *decidable* or *solvable* if and only if there is some algorithm which, given any sentences $p_1,...,p_n$ and *q* in the language for that logic, determines whether or not *q* is a semantic consequence of $p_1,...,p_n$ in the sense appropriate for that kind of logic. Validity is decidable for classical sentential logic and monadic predicate logic, though not, so far as we know, by an algorithm which is computationally tractable in all cases. However, it was proved by Alonzo Church in 1936 that the full system of first-order logic is undecidable (see Hodges §24). In fact, it suffices to add one two-place predicate to monadic predicate logic to obtain an undecidable logic. Since there is no decision procedure for validity in first-order logic, we will not extend the method of semantic tableaux to this case.

It is crucial to undecidability that there are arguments whose invalidity can only be demonstrated by interpretations with infinite domains. For if every invalid argument-form could be shown to be invalid by a finite counterexample, then even if, for any given form, there is no way of knowing in advance a size such that some interpretation of that size is guaranteed to refute the argument-form, we could still create an algorithm for deciding validity in the following way. First, it is possible to enumerate the lexicon of LFOL in an unending list such that every symbol of LFOL occurs some finite distance from the start of the list. We are given such lists for individual variables, individual constants, and for each n, for the n-place predicate symbols. We can form a single list from this infinite collection of lists by starting a new list with parentheses and logical constants and then proceeding as follows:

- Append the first individual variable, the first individual constant and the first 0-place predicate symbol.
- Append the next individual variable, the next individual constant, the next 0-place predicate symbol and the first 1-place predicate symbol.
- Append the next individual variable, the next individual constant, the next 0-place predicate symbol, the next 1-place predicate symbol and the first 2-place predicate symbol.

Continuing in this way, every symbol of the lexicon appears some finite distance from the start of the list (compare a listing of the natural numbers in the usual order 0,1,2,... versus the order 'evens first, then odds', in which each odd number occurs an infinite distance from the start of the list).

Using the lexicon list, we can list all sentences of the language LFOL so that each sentence occurs some finite distance from the start. For given any finite set of symbols from the lexicon, and a finite bound on the number of permitted occurrences of each symbol in the set, only finitely many different sentences can be formed from that set, and a finite collection of objects can always be arranged in some order. So we list the sentences of LFOL as follows:

- Take the first symbol in the lexicon list and form all sentences containing at most one occurrence of that symbol, listing them in some order.
- Take the first two symbols in the lexicon list and form all sentences containing at most two occurrences of each symbol, appending those not already on the sentence list to it in some order.
- Take the first three symbols in the lexicon list and form all sentences containing at most three occurrences of each symbol, appending those not already on the sentence list to it in some order.

Continuing in this way, every sentence of LFOL appears some finite distance from the start of the list.

Finally, we can list all NK-proofs by an analogous procedure. We begin by listing the nineteen rules of NK in some order (two rules for each logical constant plus Assumptions, DN and Df). Then:

- Take the first sentence in the sentence-list and the first rule in the rule list and form all proofs which involve at most one application of the rule, listing the resulting proofs in some order.
- Take the first two sentences in the sentence-list and the first two rules in the rule list and form all proofs which involve at most two applications of each of the rules; append to the proof list those proofs formed at this stage which are not already on it.
- Take the first three sentences in the sentence-list and the first three rules in the rule list and form all proofs which involve at most three applications of each of these rules; append to the proof list those proofs formed at this stage which are not already on it.

Continuing in this way we get a list of all possible proofs in which each proof occurs some finite distance from the start of the list.

An idealized computer could be programmed both to generate the list of proofs in the way described and also to start checking for counterexamples to any given sequent; for instance, at each stage *n*, it checks the given sequent against all possible interpretations with domain of size *n*. Suppose we are now given an arbitrary sequent of LFOL and have to determine whether or not it is valid. Then the computer will settle this for us *if every invalid sequent can be shown to be invalid by a finite interpretation*. For the sequent is either valid or invalid. If it is valid, then by the completeness of the rules of NK, a proof will eventually be generated by the computer, and if it is invalid, then, on the italicized assumption, a counterexample will eventually be generated. The computer can determine, of each proof, whether or not it is a proof of the input sequent, and of each interpretation, whether or not it is a counterexample to the input sequent. So eventually the computer will provide the answer to our question. However, as we know from Example 2.6 (page 260), the preceding italicized assumption is false: there are sequents whose invalidity can only be demonstrated by an infinite interpretation. Consequently, the computer program described would never determine whether or not such a sequent is valid, even in the absence of physical limitations on power, speed of computation, availability of memory, and the like.

8 Some limitations of first-order logic

It was easy to motivate the extension of sentential logic to monadic predicate logic, and the extension of the latter to the full system of first-order logic, since at each point we could exhibit very simple arguments which were intuitively valid but were mishandled by the system of logic developed by that point. At the point we have now reached, can we motivate a further extension of our system in the same way? With this question we venture into a vast area of lively and ongoing research and controversy. What follows is the merest scratching of the surface.

Suppose we put aside disputes, such as that between classical logicians and intuitionists, over which arguments *are* valid or invalid. Then the problem cases which remain can be divided broadly into two groups. First, there are intuitively valid English arguments which are translatable into valid LFOL arguments but only at the cost of providing translations which are rather far removed from the apparent syntactic structure of the English. Arguably, arguments with definite descriptions already illustrate this. We give two more examples, involving *second-order quantifiers* and *adverbs*. Second, there are English arguments which are intuitively valid but which cannot be translated into LFOL at all. Arguments containing *nonclassical quantifiers* provide many examples of this phenomenon.

Second-order quantification

Consider the following argument:

> A: (1) Socrates is wise and Plato is wise
> (2) ∴ There is something which Socrates is and Plato is.

(2) gives the appearance of being an existential quantification of (1), but a quantification into predicate position rather than name position. If we symbolize A, therefore, we will want the existential quantifier to be coupled not with an *individual* variable, which ranges over objects, but with a *predicate* variable, which ranges over properties. Such variables are called *second-order* variables, a quantifier coupled with one is called a *second-order* quantifier, and the extension of first-order logic which includes second-order quantifiers is called *second-order* logic.

The second-order symbolization of A would be:

> B: (1) Wa & Wb
> (2) ∴ ($\exists$F)(Fa & Fb)

which is a valid argument-form of the language of second-order logic, LSOL. Hence A is second-order valid. However, there is also a way of symbolizing A in first-order logic, by regarding properties as individual objects on a par with more ordinary objects such as people or things. Suppose we use a two-place predicate 'I_,_' to mean '_ instantiates _', and 'c' as a name of wisdom. Then it would be possible to give A the first-order symbolization

> C: (1) Iac & Ibc
> (2) ∴ ($\exists$x)(Iax & Ibx).

C correctly represents A as valid. On the other hand, the straightforward symbolization of (1) in A has been lost; we are committed to the view that 'Socrates is wise' in some sense *means* that Socrates instantiates wisdom. Among other things, this raises the specter of Plato's 'third man' argument: if it takes instantiation to relate Socrates and wisdom, then since instantiation is itself a (binary) property, what relates it to Socrates and wisdom? And whatever does this, what relates *it* to instantiation, Socrates and wisdom? And so on. So from the point of view of naturalness, the second-order formalization seems preferable. On the other hand, second-order logic has a more problematic nature than first-order logic; as we already remarked, for example, there is no second-order system of inference rules which is both sound and complete. So there are considerations to be made in favor of C, or something along similar lines, as well. For more on second-order logic, see Shapiro.

Adverbs
Consider the following argument:

> D   (1) John is running in Mary's direction
> (2) ∴ John is running.

D is certainly valid, but it is not immediately obvious how to find a valid LFOL translation of it: 'in Mary's direction' seems to be some kind of modifier, but it is certainly not a truth-functional connective, so it could not be added to the logic we have developed to this point. However, there is a well-known proposal, due to Donald Davidson, that (1) should be regarded as involving existential quantification over events, and this allows us to give an LFOL translation of it (see Davidson, Essay 6). To indicate how this would go, here is Davidsonian Loglish for (D1):

> (3) $(\exists x)(x$ is a running & x is by John & x is in Mary's direction).

Consequently, the inference from (1) to (2) in A is seen to involve not a special rule for a logic which treats 'in Mary's direction' as a logical constant, but simply our familiar &E, within an application of ∃E. There are other interesting features of (3): for instance, each of the atomic formulae in (3) could be preceded by a negation, and each of these three positions for negation seems to correspond to three different readings of the English 'John did not run in Mary's direction' according to whether the emphasis falls on 'John', 'run' or 'in Mary's direction'. For example, 'John did not *run* in Mary's direction' has the Loglish

> (4) $(\exists x)(x$ is not a running & x is by John & x is in Mary's direction).

On the other hand, one can use negation to generate variants of (3) which do not seem to have any English realization, short of quantifying over occurrences explicitly. An example is

> (5) $(\exists x)(\sim(x$ is a running & x is by John) & x is in Mary's direction)

for which the best we can do is 'Something happened in Mary's direction which did not involve both John and running'. Whether or not this is an objection to (3) raises some complicated questions about what the proper constraints on symbolization are. But at least if one is willing to countenance the very great departure from the 'surface structure' of the English of (D1) which (3) represents, Davidson's technique will enable us to accommodate all English arguments whose validity or invalidity hinges on adverbial modification.

Nonclassical quantifiers

We remarked in §2 of this chapter that there is no sentence of LFOL which means 'there are at most finitely many things such that...'. This is a consequence of the *compactness* property of the semantics of LFOL, which may be stated as follows:

> (CM): A semantics for a language $\mathcal{L}$ is *compact* if and only if, for any set of $\mathcal{L}$-sentences Γ, if for every finite subset Σ of Γ there is an interpretation which makes every sentence in Σ true, then there is an interpretation which makes every member of Γ true (for short: if Γ is finitely satisfiable then Γ is simultaneously satisfiable).

In a more advanced course it is a standard exercise on compactness, or illustration of it, to show that (CM) implies

> (IN): If Δ is a set of sentences such that, for each finite n, there is an interpretation of size n which makes every member of Δ true, then there is an infinite interpretation which makes every member of Δ true.

From (IN) we can infer that there is no way of expressing the quantifiers 'there are at most finitely many things x such that' and 'there are infinitely many things x such that'. For suppose that some LFOL sentence σ is hypothesized to mean 'there are at most finitely many x such that x = x' (i.e. 'at most finitely many things exist'). Then $\Delta = \{\sigma\}$ is a set of sentences which, for each finite n, is true in an interpretation of size n (indeed, true in every interpretation of size n). So by (IN), Δ is true in an infinite interpretation, thereby refuting the hypothesis that σ means 'at most finitely many thing exist'. It follows from this that there is no LFOL method of expressing 'there are infinitely many things x such that', since if there were, then for σ we could take $\ulcorner \sim\!\sigma' \urcorner$, where σ' is the putative LFOL-translation of 'there are infinitely many things x such that x = x'.

Consequently, it is an indisputable fact that translation into LFOL cannot establish the validity of the English argument

> E: (1) There are infinitely many rational numbers.
> (2) All rational numbers are real numbers.
> (3) ∴ There are infinitely many real numbers.

E is therefore not first-order valid, though it is intuitively valid. This is the kind of case which unarguably motivates an extension of LFOL. The simplest extension which will accommodate E involves adding a quantifier symbol to LFOL whose semantics bestows on it the meaning 'there are infinitely many things such that...'. Such a quantifier is called a *generalized* quantifier, and logic with generalized quantifiers has been widely investigated (see Westerståhl, and Partee *et al.*, Chapter 14). However, for most generalized quantifiers there is no system of rules of inference which is both sound and complete.

Adding 'there are infinitely many things such that...' to LFOL and giving the correct semantics for it is straightforward enough with the resources of modern set theory. But there are other natural-language quantifiers which are harder to deal with. Intuitively, the following argument is invalid:

> F: (1) Almost everything which is F is G.
> (2) Almost everything which is G is H.
> (3) ∴ Almost everything which is F is H.

To see F's invalidity, consider a domain of 1,000 objects, 10 of which belong to the extension of 'F', 9 of these 10 belong to the extension of 'G' but only 4 or 5 of these 9 belong to the extension of 'H'; and let all things not in the extension of 'F' be in the extensions of 'G' and 'H'. Then (1) is true, since 9 out of 10 Fs are G, and (2) is true, since there are 999 Gs and only 5 or 6 are not in the extension of 'H'. But (3) is false, since only 4 or 5 out of 10 Fs are H.

However, we cannot simply add a new quantifier symbol, '$\mathcal{A}$', say, and expect to be able to give a semantics for it which allows us to symbolize F. The semantics itself might use vague terms like 'almost', but that is not the main problem. By analogy with our treatment of 'all' and 'some', we would expect something like the following:

> ($\mathcal{A}$⊤): ⌜$(\mathcal{A}v)\phi v$⌝ is true on an interpretation $\mathcal{I}$ if and only if almost all its representative instances are true on $\mathcal{I}$.

The trouble is that without making further additions to LFOL we are still not going to be able to symbolize (F1). For example, '$(\mathcal{A}x)(Fx \,\&\, Gx)$' is not a correct symbolization of (F1), since it says that almost everything is both F and G, which can be false while (F1) is true, as our counterexample to F illustrates. And '$(\mathcal{A}x)(Fx \to Gx)$' is not a correct symbolization either, since this formula will be true if almost all its instances are true, although (F1) could be false in such a situation. This is also illustrated in our counterexample to F, where (3) is false but almost all instances of '$(\mathcal{A}x)(Fx \to Hx)$' are true, since all but nine have false antecedents, so '$(\mathcal{A}x)(Fx \to Hx)$' is true. Even after adding a new quantifier, then, (F1) still escapes symbolization. A much more radical departure from LFOL is needed, in which new modes of construction of complex formulae from simpler formulae are employed (see Westerståhl).

These few illustrations of limitations in first-order logic are cursory and indicate only a fraction of the range of phenomena which could be cited. But it would be a mistake to disparage first-order logic on the grounds of its limitations, since even if it does not go as far as might have been hoped, first-order logic is the foundation stone for the wealth of projects in philosophy, linguistics, computer science and logic which take up the challenges posed by problems that lie beyond the domain of its applicability.

❑ Exercises

I   Symbolize the following using second-order quantifiers.

(1)   Everything which Socrates is, Plato is.
*(2)   There is something which Socrates is and Plato isn't.
(3)   If there is something which Plato isn't then there is nothing which Socrates isn't.
(4)   Everyone who is something which Plato is, is everything which Socrates isn't.
*(5)   Plato and Socrates are indistinguishable.
(6)   Any two people have something in common.
(7)   Some people have nothing in common with anyone else.
(8)   If there is something which any two people are then for any two people there is something they are.

II   Let 'W' be a quantifier symbol for 'few'. Is either of the following symbolizations of 'Few Fs are G' correct? Justify your answer (study the discussion of 'almost all' in this section). Read (1) as 'for few x, x is F and x is G' and (2) as 'for few x, if x is F then x is G'.

(1) $(Wx)(Fx \,\&\, Gx)$
*(2) $(Wx)(Fx \rightarrow Gx)$

III   With 'W' as in II, which of the following are correct? Explain your verdicts of 'incorrect'.

(1) $(\forall x)Fx \vDash (Wx)Fx$
(2) $\sim(\exists x)Fx \vDash (Wx)Fx$
(3) $(\exists y)(Wx)Rxy \vDash (Wx)(\exists y)Rxy$
(4) $(Wx)(Fx \,\&\, Gx) \vDash (Wx)(Gx \,\&\, Fx)$
(5) $(Wx)(Fx \rightarrow Gx) \vDash (Wx)(\sim Gx \rightarrow \sim Fx)$

9 Summary

- An interpretation of an argument-form in LFOL consists in (i) a domain of discourse, (ii) to each individual constant in the argument-form, an assignment (perhaps implicit) of a reference drawn from the domain, and (iii) for each n-place predicate in the argument-form, an assignment to it of an extension consisting in a collection of n-tuples of objects drawn from the domain.
- Some invalid arguments can be demonstrated to be invalid only by counterexamples which have infinitely many objects in their domains.
- To construct NK-proofs in LFOL we do not need to alter the quantifier rules or add extra quantifier rules. We add two new rules for the identity symbol.

- We can use the semantics for LFOL and the inference system NK to investigate the links among structural properties of binary relations. The basic structural properties are reflexivity, symmetry and transitivity.
- NK is sound and complete: its rules suffice to prove all and only the valid arguments of LFOL. But validity for LFOL is undecidable.
- LFOL is not adequate for the symbolization of all natural-language reasoning. There are some kinds of construction which figure in logical inferences which can only be accommodated in LFOL by symbolizations which depart radically from the surface structure of the English, and there are others which cannot be accommodated in LFOL at all.

PART IV

EXTENSIONS AND ALTERNATIVES TO CLASSICAL LOGIC

9 Modal Logic

1 Modal operators

In discussing the difference between truth-functional and non-truth-functional connectives in §8 of Chapter 3, we gave the example of 'it is a matter of contingent fact that' to illustrate non-truth-functionality. This one-place connective, syntactically of the same type as 'it is not the case that', is an example of a *modal* operator. Other examples are 'it is impossible that', 'it is possible that' and 'it is necessary that'. Their meanings may be explained in terms similar to those in which we explained the meaning of 'it is a matter of contingent fact that'. Thus it is possible that *p* if and only if there is a way things could have gone in which *p* is true; in particular, everything that is actually true is possible (something could hardly be actually true and at the same time impossible!) and in addition everything which is not necessarily false is possible. So it is possible that Margaret Thatcher leads the British Conservative party in four consecutive election campaigns, since there is a way things could have gone in which she holds the party leadership through four consecutive elections. Correspondingly, something is necessary if and only if it holds in every way things could have gone. Mathematical truths are the standard examples of necessary truths; for example, there is no way things could have gone in which 2 + 2 is anything other than 4. In modal logic, ways things could have gone are called *possible worlds*, the 'world' emphasizing that we are interested in *total* courses of events. This terminology traces to the philosopher and mathematician Leibniz (1646–1716) who was one of the codiscoverers of calculus.

These senses of 'it is possible that' and 'it is necessary that' are sometimes called the *broadly logical* or *metaphysical* senses, by contrast with the *epistemic* senses. In the epistemic sense, 'it is possible that' means something like 'it does not contradict anything we know that'. So while it is metaphysically possible that Margaret Thatcher leads the British Conservative party in four consecutive election campaigns, it is not epistemically possible, since we know that she lost the party leadership after three. In this chapter, 'it is possible' is always understood metaphysically.

In modal logic, 'it is possible that' and 'it is necessary that' are usually taken to be the primitive operators. 'It is possible that' is symbolized by '◇' and 'it is necessary that' by '□'. In fact, it suffices to take just one of these two as primitive, since the other can be defined using it and negation (just as either of '∃' and '∀' can be taken as primitive and used to define the other). For example, if a statement is possibly true, then its negation is not necessarily true. So if we have '□' as primitive, we can define '◇' by:

(Df. ◇): ◇p ≝ ~□~p.

But equally, if we have '◇' as primitive, then since a statement *p* is necessarily true if and only if it is not possible that its negation ⌜~p⌝ is true, we can define '□' by:

(Df. □): □p ≝ ~◇~p.

However, we will take both operators as primitive and reflect these interdefinabilities in interderivabilities. Thus we form the language of sentential modal logic, LSML, by adding '□' and '◇' to the lexicon of LSL, subject to the formation rule

(*f-m*): If *p* is a wff of LSML, then so are ⌜□p⌝ and ⌜◇p⌝.

Thus '□A → ◇(B & C)', '□□A → ◇□◇A' and '□(A ↔ ~◇(B & ~C))' are examples of LSML wffs; note that like (*f-~*), (*f-m*) does not introduce parentheses, and is subject to an analogous scope convention, in that it takes smallest possible scope. Hence the string '◇A & B' is understood as '(◇A) & B', not '◇(A & B)'.

2 A definition of semantic consequence for sentential modal logic

Many of the central questions of philosophy involve modal concepts crucially. Perhaps the best-known example is the problem of free will. When a person acts in a certain way, we can raise the question whether, given all the conditions leading up to her action, it was possible for her to have acted differently, and if not, whether this shows that the sense we sometimes have of acting freely, from our own choice, is illusory. Nowadays, intelligent discussion of this issue and others like it requires, among other things, a grasp of the logic of possibility and necessity, which is what modal logic seeks to provide.

We begin by extending the semantics for LSL to LSML, which demands that we extend our notion of an interpretation to accommodate the non-truth-functional operators '□' and '◇'. What is the new notion of interpretation to be, and what evaluation clauses should '□' and '◇' receive? The evaluation clauses we use treat '□' and '◇' as quantifiers over possible worlds: '□' means 'in *all* possible worlds' while '◇' means 'in *some* possible world'. So, for example, the formula '◇(A & B)' will be true if there is some possible world where 'A & B' is true. To evaluate a sentence such as '◇(A & B)' is to evaluate it for truth at the actual world. But the presence of '◇' in '◇(A & B)' means that we need to know the truth-values of sentence-letters at *every* world, so as to know whether or not there *is* a world where 'A & B' is true. Thus an interpretation has to say what worlds there are, which one is the actual world, and what the truth-values of the sentence-letters of the language are at each world. Corresponding to the idea that a possible world is a *complete* alternative way things could have been, each world has to assign a truth-value to *every* sentence-letter of the appropriate language. Indeed, though this is slightly unusual, it is well motivated to *identify* a world with the assignment it makes to sentence-letters. In most treatments of modal logic, the worlds are treated as structureless points and an assignment to sentence-letters is associated with each world. This makes good sense if we interpret worlds as instants of time rather than ways things could have been, since we can distinguish between a time and what happens at it (what truth-values it assigns to sentence-letters); for example, some things which happened in 1707 might not have happened at all, so the year 1707 is one thing and the course of events which unfolds in it another. But there is no analogous distinction for worlds, between, say, the actual world and what happens at it: what happens at a world is *constitutive* of that world's identity, so that if things had gone differently, another world would have been actual in place of this one. We therefore make the following definition:

> An *interpretation* of an LSML sequent is a set W of possible worlds; each $w \in W$ is an assignment of truth-values to the sentence-letters which occur in the sequent, every w assigning $\perp$ to 'λ'; and one $w \in W$ is designated the actual world, conventionally written 'w^*'.

As usual, an argument's validity turns on whether it has an interpretation that makes its premises true and conclusion false. That is,

> For any formulae $p_1,...,p_n$ and q of LSML, $p_1,...,p_n \vDash q$ if and only if there is no interpretation W of the sentence-letters in $\{p_1,...,p_n,q\}$ under which $p_1,...,p_n$ are all true and q is false.

To apply this, we need to know how to decide whether or not a formula is true in an interpretation of an LSML argument-form. By what was said above, we identify truth with truth in the actual world and the operators with quantifiers over possible worlds. However, there is one further complication. When evaluating a formula for truth at the actual world, we have seen that the occurrence of a modal operator in it can require us to evaluate a subformula at some nonactual world. So the general notion of truth we must work with is that of truth at an arbitrary world, actual truth being a special case of this. The definitions are:

(*Eval At*): A sentence-letter π is true (false) in an interpretation W at a world $w \in W$ if and only if w assigns π the value $\top$ ($\perp$).

(*Eval □*): A formula of the form $\ulcorner \Box p \urcorner$ is true in an interpretation W at a world $w \in W$ if and only if, for all worlds $u \in W$, p is true at u.

(*Eval* ◇): A formula of the form ⌜◇p⌝ is true in an interpretation W at a world w ∈ W if and only if there is at least one world u ∈ W such that p is true at u.

(*Truth*): A formula p of LSML is true in an interpretation W if and only if p is true at w* ∈ W.

This completes the explanation of all the terms in the definition of semantic consequence, $p_1,...,p_n \vDash q$. We could call the system of modal logic determined by these definitions *standard* modal logic. However, systems of modal logic, of which this is only one (though perhaps the most natural), have traditional names bestowed on them by, or after, the modern originators of the subject, and our standard system is called *S5* in this nomenclature. Though we have no intention of discussing S0–S4, we will use the traditional name for our system.

To illustrate the apparatus, we establish some simple invalidities of S5. We describe an interpretation by listing its worlds and specifying what truth-value assignments each world makes. We also depict the interpretations in an obvious way. Here are some useful abbreviations: when a world w assigns ⊤ (⊥) to a sentence-letter π, we write, '*w: π ↦ ⊤*' ('*w: π ↦ ⊥*'); and when a formula p is true (false) at a world w, we write '*w[p] = ⊤*' ('= ⊥'). Rewriting our four clauses with such abbreviations produces:

(*EA*): ∀w ∈ W, w[π] = ⊤ if and only if w: π ↦ ⊤;
(*E□*): ∀w ∈ W, w[□p] = ⊤ if and only if (∀u ∈ W) u[p] = ⊤;
(*E◇*): ∀w ∈ W, w[◇p] = ⊤ if and only if (∃u ∈ W) u[p] = ⊤;
(*Tr*): W[p] = ⊤ if and only if $w*_w[p]$ = ⊤.

The clause (*Tr*) says that an interpretation W makes the formula p true if and only if p is true at the actual world w* of W.

Example 1: Show ◇A & ◇B ⊭ ◇(A & B).

Intuitively, 'A' and 'B' might be contrary statements (statements which cannot both be true), neither of which is impossible by itself. But because they are contrary, their *conjunction* is impossible—there is no one world where both conjuncts are true. Formally, an interpretation embodying this is W = {w*, u}, w*: A ↦ ⊥, B ↦ ⊤, u: A ↦ ⊤, B ↦ ⊥, or as pictured below:

$$\overset{\bullet}{w*} \qquad \overset{\bullet}{u}$$

A ↦ ⊥   A ↦ ⊤
B ↦ ⊤   B ↦ ⊥

Explanation: w*[◇A] = ⊤ by (*E◇*) since u[A] = ⊤ by (*EA*), and w*[◇B] = ⊤ by (*E◇*) since w*[B] = ⊤ by (*EA*). Hence w*[◇A & ◇B] = ⊤ and so by (*Tr*), we find that in this W, '◇A & ◇B' is true. However, since w*[A & B] = ⊥ and u[A & B] = ⊥, we also have w*[◇(A & B)] = ⊥ by (*E◇*), and so by (*Tr*), '◇(A & B)' is false in W.

Example 2: Show A, $\Box$(A → B) $\not\vDash$ $\Box$B.

This example is related to a philosophical view known as *fatalism*, according to which there is a certain inevitability about how the future will turn out. This inevitability is usually expressed in some such thought as 'If I am going to be killed in an air crash then I am going to be killed in an air crash, and there is no avoiding it'. It is quite likely that the superficial appeal of fatalism lies in a confusion between conditionals of the forms 'A → $\Box$A' and '$\Box$(A → A)' respectively. Is the *it* which cannot be avoided the truth of 'A' or the truth of 'A → A'? The truth of the conditional 'A → A' is of course unavoidable, since it is a simple tautology. But its unavoidability does not impute any kind of *unavoidable* truth to 'A' itself, even if 'A' is *in fact* true ('in fact' is weaker than 'unavoidable'). Our example brings out the mechanics of how this can be. *Interpretation:* W = {w*, u}, w*: A ↦ ⊤, B ↦ ⊤, u: A ↦ ⊥, B ↦ ⊥, or as pictured below:

$$
\begin{array}{cc}
\bullet & \bullet \\
w^* & u \\
A \mapsto \top & A \mapsto \bot \\
B \mapsto \top & B \mapsto \bot
\end{array}
$$

Explanation: w*[A] = ⊤ by (*EA*), and since w*[A → B] = ⊤, u[A → B] = ⊤, we have w*[$\Box$(A → B)] = ⊤. So by (*Tr*), both premises hold in W. But u[B] = ⊥, so by (*E$\Box$*), w*[$\Box$B] = ⊥. Thus '$\Box$B' is false in W.

By now the reader may have begun to sense a certain similarity between S5 and monadic predicate logic. We can bring this more into the open with the following example, which should be compared with Example 6.2.2 (page 181).

Example 3: Show ◊(A & B), ◊(A & C), $\Box$(B → ~C) $\not\vDash$ $\Box$(A ↔ (B ∨ C)).

$$
\begin{array}{ccc}
\bullet & \bullet & \bullet \\
w^* & u & v \\
A \mapsto \top & A \mapsto \top & A \mapsto \top \\
B \mapsto \bot & B \mapsto \top & B \mapsto \bot \\
C \mapsto \bot & C \mapsto \bot & C \mapsto \top
\end{array}
$$

Interpretation: W = {w*,u,v}, w*: A ↦ ⊤, B ↦ ⊥, C ↦ ⊥, u: A ↦ ⊤, B ↦ ⊤, C ↦ ⊥, v: A ↦ ⊤, B ↦ ⊥, C ↦ ⊤. *Explanation:* w*[A ↔ (B ∨ C)] = ⊥, so w*[$\Box$(A ↔ (B ∨ C))] = ⊥ as well. u[A & B] = ⊤ and v[A & C] = ⊤, therefore w*[◊(A & B)] = ⊤ and also w*[◊(A & C)] = ⊤. Finally, w*[B → ~C] = ⊤, u[B → ~C] = ⊤ and v[B → ~C] = ⊤, so w*[$\Box$(B → ~C)] = ⊤.

On a similar theme, compare the following example with the monadic predicate logic problem (∀x)Fx → (∀x)Gx $\not\vDash$ (∀x)(Fx → Gx):

Example 4: Show $\Box$A → $\Box$B $\not\vDash$ $\Box$(A → B).

Interpretation: W = {w*,u}, w*: A ↦ ⊥, B ↦ ⊤, u: A ↦ ⊤, B ↦ ⊥, or as pictured at the top of the next page:

$$\overset{\bullet}{w^*} \qquad \overset{\bullet}{u}$$

$$A \mapsto \bot \qquad A \mapsto \top$$
$$B \mapsto \top \qquad B \mapsto \bot$$

Explanation: Since $w^*[A] = \bot$, $w^*[\Box A] = \bot$, therefore $w^*[\Box A \rightarrow \Box B] = \top$. But $u[A \rightarrow B] = \bot$, so $w^*[\Box(A \rightarrow B)] = \bot$.

These examples indicate that from the point of view of the semantics, (i) '$\Box$' and '$\diamond$' behave like '$\forall$' and '$\exists$', and (ii) a sentence-letter of the modal language LSML behaves like a one-place predicate of monadic predicate logic.[1] However, for reasons which will become apparent in the next section, iteration of modal operators is less straightforwardly comparable. One example of an invalidity involving iterated operators is the following:

Example 5: Show $\Box\diamond A \not\models \diamond\Box A$.

Interpretation: $W = \{w^*, u\}$, w^*: $A \mapsto \bot$, u: $A \mapsto \top$, or as pictured below:

$$\overset{\bullet}{w^*} \qquad \overset{\bullet}{u}$$

$$A \mapsto \bot \qquad A \mapsto \top$$

Explanation: since $u[A] = \top$, by $(E\diamond)$ $u[\diamond A] = \top$ and $w^*[\diamond A] = \top$. So '$\diamond A$' is true at all worlds, and hence by $(E\Box)$, $w^*[\Box\diamond A] = \top$. But since $w^*[A] = \bot$, then by $(E\Box)$, both $w^*[\Box A] = \bot$ and $u[\Box A] = \bot$. Hence by $(E\diamond)$, $w^*[\diamond\Box A] = \bot$. Notice how the iteration of operators in the premise and conclusion requires the iterated application of the Eval rules to explain the truth-values of the premise and the conclusion.

We identify a modal system with its valid sequents, and to indicate that we are talking about the semantic-consequence relation defined in this section, we write '$p_1,...,p_n \models_{S5} q$' rather than just '$p_1,...,p_n \models q$'. Thus the system S5 is the collection of all semantically correct sequents $p_1,...,p_n \models_{S5} q$, where $p_1,...,p_n$ and q are sentences of LSML. This covers only sequents with finitely many premises, but it is true of the best-known modal systems S that if $\Sigma \models_S q$, then there is a finite subset $\{p_1,...,p_n\}$ of Σ such that $p_1,...,p_n \models_S q$. So identifying a system with its finite sequents, though incorrect in general, is adequate for our purposes.

---

[1] In predicate logic, one thinks of the predicate-letters as standing for properties of objects. An analogous viewpoint about sentence-letters in modal logic is that they stand for properties of worlds.

❑ Exercises

Show the following, with explanations:

(1) ◇~A ⊭$_{S5}$ ~◇A
(2) □(A ∨ B) ⊭$_{S5}$ □A ∨ □B
(3) ~(A → □B) ⊭$_{S5}$ □(A → ~B)
(4) A ↔ □B ⊭$_{S5}$ □(A ↔ B)
(5) ~◇A ∨ ~◇B ⊭$_{S5}$ ~◇(A ∨ B)
*(6) ◇(A → B) ⊭$_{S5}$ ◇A → ◇B
(7) ◇(A ↔ B) ⊭$_{S5}$ ◇(A ∨ B)
(8) □(A → ~B), □(B → C) ⊭$_{S5}$ □(A → ~C)
(9) ◇(A → ⅄), ◇(⅄ → A) ⊭$_{S5}$ □(A ↔ ⅄)
(10) □(A → B) ⊭$_{S5}$ ~□(A → ~B)
*(11) □A → ◇B ⊭$_{S5}$ □(A → B)
(12) □(A → B) → □(C → D) ⊭$_{S5}$ ◇(A & B) → □(C → D)
(13) ◇(A → □B) ⊭$_{S5}$ ◇A → □B
(14) □(A → ◇B) ⊭$_{S5}$ ◇(A → □B)
(15) □◇(A → B) ⊭$_{S5}$ □(◇A → B)
(16) ◇□(A → B) ⊭$_{S5}$ ◇A → □B
(17) □(◇A → ◇B) ⊭$_{S5}$ □◇A → ◇□B
(18) ◇(□A ∨ ◇B) ⊭$_{S5}$ ~◇A → □B

3 The canonical translation

In this section we will make the analogy between the modal language LSML and the language of monadic predicate logic more precise. We do this by explaining how to translate modal formulae into monadic formulae. However, the language we translate into is not LMPL as defined in §4 of Chapter Five, since translation into this language would obscure certain important features of the modal language. Rather, we introduce a new monadic language, the Language of Canonical Translation (LCT), as the language into which LSML is translated. LCT differs from LMPL in certain ways. For each sentence-letter π of LSML define the corresponding monadic predicate-letter λ_π to be the result of writing π followed by the prime symbol "'". For instance, the sentence-letter 'A' of LSML corresponds to the monadic predicate 'A'' (strictly, we should use subscripts to differentiate the monadic predicate 'A'' of LCT from the sentence-letter 'A'' of LSML, but instead we avoid using primed sentence-letters in LSML).

The lexicon of LCT:

One individual variable 'w'; no individual constants; a sentence-letter '⅄'; for each sentence-letter π of LSML except '⅄', the corresponding monadic predicate-letter λ_π; sentential connectives, quantifier symbols '∀' and '∃', and parentheses.

The formation rules of LCT:

(*f-at*): 'ʌ' is an atomic wff, and for any predicate-letter λ, ⌜λw⌝ is an atomic wff.

(*f-con*): If ϕ and ψ are wffs (atomic or otherwise) so are ⌜~ϕ⌝, ⌜(ϕ & ψ)⌝, ⌜(ϕ ∨ ψ)⌝, ⌜(ϕ → ψ)⌝ and ⌜(ϕ ↔ ψ)⌝.

(*f-q*): If ϕ is a wff, then ⌜(∃w)ϕ⌝ and ⌜(∀w)ϕ⌝ are wffs.

(*f!*): Nothing is a wff unless it is certified as such by the previous rules.

There are therefore some significant differences between LCT and the standard language of monadic predicate logic LMPL, in particular the following three: (a) LCT has only one sentence-letter, 'ʌ'; (b) LCT has only one individual variable and no individual constants; (c) in the syntax clause for the quantifiers, there is no requirement that ϕ has any free occurrences of 'w', so redundant quantification is permitted, and there is no requirement that ϕ has no bound occurrences of 'w', so double-binding is permitted. The motivation for these points of difference will be apparent when the method of translating from LSML into LCT is set out.

The goal of translation is to match an expression of one language with an 'equivalent' expression of another. So we have to decide what equivalence should mean in this context and how the translation should go. We begin with the second task. The idea is to describe a procedure for step-by-step transcription of an LSML wff into an LCT wff, a procedure that processes the LSML wff connective by connective, working inward from the main connective. It may seem that there will be a problem: modal operators are to be rendered as quantifiers, and LSML formulae can contain more than one modal operator; but there is only one variable in LCT to which to attach quantifiers. How then is the meaning of the LSML formula to be preserved? We can see how this difficulty is avoided by inspecting some target translations. For example, the translation of '◇(A → □B)' is '(∃w)(A′w → (∀w)B′w)', in which the last occurrence of 'w' is within the scope of two quantifiers corresponding to the '◇' and '□' of the LSML formula. However, the effect of the semantics for LCT will be to ensure that it is the *closest* quantifier which binds the last occurrence of 'w', just as, in the modal formula, it is the closest modal operator that determines the world of evaluation for what is within its scope. A similar problem arises with LSML formulae which have subformulae not within the scope of any modal operator, such as the antecedent of 'A → □B'. In the semantics of the previous section, such subformulae are interpreted as making a statement about w^*, the actual world. But there is no 'w^*' in the lexicon of LCT. However, the translation of 'A → □B' will be 'A′w → (∀w)B′w', in which the first occurrence of 'w' is free. We will arrange that free occurrences of 'w' in LCT formulae are interpreted as standing for the actual world. These two points, about double-binding and free occurrences of 'w', bring out why it is that one variable is all we require for translation of LSML.

The translation procedure is given by a set of rules, one for each of the connectives occurring as the main connective of the expression to be translated. The rules are:

(*Trans-at*): (a) *Trans*[⅄] = '⅄'; (b) if π is a sentence-letter of LSML other than '⅄', then *Trans*[π] = ⌜λ_πw⌝, where λ_π is the primed predicate corresponding to the sentence-letter π;

(*Trans-~*): *Trans*[~φ] = ⌜~*Trans*[φ]⌝;

(*Trans-&*): *Trans*[φ & ψ] = ⌜(*Trans*[φ]) & (*Trans*[ψ])⌝;

(*Trans-∨*): *Trans*[φ ∨ ψ] = ⌜(*Trans*[φ]) ∨ (*Trans*[ψ])⌝;

(*Trans-→*): *Trans*[φ → ψ] = ⌜(*Trans*[φ]) → (*Trans*[ψ])⌝;

(*Trans-↔*): *Trans*[φ ↔ ψ] = ⌜(*Trans*[φ]) ↔ (*Trans*[ψ])⌝;

(*Trans-□*): *Trans*[□φ] = ⌜(∀w)*Trans*[φ]⌝;

(*Trans-◇*): *Trans*[◇φ] = ⌜(∃w)*Trans*[φ]⌝.

We can verify each sample translation given above by displaying the *transformation sequence* of the relevant LSML formula, that is, the sequence of applications of *Trans* rules which produces the LCT translation. For '◇(A → □B)' we have:

Trans[◇(A → □B)]	= (∃w)*Trans*[(A → □B)]	(*Trans-◇*)
	= (∃w)(*Trans*[A] → *Trans*[□B])	(*Trans-→*)
	= (∃w)(A'w → *Trans*[□B])	(*Trans-at*)
	= (∃w)(A'w → (∀w)*Trans*[B])	(*Trans-□*)
	= (∃w)(A'w → (∀w)B'w)	(*Trans-at*)

and

Trans[A → ◇B]	= *Trans*[A] → *Trans*[◇B]	(*Trans-→*)
	= A'w → *Trans*[◇B]	(*Trans-at*)
	= A'w → (∃w)*Trans*[B]	(*Trans-◇*)
	= A'w → (∃w)B'w	(*Trans-at*)

Strictly speaking, corners should be employed at all intermediate stages of the translation of a formula. For instance, at the penultimate line of the second translation, we should have ''⌜A'w → (∃w)*Trans*[B])⌝'' abbreviating "the result of writing 'A'w → (∃w)' followed by the translation of 'B'". But we reduce clutter by not displaying the corners.

Every sentence of LSML has a translation into LCT, and it is not difficult to see that every sentence of LCT is the translation of an LSML sentence. Thus the translation scheme determines a one-one correspondence between the sentences of the two languages, which is why LCT is called the *canonical* language for the translation. If we add just one more individual variable, say 'u', to LCT, there will then be sentences which do not translate any LSML sentence, for instance '(∃u)(∀w)(A'w → B'u)'. This formula is not the *Trans*-translation of any LSML formula; in particular, it is not the translation of '◇□(A → B)', for there is no analogue of binding by which the '◇' can be related to the 'B'. In LSML, the operator that determines world of evaluation is always the closest, but when we have two variables in the monadic language, the semantically relevant quantifier need not be the closest, as the example '(∃u)(∀w)(A'w → B'u)' illustrates. The same point arises if we add an individual constant 'w*' to LCT, for then we have such LCT formulae as '$(\forall w)(A'w \rightarrow B'w^*)$', which cannot be the *Trans*-translations of any LSML formulae; for '$B'w^*$' is within the scope of '$(\forall w)$', yet the quantifier does not affect the meaning of '$B'w^*$', while by contrast the closest modal operator within whose scope a sentence-letter lies always determines the world of evaluation for it; thus '$\Box(A \rightarrow B)$' is not the *Trans*-translation of '$(\forall w)(A'w \rightarrow B'w^*)$'. However, the existence of sentences in these expansions of LCT which do not translate any LSML sentences is consistent with its being the case that for each such sentence there is a logically equivalent sentence which *is* the translation of some LSML sentence. This amounts to its being the case that for each sentence in either of these LCT expansions there is a logically equivalent sentence in LCT. In fact this is so, though we will not pursue the matter at this point except to mention that '$(\forall w)(A'w \rightarrow B'w^*)$' is logically equivalent to '$(\exists w)A'w \rightarrow B'w$', which is the translation of '$\Diamond A \rightarrow B$'.

It seems intuitively clear that the translation scheme is correct, but we can be precise about what we mean by 'correct'. If W is an interpretation of sentences of LSML, define I_W, the monadic interpretation corresponding to W, as the interpretation I whose domain D is W and which, for each sentence-letter π assigned a truth-value by the members of W, assigns to the corresponding predicate-letter λ_π the extension consisting in exactly those $w \in W$ such that $w(\pi) = \top$.

Example 1: Given the LSML interpretation $W = \{w^*, u\}$, w^*: A $\mapsto \bot$, B $\mapsto \top$, u: A $\mapsto \top$, B $\mapsto \bot$, the corresponding monadic interpretation I_W is D = $\{w^*, u\}$, Ext(A') = $\{u\}$ and Ext(B') = $\{w^*\}$.

We have the convention that we do not ever use the variable 'w' in the metalanguage to specify W or the domain D of I_W, which allows us to make the following stipulation:

- For every LSML interpretation W, if I_W is the corresponding monadic interpretation, then Ref(w) in I_W is w^*_w.

In evaluating LCT-formulae, the rules for the quantifiers are the same as those in the semantics for LMPL in §1 of Chapter 6: existential formulae are true if and only if they have at least one true instance, and universal formulae are true if and only if they have only true instances. However, there is a problem about forming the instances. For example, if D = $\{w^*, u, v\}$ and we want to evaluate the formula '$A'w \rightarrow (\exists w)B'w$', then if Ref(w) (i.e., w^*) is in Ext(A'), the conditional is true if and only if at least one of three representative instances of '$(\exists w)B'w$' is true. But we have no names in LCT with which to form instances; LCT has only the individual variable 'w'. But this is not a real problem, since for each collection of terms $\{t_1,\ldots,t_n,\ldots\}$ used in the metalanguage to list the worlds of an LSML interpretation W, we can form an *expansion* of the object language LCT, LCT+, which is the same as LCT except that we add corresponding terms $\{s_1,\ldots,s_n,\ldots\}$ to the lexicon and adjust the clause (*f-at*) of the syntax accordingly. Applying the evaluation rules for the quantifiers to quantified sentences of LCT, the instances in a particular case are understood to be the instances in the expanded language LCT⁺ for that case. We need some conventional method of associating metalanguage terms with object language ones, and to avoid further notational complexity we will simply use the same letters in object language expansions as in the metalanguage. Styling as roman or italic will serve to keep the manners of occurrence distinguished.

Example 2: Using the LSML interpretation of Example 1, $W = \{w^*, u\}$, $w^*: A \mapsto \bot$, $B \mapsto \top$, $u: A \mapsto \top$, $B \mapsto \bot$, we evaluate *Trans*[□(◇A → B)] in the corresponding LCT⁺ interpretation $\mathcal{I}_W$, which has D = $\{w^*, u\}$, Ext(A′) = $\{u\}$ and Ext(B′) = $\{w^*\}$. *Trans*[□(◇A → B)] = '(∀w)((∃w)A′w → B′w)', which is true if and only if its instances (i) '(∃w)A′w → B′w*' and (ii) '(∃w)A′w → B′u' are both true. But (ii) is false, since '(∃w)A′w' is true (because 'A′u' is true since $u \in$ Ext(A′)) and 'B′u' is false, because $u \notin$ Ext(B′). Hence *Trans*[□(◇A → B)] is false in $\mathcal{I}_W$, and it can easily be verified that '□(◇A → B)' is itself false in W.

Example 2 illustrates the sense in which *Trans* associates each LSML formula with an 'equivalent' formula of LCT. If ϕ is a formula of LSMT and ψ is a formula of LCT, say that ϕ and ψ are *semantically equivalent* if and only if for every LSML interpretation W, $W[\phi] = \top$ if and only if ψ is true in $\mathcal{I}_W$. Inspection of the translation scheme *Trans* should be sufficient to convince the reader that for any formula ϕ of LSMT, *Trans*[ϕ] is semantically equivalent to ϕ.

The association between LCT and LSML allows us to transfer certain results about LCT over to LSML. Of these, the main one available at this point is decidability. At the end of §6 of Chapter 6, we remarked that monadic predicate logic is known to be decidable (6.7 contains a decision procedure). Since this result is independent of whether redundant quantification and double-binding are allowed in the language, and since the semantically correct sequents of LCT are a subset of those of a version of monadic predicate logic which allows redundant quantification and double-binding, it follows that this subsystem is decidable as well. Hence S5 is also decidable, by translation into LCT.

❏ Exercises

I   Give the LCT-translations of the following sentences, exhibiting the step-by-step transformation sequences you use:

(1) □(A ∨ B)   (2) □A ∨ □B
(3) □□A & ~◇◇~A   *(4) □(A → ◇⅄)
(5) ◇(A → □⅄)

II   Recover the sentences of LSML which are translated by the following LCT formulae.

(1) (∀w)(A′w ↔ (∃w)(B′w ∨ ⅄))   *(2) (∃w)(∀w)A′w → A′w
(3) (∀w)((∃w)A′w ∨ (∃w)B′w)   (4) A′w → ~(∃w)~(∀w)⅄
(5) (∀w)(⅄ → (∃w)(A′w & ⅄))

4 Natural deduction in S5

We are expounding modal logic in the same sequence as the various systems of classical logic already covered: first the syntax of the language, then the definition of semantic consequence and finally the definition of deductive consequence. So we now want to formulate introduction and elimination rules for the modal operators, and having seen the canonical translation of LSML, we have some idea of what these rules should be like.

In fact, the rules of □E and ◇I are obvious in their own rights. If a statement is necessary, then it is true, and if it is true, then it is possible. So we have these two rules:

> *Rule of □-Elimination:* For any sentence $\ulcorner \Box p \urcorner$, if $\ulcorner \Box p \urcorner$ occurs at a line j in a proof, then at line k we may infer p, labeling the line 'j □E' and writing on its left the same numbers as appear on the left at line j. Schematically,

$$a_1,...,a_n \quad (j) \qquad \Box p$$
$$\vdots$$
$$a_1,...,a_n \quad (k) \qquad p \qquad\qquad\qquad j\ \Box E$$

> *Rule of ◇-Introduction:* For any sentence p, if p occurs at a line j in a proof, then at line k we may infer $\ulcorner \Diamond p \urcorner$, labeling the line 'j ◇I' and writing on its left the same numbers as occur on the left of line j. Schematically,

$$a_1,...,a_n \quad (j) \qquad p$$
$$\vdots$$
$$a_1,...,a_n \quad (k) \qquad \Diamond p \qquad\qquad\qquad j\ \Diamond I$$

Adding these two rules and the other rules to come in this section, we extend the system NK of classical natural deduction to the modal logical system S5. With just the two rules already given, we can prove such simple S5 sequents as the following:

Example 1: Show $\Box(A\ \&\ B) \vdash_{S5} \Diamond B$.

1	(1)	□(A & B)	Premise
1	(2)	A & B	1 □E
1	(3)	B	2 &E
1	(4)	◇B	3 ◇I ◆

But we cannot prove anything very interesting with just these two rules. We need at least a rule of □I, and here we face a problem comparable to the one that arose over ∀I in monadic predicate logic. The issue there was how we could justify moving from a statement that such and such an object satisfies a certain condition to the statement that every object satisfies it. In the present context the question is how, having established that something is actually the case (is true at the actual world), we are entitled to conclude that the same thing is necessary (is true at every world). In the case of predicate logic, the solution is to allow the introduction of the universal quantifier when and only when the statement about the particular object (the instance) has been derived by steps which would suffice for the derivation of any other instance of the target universal sentence depending on the same premises and assumptions. A simple syntactic criterion is available for when a particular proof meets this condition: the relevant individual constant in the instance which has been derived must not occur in any of the premises or assumptions on which the instance depends. However, this idea cannot be straightforwardly extended to S5, since in LSML there is no analogue of the individual constants which appear in the LMPL sentences to which ∀I is applied and so there is no direct analogue of the idea of 'any other instance'.

What we have to consider is whether, if things had been different in any way, the premises and undischarged assumptions used in deriving the statement p to which we wish to apply □I to obtain $\ulcorner \Box p \urcorner$ would still have held. If those premises and assumptions would still have held, then since their consequences would also be unchanged, p would continue to hold. Thus, as things are, the step to $\ulcorner \Box p \urcorner$ is justifiable. Statements which would hold no matter how things might have varied are necessary. So to put it briefly, p is necessary if p depends only on necessary truths. Hence we can give a syntactic criterion for □I if we can say when a premise or assumption is necessary. It would not be correct to say that a premise is necessary if and only if it has '□' as main connective. We have seen in the semantics that whenever '◇A' is true at one world it is true at all; and similarly, when such a formula as '□A & □B' is true at one world it is true at all. So something proved on the basis of premises and assumptions like these, *implicitly* necessary premises, should also permit □I.

A syntactic criterion which picks out the right class can be gleaned by considering the LCT translations of, say, '◇A' and '□A & □B', which are respectively '(∃w)A′w' and '(∀w)A′w & (∀w)B′w'. These two formulae contain no free occurrences of 'w', and consequently, prefixing them with '∀w' is to prefix them with a redundant quantifier, so that their truth-values remain unchanged. In general, if p is a wff of LCT in which every occurrence of 'w' is within the scope of a quantifier, then prefixing p with '∀w' is redundant in not merely the syntactic sense, but in the semantic sense that it cannot change p's truth-value. If p is a wff of LSML which says something necessary, then prefixing p with '□' should also have no effect. Since the redundancy of prefixing a quantifier to ϕ in LCT is guaranteed if every occurrence of 'w' in ϕ is already within the scope of a quantifier, redundancy of prefixing a modal operator in LSML is guaranteed by the modal analogue of this condition. A little reflection on the canonical translation indicates that the analogous condition is that every sentence-letter in p other than 'λ' should occur within the scope of a '□' or '◇', since it is in translating sentence-letters that occurrences of 'w' are introduced, and a sentence-letter in p not within the scope of a modal operator will produce a free occurrence of 'w' in p's canonical translation. Hence the correct restriction on an application of □I to a formula p at a line j is that at j, p should depend on no premise or assumption in which there is a sentence-letter other than '⋏' that is not within the scope of a modal operator, '□' or '◇'.

An LSML-wff in which every occurrence of a sentence-letter other than '⋏' is within the scope of a modal operator is said to be *fully modalized*. We can now state the rule of □I as follows:

> **Rule of □-Introduction:** For any sentence p, if p occurs at a line j in a proof, then at line k we may infer $\ulcorner \Box p \urcorner$, labeling the line 'j □I' and writing on its left the same numbers as occur on the left of line j, provided that the premises and assumptions corresponding to these numbers are all fully modalized. Schematically,

$$a_1,...,a_n \quad \text{(j)} \qquad p$$
$$\vdots$$
$$a_1,...,a_n \quad \text{(k)} \qquad \Box p \qquad\qquad \text{j □I}$$

where the formulae at lines $a_1,...,a_n$ are fully modalized.

With this rule we can prove a wider variety of sequents, such as the following:

Example 2: Show $\Box(A \to B) \vdash_{S5} \Box A \to \Box B$.

1	(1)	□(A → B)	Premise
2	(2)	□A	Assumption
1	(3)	A → B	1 □E
2	(4)	A	2 □E
1,2	(5)	B	3,4 →E
1,2	(6)	□B	5 □I
1	(7)	□A → □B	2,6 →I ◆

Recalling the point that '◇' can be regarded as an abbreviation of '~□~' or '□' as an abbreviation of '~◇~', we would expect to be able to show two interderivabilities analogous to QS, which we will label MS (modality shift):

MS: $\sim\!\Diamond A \dashv\vdash_{S5} \Box\!\sim\!A$
    $\sim\!\Box A \dashv\vdash_{S5} \Diamond\!\sim\!A$

Here is the left-to-right direction of $\sim\!\Diamond A \dashv\vdash_{S5} \Box\!\sim\!A$:

Example 3: Show $\sim\!\Diamond A \vdash_{S5} \Box\!\sim\!A$.

1	(1)	~◇A	Premise
2	(2)	A	Assumption
2	(3)	◇A	2 ◇I
1,2	(4)	⋏	1,3 ~E
1	(5)	~A	2,4 ~I
1	(6)	□~A	5 □I ◆

□I at (6) is legitimate since (5) depends only on (1) and (1) is fully modalized, since its only sentence-letter, 'A', is within the scope of the '◇'.

The remaining rule for the modal operators is ◇E. We use the canonical translation from LSML into LCT, the analogy between '◇' and '∃', and considerations about ∃E in LCT deductions to arrive at the form of the rule. In LMPL one may infer a formula q from an existential sentence $\ulcorner(\exists v)\phi v\urcorner$ if q has been derived from an instance ϕt, provided there is no occurrence of the term t (i) in $\ulcorner(\exists v)\phi v\urcorner$, or (ii) in q or (iii) in any assumption or premise used to derive q from ϕt. In LCT we only have the term 'w', which functions in unbound occurrences as an individual constant for the actual world. So in forming an instance of an existential sentence, we must perforce use 'w' again if we are to remain within LCT. Suppose, for example, that $\ulcorner(\exists v)\phi v\urcorner$ is '(∃w)(A'w → (∃w)B'w)'. Then we have no option but to assume the instance 'A'w → (∃w)B'w'. So if we are to infer a formula q from this and discharge the assumption justifiably, we must ensure that the derivation of q does not exploit the fact that the same term occurs free in the assumption as occurs in any premises or other assumptions there may be which have a free variable. Moreover, we may subsequently in the proof obtain '(∃w)B'w' from 'A'w → (∃w)B'w' by →E and have to assume an instance of it with a view to using ∃E again. As before, the instance has to be 'B'w', so if we wish to derive something from it and then discharge 'B'w', our reasoning must not exploit the fact that there are free occurrences of the same term in earlier premises or assumptions that have a free variable. Of course, we could avoid this difficulty by adding individual constants to LCT, but then the strict parallel with '◇' in LSML would be lost. Fortunately, there is a simple restriction that keeps applications of ∃E in LCT justifiable in situations like these: if we have proved q from $\ulcorner\phi w\urcorner$ then we may only discharge $\ulcorner\phi w\urcorner$ and assert (by adjusting premise and assumption numbers on the left) that q follows from $\ulcorner(\exists w)\phi w\urcorner$ if none of the other premises or assumptions used in deriving q contain free occurrences of 'w'. This is essentially restriction (iii) as listed above in the standard ∃E rule. As for analogues of restrictions (i) and (ii), (i) lapses since there cannot be a free occurrence of 'w' in the existential formula: the main connective of this formula is '(∃w)', and so all subsequent occurrences of 'w' are bound either by this quantifier or by ones within its scope, as in our example '(∃w)(A'w → (∃w)B'w)'. But we still need a version of (ii): q should not be allowed to contain free occurrences of 'w' since on discharging the assumption we would be left with a statement specifically about the actual world, while it may be that nothing follows from our premises about any *particular* world.

As we already remarked, the condition that an LCT formula has no free occurrences of 'w' is matched in LSML by the condition that an LSML formula has all occurrences of sentence-letters other than 'ʌ' within the scope of a modal operator, that is, the condition that the LSML formula is fully modalized. Thus the rule of ◇E should say that q may be inferred from $\ulcorner\diamond p\urcorner$ if q has been derived from p, *provided* q is fully modalized and all other premises and assumptions used to obtain it are fully modalized. Intuitively, this corresponds to the idea that we can infer q from the mere fact that p is possible only if we restrict ourselves to ancillary premises and assumptions that would be the case no matter how things are (premises and assumptions about what is necessary, possible and impossible), and if q itself would be the case, given p, no matter how things might be in other respects.

> **Rule of ◇-Elimination:** If a sentence $\ulcorner ◇p \urcorner$ occurs at a line i and depends on lines $a_1,\ldots,a_n$, and p is assumed at a line j, then if q is inferred at line k depending on lines $b_1,\ldots,b_u$, then at line m we may infer q, labeling the line 'i,j,k ◇E' and writing on its left the line numbers X = $\{a_1,\ldots,a_n\} \cup \{b_1,\ldots,b_u\}/j$, *provided* (i) all formulae at lines $b_1,\ldots,b_u$ other than j are fully modalized and (ii) q is fully modalized. Schematically,

$$
\begin{array}{llll}
a_1,\ldots,a_n & \text{(i)} & ◇p & \\
& & \vdots & \\
j & \text{(j)} & p & \text{Assumption} \\
& & \vdots & \\
b_1,\ldots,b_u & \text{(k)} & q & \\
& & \vdots & \\
X & \text{(m)} & q & \text{i,j,k ◇E}
\end{array}
$$

where all the formulae at lines $b_1,\ldots,b_u$ other than j are fully modalized and q is fully modalized.

As usual, we allow the case where j $\notin \{b_1,\ldots,b_u\}$, that is, the case where p is not used in the derivation of q. In that case, X = $\{a_1,\ldots,a_n\} \cup \{b_1,\ldots,b_u\}$.

As a simple illustration of ◇E we establish the following:

Example 4: Show ◇A & ◇B ⊢$_{S5}$ ◇(A & ◇B).

1	(1)	◇A & ◇B	Premise
1	(2)	◇A	1 &E
3	(3)	A	Assumption
1	(4)	◇B	1 &E
1,3	(5)	A & ◇B	3,4 &I
1,3	(6)	◇(A & ◇B)	5 ◇I
1	(7)	◇(A & ◇B)	2,3,6 ◇E ◆

The target formula at (6) is fully modalized, and apart from (3) the only other premise or assumption on which (6) depends is (1), which is also fully modalized. Hence the application of ◇E at (7) is correct.

We can also prove the converse of Example 3, completing the proof of one of the MS interderivabilities (the other is an exercise).

Example 5: Show □~A ⊢$_{S5}$ ~◇A.

1	(1)	□~A	Premise
2	(2)	◇A	Assumption
3	(3)	A	Assumption

1	(4)	~A	1 □E
1,3	(5)	⅄	4,3 ~E
1,2	(6)	⅄	2,3,5 ◇E
1	(7)	~◇A	2,6 ~I ◆

Our treatment of '⅄' as fully modalized in itself means that we take what is absurd to be necessarily absurd.

Constructing certain proofs in S5 has a slightly different flavor from constructing them in nonmodal logic. In nonmodal proofs one proceeds, very generally, by applying elimination rules to premises and assumptions to obtain certain subformulae of these premises and assumptions, and then applying introduction rules to these subformulae to obtain the conclusion. However, the 'full modalization' condition on the rules of □I and ◇E can block the corresponding way of proceeding in modal logic. For instance, suppose we have the following problem:

Example 6: Show A ⊢$_{S5}$ □◇A.

The following is an *incorrect* proof of this sequent:

1	(1)	A	Premise
1	(2)	◇A	1 ◇I
1	(3)	□◇A	2 □I

Line 3 is wrong, because □I is applied to a formula, (2), which depends on a premise that is not fully modalized. One can get around this difficulty by using ~I, MS and DN, but it seems unsatisfactory that a proof of this sequent should involve the negation rules: intuitively, we know that if A is possible, we can infer that it is necessary that it is possible, and we also know that if A holds, then we can infer that A is possible. We would like to be able to put these two facts together in a straightforward proof. But the closest we can get to the intuitive reasoning is the following:

1	(1)	A	Premise
1	(2)	◇A	1 ◇I
3	(3)	◇A	Assumption
3	(4)	□◇A	3 □I
	(5)	◇A → □◇A	3,4 →I
1	(6)	□◇A	2,5 →E ◆

The application of □I is now legal, since the line to which it is being applied, (3), depends on an assumption (itself) which *is* fully modalized. But contriving this gives the proof the strange feature that the conditional at (5) is inferred by →I and is then used in →E. A formula which is derived by an I-rule for a connective *c* and then has the E-rule for *c* applied to its main connective (the just-introduced occurrence of *c*) is called a *maximum* formula. In nonmodal logic there is never any need to reason through a maximum formula (or to use the negation rules simply to avoid having to do this); but because of the way the restrictions on □I and ◇E work, S5 has this awkward feature.[2]

This completes our account of natural deduction in S5, except to say that we can employ Sequent Introduction with the same freedom as in nonmodal sentential logic, using exactly the same definition of 'substitution-instance' as in §8 of Chapter 4. This means in particular that we can substitute formulae with modal operators for sentence-letters in nonmodal sequents; for instance, □A ⊢$_{S5}$ ~~□A is a substitution-instance of DN⁺. The modal-shift interderivabil-ities ~◇A ⊣⊢$_{S5}$ □~A and ~□A ⊣⊢$_{S5}$ ◇~A are especially useful for SI.

In §2 and §4 of this chapter we have introduced two modal systems, one defined by a semantic-consequence relation as the collection of all semantically correct sequents $p_1,...,p_n$ ⊨$_{S5}$ q, and the other defined by a deductive-conse-quence relation as the collection of all provable sequents $p_1,...,p_n$ ⊢$_{S5}$ q. An obvious question concerns the relationship between the two consequence rela-tions, ⊨$_{S5}$ and ⊢$_{S5}$, and we state without proof the fact that they are the same (our notation would hardly be justifiable otherwise!). That is, for any LSML wffs $p_1,...,p_n$ and q, whenever $p_1,...,p_n$ ⊢$_{S5}$ q then also $p_1,...,p_n$ ⊨$_{S5}$ q, *and* vice versa. In the terminology of §11 of Chapter 4, the first part of this claim, that if $p_1,...,p_n$ ⊢$_{S5}$ q then $p_1,...,p_n$ ⊨$_{S5}$ q, states that the rules of inference which define S5 deductive consequence are *sound* with respect to the semantics which defines S5 semantic consequence. For whenever q is derivable from $p_1,...,p_n$ by the S5 rules, then q is indeed a semantic consequence of $p_1,...,p_n$ by the S5 semantics. The second part, which states that if $p_1,...,p_n$ ⊨$_{S5}$ q then $p_1,...,p_n$ ⊢$_{S5}$ q, means that the S5 rules are *complete* with respect to the semantics which defines S5, for whenever q is a semantic consequence of $p_1,...,p_n$ according to the S5 semantics, then q is derivable from $p_1,...,p_n$ by the S5 rules. The proofs of these result are given in more advanced texts.

❑ Exercises

Show the following.

(1) ◇A → □B ⊢$_{S5}$ □A → □B
(2) □B ⊢$_{S5}$ □(A → B)
(3) □~A ⊢$_{S5}$ □(A → B)
(4) ◇(A & ◇B) ⊢$_{S5}$ ◇A & ◇B
(5) ⊢$_{S5}$ ◇□(◇A → □◇A)
*(6) □(A & ◇B) ⊢$_{S5}$ ◇(A & B)
(7) □A ⊢$_{S5}$ ◇~A → B
(8) ◇A ∨ ◇B ⊢$_{S5}$ ◇(A ∨ B)
(9) □(◇A ∨ ◇B) ⊢$_{S5}$ □◇A ∨ □◇B
(10) □(A → ⋏) ⊢$_{S5}$ ~◇A
(11) ~□A ⊣⊢$_{S5}$ ◇~A (without using MS)

---

[2] A way of altering the rules to get around this is given in Prawitz, Chapter 6. A different approach is to add a rule of Cut, as in *MacLogic*. But these strategies appear to me to obscure what is really going on.

(12) A ⊢$_{S5}$ □◇A (without any maximum formulae)
(13) A → □B, □(B → C) ⊢$_{S5}$ A → □C
(14) ◇□A ⊢$_{S5}$ A
(15) ⊢$_{S5}$ ◇□A → □◇A
(16) ⊢$_{S5}$ □(□A → □B) ∨ □(□B → □A)
(17) □◇A → ◇□A ⊢$_{S5}$ ◇A → □A
(18) □◇A → ◇□A ⊢$_{S5}$ ◇A → A
*(19) □(A → □B) ⊢$_{S5}$ □(◇A → B)
(20) □(□A ∨ B) ⊢$_{S5}$ □A ∨ □B
(21) ⊢$_{S5}$ □(A → ◇B) ∨ □(B → ◇A)
(22) ◇A → ◇B ⊢$_{S5}$ ◇(A → B)

5 First-order modal logic: S5 semantics

We turn now to extending sentential modal logic by adding quantifiers, predicates and identity. From the grammatical point of view, we form the language of first-order modal logic, LFOML, simply by adding '□' and '◇' to the lexicon of LFOL and augmenting the syntax formation rules with

(f-m): If ϕ is a wff of LFOML, then so are ⌜□ϕ⌝ and ⌜◇ϕ⌝.

Many interesting new issues arise in extending S5 from sentential to first-order logic. Consider, for example, the difference between

(1) □(∀x)Mx

and

(2) (∀x)□Mx.

There is a philosophical view known as *materialism*, according to which, roughly speaking, everything is material, that is, made of matter: '(∀x)Mx'. However, two varieties of materialism may be distinguished. According to the first, it is a contingent truth that everything is made of matter: though there are in fact no irreducibly mental or spiritual entities, there could have been. According to the second variety, materialism is a necessary truth: no matter how things might have gone, everything would have been made of matter. This is the view expressed by (1), which says that in every possible world, everything is made of matter. A position according to which materialism is contingent is the negation of (1), '~□(∀x)Mx', or equivalently, '◇(∃x)~Mx'. But (2) says something different from (1). According to (2), each thing is such that necessarily it is made of matter. This is a claim with which the contingent materialist can agree, for he can hold that the various material things which in fact exist could not *themselves* have been immaterial. Rather, though there could have been immaterial entities, these would be things quite different from any that actually exist. For example, the contingent materialist may hold that God, an immaterial being, does not exist although He could have existed. This view is consistent with (2), since to hold that there could have existed an immaterial being does not require holding that something which does exist, and is therefore material, could have been immaterial. The contingent materialist who allows that God could have existed is unlikely to say that some actual material thing—the smallest of Jupiter's moons, say—could have been God.

Thus there is a difference in meaning between (1) and (2) which we would expect to be articulated by semantics for first-order modal language. Specifically, in an S5 semantics for the language of first-order modal logic LFOML we would expect to be able to establish

(3) $(\forall x)\Box Mx \nvDash_{S5} \Box(\forall x)Mx$.

For the formula on the left in (3) is something the contingent materialist can allow to be true, while the formula on the right is one she must say is false.

Recall that an S5 interpretation in sentential modal logic is a set W with a distinguished element w^*, where each w in W is itself an interpretation for non-modal sentential logic, that is, an assignment of truth-values to sentence-letters. Correspondingly, an S5 interpretation in first-order modal logic is essentially just a collection of worlds, each of which is an interpretation for non-modal first-order logic—a domain of discourse and an assignment of extensions to predicates and references to names. However, some complications arise from interrelationships among worlds, so this is not exactly how we describe first-order interpretations.

> An interpretation I of an LFOML sequent is a pair of sets, W (possible worlds) and D (possible objects) that is, $I = (W,D)$. Each $w \in W$ selects a subset $w(D)$ of D and also assigns (i) a truth-value to each sentence-letter which occurs in the sequent ($w[\lambda] = \bot$), (ii) a reference $w[c]$ in D to each individual constant c in the sequent, and (iii) an extension $w[\lambda]$ to each n-place predicate which occurs in the sequent; and one $w \in W$ is designated the actual world, conventionally written 'w^*'.

$w(D)$ is the set of things which exist at w, and D is the set of all possible objects—the things which exist at some world or other. When giving an interpretation, we retain the '$\mapsto$' notation for sentence-letters, but otherwise we use parentheses and brackets.

We make certain stipulations about interpretations.

(NED): D must be nonempty.

($\Diamond\exists$): For each x in D there is at least one w in W with x in $w(D)$.

(RD): For each individual constant c, all worlds in W assign c the *same* reference (*r*igid *d*esignation).

(EN): For each n-place predicate λ in the sequent, for each w in W, $w[\lambda]$ is a collection of n-tuples of elements drawn from D, that is, $w[\lambda]$ is not restricted to $w(D)$ (*e*xistence *n*eutrality).

It is a consequence of (*NED*) and ($\diamond \exists$) that for at least one world w in W, $w(D)$ is nonempty. It is also a consequence that D is the union of all the $w(D)$ for w in W, so every possible object *possibly exists*—hence the abbreviation '$\diamond \exists$'. In specifying an interpretation $\mathcal{I}$ we will only display the domains of the individual worlds, implicitly understanding D to be their union.

A consequence of (*RD*) is that an individual constant can refer at a world to an object which does not exist at that world. No doubt it would be a neat trick for persons who *exist in* a counterfactual situation in which Socrates does not exist to use 'Socrates' to refer to Socrates. But *we* can use 'Socrates' to talk about a counterfactual situation which we hypothesize, such as one in which Socrates does not exist, and we can say that about it. Thus (*RD*) means only that in discourse about counterfactual situations, names continue to refer to whatever they actually refer to.

Lastly, we do not require that the atomic predicates be true at a world only of things which exist at that world. This permissiveness is more controversial, but logic should not prejudge the controversy. An example of an existence-neutral atomic predicate might be the set-membership relation; arguably, 'Socrates $\in$ {Socrates}' is true even at worlds where Socrates does not exist.

As (3) indicates, the interaction of modal operators and quantifiers is an important feature of first-order modal logic, so we should state our evaluation rules for the connectives with some care. The rules for sentence-letters, the truth-functional connectives and the modal operators are the same as in the standard semantics for sentential modal logic, but we need new rules for atomic sentences containing predicates and individual constants, and for the quantifiers. The case of atomic sentences is easy, since we can use the same approach as in nonmodal first-order logic. For example, if 'R' is a two-place atomic predicate and 'b' and 'd' are two individual constants, then whether or not 'Rbd' is true at w depends on whether or not the pair of objects consisting in the reference of 'b' at w and the reference of 'd' at w (in that order) belongs to the extension of 'R' at w: $w[Rbd] = \top$ if and only if $\langle w[b], w[d] \rangle \in w[R]$. Clause (*EA*) on page 316 generalizes this.

For the quantifiers we have a choice, because there are *two* relevant domains, $w(D)$ and D. Should '($\exists$x)Fx' be true at a world w in an interpretation $\mathcal{I} = (W,D)$ if and only if at least one instance $\ulcorner Ft \urcorner$ is true at w, where t denotes something in D? Or should we have the more restrictive condition that '($\exists$x)Fx' is true at a world w in an interpretation $\mathcal{I} = (W,D)$ if and only if at least one instance $\ulcorner Ft \urcorner$ is true at w, where t denotes something in $w(D)$? The second account is the natural one. The '$\exists$' of '($\exists$x)Fx' is meant to express *existence*, and $w(D)$ is the set of things which exist at w; consequently, only they should be relevant to whether or not 'there is (exists) at least one F' is true at w. Similarly, we should say that '($\forall$x)Fx' is true at a world w in an interpretation $\mathcal{I} = (W,D)$ if and only if every instance $\ulcorner Ft \urcorner$, where t denotes something in $w(D)$, is true at w. After all, the materialist who asserts that *everything* is made of matter is not bothered by things which could have existed but do not, and which consequently are not made of matter; her 'everything' is not meant to include these *merely possible* objects.

We therefore arrive at the following evaluation clauses:

(EA): $w[\lambda t_1...t_n] = \top$ if and only if the n-tuple of objects $\langle w[t_1],...,w[t_n]\rangle$ belongs to $w[\lambda]$, the extension of λ at w. When $n = 1$, that is, if λ is monadic, we dispense with pointed brackets and simply require $w[t] \in w[\lambda]$. If $n = 0$, then λ is a sentence-letter and $w[\lambda] = \top$ if and only if $w: \lambda \mapsto \top$.

(E=): $w[t = t'] = \top$ if and only if $w[t] = w[t']$.

(ECon): $w[{\sim}\phi] = \top$ if and only if $w[\phi] = \bot$; $w[\phi \mathbin{\&} \psi] = \top$ if and only if $w[\phi] = \top$ and $w[\psi] = \top$; $w[\phi \vee \psi] = \top$ if and only if $w[\phi] = \top$ or $w[\psi] = \top$; $w[\phi \rightarrow \psi] = \top$ if and only if, $w[\psi] = \top$ if $w[\phi] = \top$; $w[\phi \leftrightarrow \psi] = \top$ if and only if $w[\phi] = \top$ and $w[\psi] = \top$ or $w[\phi] = \bot$ and $w[\psi] = \bot$.

(E□): $w[\Box\phi] = \top$ if and only if $(\forall u \in W)\ u[\phi] = \top$.

(E◊): $w[\Diamond\phi] = \top$ if and only if $(\exists u \in W)\ u[\phi] = \top$.

(E∀): $w[(\forall v)\phi v] = \top$ if and only if for every instance ϕt with $w(t) \in w(D)$, $w[\phi t] = \top$.

(E∃): $w[(\exists v)\phi v] = \top$ if and only if there is at least one instance ϕt with $w(t) \in w(D)$ such that $w[\phi t] = \top$.

(Tr): A sentence ϕ of LFOML is true in an interpretation $\mathcal{I} = (W,D)$ if and only if ϕ is true at w^* in W.

It is also common in first-order modal logic to use a monadic predicate 'E_' for '_ exists'. We could treat 'E_' as a new logical constant, providing it with an evaluation clause here and rules of inference in the next section, but it is simpler to take it just as a device of abbreviation: in the context of any formula ϕ there is a variable v such that ⌜Et⌝ abbreviates the subformula ⌜$(\exists v)(v = t)$⌝ of ϕ; v may be any variable such that ⌜Et⌝ is not already within the scope of some ⌜$(\exists v)$⌝ or ⌜$(\forall v)$⌝ in ϕ. By evaluation clauses (E=) and (E∃), this means that at any world w the predicate 'E' applies to exactly the elements of $w(D)$, the things which exist at w.

The definition of semantic consequence for first-order modal logic is essentially as it was for sentential modal logic.

> For any wffs $p_1,...,p_n$ and q of LFOML, $p_1,...,p_n \vDash_{S5} q$ if and only if there is no interpretation $\mathcal{I}$ of the sequent under which $p_1,...,p_n$ are all true and q is false.

We turn now to illustrating these concepts by establishing some invalidities, beginning with (3), given earlier.

Example 1: Show $(\forall x)\Box Mx \nvDash_{S5} \Box(\forall x)Mx$.

The intuitive idea embodied in the example of contingent materialism is that while each actual object may be material at every world, there may be objects at other worlds which are immaterial; these objects would be things which could have existed, but in fact do not. Thus we have the following interpretation $\mathcal{I}$: $W = \{w^*, u\}$, $w^*(D) = \varnothing$, $u(D) = \{\alpha\}$, $w^*[M] = \varnothing$, $u[M] = \varnothing$, as pictured at the top of the next page. We maintain the alphabetic convention, so 'a' denotes α. Now let us step through the evaluation of Example 1's premise and conclu-

•   •
w^* u
∅ $\{\alpha\}$
M: ∅   M: ∅

sion in detail. Beginning with the conclusion, we see that 'Ma' is an instance of '$(\forall x)Mx$'; also $u[Ma] = \bot$, since $u[a]$, namely, α, is not in Ext(M) at u. So by ($E\forall$), $u[(\forall x)Mx] = \bot$. So by ($E\Box$), $w^*[\Box(\forall x)Mx] = \bot$. On the other hand, every instance of $\ulcorner \Box Mt \urcorner$ with $w^*[t] \in w^*(D)$ is true at w^* (since there are no such instances), and so $w^*[(\forall x)\Box Mx] = \top$ as required.

This particular example illustrates the general point that when the domain of a world is empty, all universal sentences are true at that world. This is because 'everything...' in quantified modal logic has the force of 'for every x, if x exists then...'. Hence all instances of this universally quantified conditional formed by using a name t of an element of D are true at an empty world since they all have false antecedents of the form $\ulcorner t \text{ exists} \urcorner$. Concomitantly, an existential sentence must be false at a world with an empty domain, no matter what follows the initial existential quantifier; for if $w(D)$ is empty, there can be no true instance of the existential sentence formed using an individual constant whose reference is in $w(D)$, since there are no such constants.

We can also establish the converse of the previous problem. Just as that problem's solution turned on the presence of a world containing an object which does not exist at the actual world, so the solution to its converse turns on the presence of a world *lacking* an object which *does* exist at the actual world.

Example 2: Show $\Box(\forall x)Mx \nvDash_{S5} (\forall x)\Box Mx$.

Interpretation: $W = \{w^*, u\}$, $w^*(D) = \{\alpha\}$, $u(D) = \emptyset$, $w^*[M] = \{\alpha\}$, $u[M] = \emptyset$, as pictured immediately below. *Explanation:* $w^*[(\forall x)Mx] = \top$ because 'Ma' is the only

•   •
w^* u
$\{\alpha\}$ ∅
M: $\{\alpha\}$ M: ∅

instance of '$(\forall x)Mx$' whose individual constant denotes something in w^* and $w^*[Ma] = \top$; also, $u[(\forall x)Mx] = \top$ because $u(D)$ is empty (see the preceding discussion of empty worlds). Hence by ($E\Box$), $w^*[\Box(\forall x)Mx] = \top$. However, again by ($E\Box$), $w^*[\Box Ma] = \bot$, because $u[Ma] = \bot$. Therefore $w^*[(\forall x)\Box Mx] = \bot$, because '$\Box Ma$' is a false instance formed with an individual constant whose reference is in $w^*(D)$.

This example brings out an infelicity in our discussion of contingent materialism. We suggested that a contingent materialist could accept '$(\forall x)\Box Mx$' on the grounds that though there could have been immaterial entities, they would have been different things from any actual entity. But this view does not motivate '(∀x)□Mx', which says that each actual entity is material at every world. Since material things exist contingently, there are worlds at which they do not exist, and we would hardly want to say that at *those* worlds they are material—there may be existence-neutral atomic predicates, but 'is material' is not one of them. The view which should really be endorsed by the contingent materialist who doubts that actual material things could have been purely mental or spiritual is that any actual entity is material at every world *at which it exists*. This would be formalized as '(∀x)□(Ex → Mx)' and is a weaker claim, as the following example establishes (to avoid monotony, we do not use empty worlds this time, though we could).

Example 3: Show (∀x)□(Ex → Mx) ⊮$_{S5}$ (∀x)□Mx.

Interpretation: W = {w^*,u}, $w^*(D)$ = {α,β}, $u(D)$ = {α}, w^*[M] = {α,β}, u[M] = {α}, as pictured below:

$$\bullet \qquad \bullet$$
$$w^* \qquad u$$
$$\{\alpha,\beta\} \qquad \{\alpha\}$$
$$\text{M: } \{\alpha,\beta\} \quad \text{M: } \{\alpha\}$$

Explanation: w^*[Ea → Ma] = ⊤ and u[Ea → Ma] = ⊤, so w^*[□(Ea → Ma)] = ⊤. Also, w^*[Eb → Mb] = ⊤ and u[Eb → Mb] = ⊤ (since u[Eb] = ⊥) so w^*[□(Eb → Mb)] = ⊤. Since '□(Ea → Ma)' and '□(Eb → Mb)' are all the representative instances of '(∀x)□(Ex → Mx)' formed from individual constants whose referents are in $w^*(D)$, w^*[(∀x)□(Ex → Mx)] = ⊤. But w^*[(∀x)□Mx] = ⊥ since w^*[□Mb] = ⊥, since u[Mb] = ⊥.

A weaker claim still, one which is consistent with holding that there *are* immaterial beings, is the claim that, if something is in fact material, then necessarily it is so if it exists at all: '(∀x)(Mx → □(Ex → Mx))'. It is an exercise to show that this thesis does not entail '(∀x)□(Ex → Mx)'. When a predicate is said to apply necessarily to a thing if the thing exists, the property for which the predicate stands is said to be an *essential* property of the thing, or *essential to* it. So according to '(∀x)□(Ex → Mx)', being material is essential to everything. We can also predicate an essential property of things of a certain kind, which is to say that the property is essential to things of that kind. Thus '(∀x)(Mx → □(Ex → Mx))' means that being material is essential to material things. It is a matter of considerable controversy in contemporary philosophy whether or not there are 'nontrivial' examples of essential properties, such as materiality (the definitions make existence an essential property, but this is trivial).

Returning to more technical issues, our semantic framework allows us to make good sense of the idea of a nonexistent object: the nonexistent objects or *mere possibles* of a world w are the things which exist at other worlds but not at w. However, the quantifiers express the notion of existence. This means that *being within the domain of quantification* and *being an object* come apart in first-order modal logic in a way that they do not in nonmodal logic. The contrast with nonmodal logic is strikingly illustrated by the following very simple

❑ Exercises

Show the following, with explanations:

(1) ◇~A ⊮$_{S5}$ ~◇A
(2) □(A ∨ B) ⊮$_{S5}$ □A ∨ □B
(3) ~(A → □B) ⊮$_{S5}$ □(A → ~B)
(4) A ↔ □B ⊮$_{S5}$ □(A ↔ B)
(5) ~◇A ∨ ~◇B ⊮$_{S5}$ ~◇(A ∨ B)
*(6) ◇(A → B) ⊮$_{S5}$ ◇A → ◇B
(7) ◇(A ↔ B) ⊮$_{S5}$ ◇(A ∨ B)
(8) □(A → ~B), □(B → C) ⊮$_{S5}$ □(A → ~C)
(9) ◇(A → λ), ◇(λ → A) ⊮$_{S5}$ □(A ↔ λ)
(10) □(A → B) ⊮$_{S5}$ ~□(A → ~B)
*(11) □A → ◇B ⊮$_{S5}$ □(A → B)
(12) □(A → B) → □(C → D) ⊮$_{S5}$ ◇(A & B) → □(C → D)
(13) ◇(A → □B) ⊮$_{S5}$ ◇A → □B
(14) □(A → ◇B) ⊮$_{S5}$ ◇(A → □B)
(15) □◇(A → B) ⊮$_{S5}$ □(◇A → B)
(16) ◇□(A → B) ⊮$_{S5}$ ◇A → □B
(17) □(◇A → ◇B) ⊮$_{S5}$ □◇A → ◇□B
(18) ◇(□A ∨ ◇B) ⊮$_{S5}$ ~◇A → □B

3 The canonical translation

In this section we will make the analogy between the modal language LSML and the language of monadic predicate logic more precise. We do this by explaining how to translate modal formulae into monadic formulae. However, the language we translate into is not LMPL as defined in §4 of Chapter Five, since translation into this language would obscure certain important features of the modal language. Rather, we introduce a new monadic language, the Language of Canonical Translation (LCT), as the language into which LSML is translated. LCT differs from LMPL in certain ways. For each sentence-letter π of LSML define the corresponding monadic predicate-letter λ_π to be the result of writing π followed by the prime symbol ''. For instance, the sentence-letter 'A' of LSML corresponds to the monadic predicate 'A''' (strictly, we should use subscripts to differentiate the monadic predicate 'A''' of LCT from the sentence-letter 'A''' of LSML, but instead we avoid using primed sentence-letters in LSML).

The lexicon of LCT:

One individual variable 'w'; no individual constants; a sentence-letter 'λ'; for each sentence-letter π of LSML except 'λ', the corresponding monadic predicate-letter λ_π; sentential connectives, quantifier symbols '∀' and '∃', and parentheses.

The formation rules of LCT:

(f-at): 'ʎ' is an atomic wff, and for any predicate-letter λ, ⌜λw⌝ is an atomic wff.

(f-con): If ϕ and ψ are wffs (atomic or otherwise) so are ⌜~ϕ⌝, ⌜(ϕ & ψ)⌝, ⌜(ϕ ∨ ψ)⌝, ⌜(ϕ → ψ)⌝ and ⌜(ϕ ↔ ψ)⌝.

(f-q): If ϕ is a wff, then ⌜(∃w)ϕ⌝ and ⌜(∀w)ϕ⌝ are wffs.

(f!): Nothing is a wff unless it is certified as such by the previous rules.

There are therefore some significant differences between LCT and the standard language of monadic predicate logic LMPL, in particular the following three: (a) LCT has only one sentence-letter, 'ʎ'; (b) LCT has only one individual variable and no individual constants; (c) in the syntax clause for the quantifiers, there is no requirement that ϕ has any free occurrences of 'w', so redundant quantification is permitted, and there is no requirement that ϕ has no bound occurrences of 'w', so double-binding is permitted. The motivation for these points of difference will be apparent when the method of translating from LSML into LCT is set out.

The goal of translation is to match an expression of one language with an 'equivalent' expression of another. So we have to decide what equivalence should mean in this context and how the translation should go. We begin with the second task. The idea is to describe a procedure for step-by-step transcription of an LSML wff into an LCT wff, a procedure that processes the LSML wff connective by connective, working inward from the main connective. It may seem that there will be a problem: modal operators are to be rendered as quantifiers, and LSML formulae can contain more than one modal operator; but there is only one variable in LCT to which to attach quantifiers. How then is the meaning of the LSML formula to be preserved? We can see how this difficulty is avoided by inspecting some target translations. For example, the translation of '◇(A → □B)' is '(∃w)(A′w → (∀w)B′w)', in which the last occurrence of 'w' is within the scope of two quantifiers corresponding to the '◇' and '□' of the LSML formula. However, the effect of the semantics for LCT will be to ensure that it is the *closest* quantifier which binds the last occurrence of 'w', just as, in the modal formula, it is the closest modal operator that determines the world of evaluation for what is within its scope. A similar problem arises with LSML formulae which have subformulae not within the scope of any modal operator, such as the antecedent of 'A → □B'. In the semantics of the previous section, such subformulae are interpreted as making a statement about w^*, the actual world. But there is no 'w^*' in the lexicon of LCT. However, the translation of 'A → □B' will be 'A′w → (∀w)B′w', in which the first occurrence of 'w' is free. We will arrange that free occurrences of 'w' in LCT formulae are interpreted as standing for the actual world. These two points, about double-binding and free occurrences of 'w', bring out why it is that one variable is all we require for translation of LSML.

The translation procedure is given by a set of rules, one for each of the connectives occurring as the main connective of the expression to be translated. The rules are:

q is a theorem. (Snd) and (Comp) are sometimes known as *finitary* soundness and *finitary* completeness respectively, because they are restricted to the cases where there are only finitely many premises. In first-order logic, cases where there are infinitely many premises are of considerable interest, so that soundness and completeness are often given their *strong* formulations: a system S is said to be *strongly sound* if and only if, for any set of LFOL-sentences Γ, *if* $\Gamma \vdash_S q$ then $\Gamma \vDash q$, and *strongly complete* if and only if, for any set of LFOL-sentences Γ, if $\Gamma \vDash q$ then $\Gamma \vdash_S q$. In these latter definitions, Γ may be either a finite or an infinite set of sentences. NK is strongly sound and strongly complete, though this is not implied merely by its finitary soundness and finitary completeness. These topics are pursued in more advanced texts.

In discussing sentential NK, we remarked that the soundness of the system of rules is obvious, while their completeness is not obvious, since although one can tell by inspecting the list of sentential rules that none are fallacious, one cannot tell by inspection that every rule needed in the proof of a valid sequent of LSL has been included on the list. In first-order logic, completeness is certainly at least as unobvious as in sentential logic, but soundness (finitary or strong) is no longer obvious, because of the need for restrictions on ∀I and ∃E. We motivated each restriction we imposed by indicating a problem that would arise without it, but this is no guarantee that we have thought of *all* the restrictions we need; perhaps some crucial restriction has been omitted, and as a result some invalid sequent has an NK-proof. In fact this is not so, but it requires a nontrivial proof to demonstrate that the rules are sound.

These remarks about the full system of first-order logic hold good of monadic predicate logic as well. The soundness of the rules for the full system entails the soundness of the rules for the monadic system, since the latter's rules are a subset of the former's. Consequently, a proof of an invalid sequent in LMPL would *ipso facto* be a proof of an invalid sequent in LFOL, and as there are no provable invalidities in first-order logic, it follows that there are none in monadic predicate logic. However, the completeness of the rules for the full system does not immediately entail the completeness of the rules for the monadic system, since it is conceivable that there is some valid sequent whose premises and conclusion contain only sentence-letters or one-place predicates, yet any NK-proof of it requires the introduction of some formula containing a many-place predicate. But there are results in a branch of logic known as *proof theory* which rule out this possibility (see Sundholm).

There remains the question of decidability. Recall from §6 of Chapter 6 that the validity problem for a kind of logic is *decidable* or *solvable* if and only if there is some algorithm which, given any sentences $p_1,...,p_n$ and q in the language for that logic, determines whether or not q is a semantic consequence of $p_1,...,p_n$ in the sense appropriate for that kind of logic. Validity is decidable for classical sentential logic and monadic predicate logic, though not, so far as we know, by an algorithm which is computationally tractable in all cases. However, it was proved by Alonzo Church in 1936 that the full system of first-order logic is undecidable (see Hodges §24). In fact, it suffices to add one two-place predicate to monadic predicate logic to obtain an undecidable logic. Since there is no decision procedure for validity in first-order logic, we will not extend the

For sequent-to-sequent proofs, the identity rules have the following form:

Rule of =I: At any point in a proof we may include a line of the form $\vdash t = t$.

Rule of =E: From the sequents $\Gamma \vdash t_1 = t_2$, $\Sigma \vdash \phi t_1$, we may infer the sequent $\Gamma, \Sigma \vdash \phi t_2$.

So Example 1 may be reformatted as a sequent-to-sequent proof as follows:

(1)	$(\exists x)(Fx \ \& \ x = a) \vdash (\exists x)(Fx \ \& \ x = a)$	Premise
(2)	$Fb \ \& \ b = a \vdash Fb \ \& \ b = a$	Assumption
(3)	$Fb \ \& \ b = a \vdash Fb$	2 &E
(4)	$Fb \ \& \ b = a \vdash b = a$	2 &E
(5)	$Fb \ \& \ b = a \vdash Fa$	3,4 =E
(6)	$(\exists x)(Fx \ \& \ x = a) \vdash Fa$	1,2,5 ∃E ◆

❑ Exercises

(1) Arrange the remaining examples in §4 in tree format.
(2) Arrange the remaining examples in §4 in sequent-to-sequent format.

7 Semantic consequence, deductive consequence and decidability

The same questions arise about the strength of our rules of inference for first-order logic as arose about the rules for the truth-functional connectives. We repeat the definitions here:

- A system S of rules of inference is said to be *sound* if and only if every S-provable argument-form is valid, that is, if and only if whenever $p_1, ..., p_n \vdash_S q$, then $p_1, ..., p_n \vDash q$.
- A system S of rules of inference is said to be *complete* if and only if every valid argument-form is S-provable, that is, if and only if whenever $p_1, ..., p_n \vDash q$, then $p_1, ..., p_n \vdash_S q$.

Our system of rules NK for first-order logic is both sound and complete. In other words, for any closed sentences $p_1, ..., p_n$ and q of LFOL, the following statements are true:

(Snd): If $p_1, ..., p_n \vdash_{NK} q$ then $p_1, ..., p_n \vDash q$.
(Comp): If $p_1, ..., p_n \vDash q$ then $p_1, ..., p_n \vdash_{NK} q$.

This includes the case where $n = 0$; that is, by (Snd), every theorem q is logically valid (true in all interpretations), and by (Comp), every logically valid sentence example, which does not even involve modal operators:

Example 4: Show Fa $\nVDash_{S5}$ ($\exists$x)Fx.

Interpretation: $W = \{w^*, u\}$, $w^*(D) = \varnothing$, $u(D) = \{\alpha\}$, $w^*[F] = \{\alpha\}$, $u[F] = \varnothing$ (a solution could also be given with $u[F] = \{\alpha\}$). *Explanation:* We have $w^*[Fa] = \top$

$$
\begin{array}{cc}
\bullet & \bullet \\
w^* & u \\
\varnothing & \{\alpha\} \\
\text{F: }\{\alpha\} & \text{F: }\varnothing
\end{array}
$$

because $w[a] \in w[F]$, but $w^*[(\exists x)Fx] = \bot$ because $w^*(D)$ is empty, or in more detail, because '($\exists$x)Fx' has no true instance formed with an individual constant whose reference is in $w^*(D)$. Note that the role of the second world u in this interpretation is to satisfy condition ($\diamond\exists$) on interpretations, that every possible object exists in at least one world, and we cannot use w^* if we want the conclusion of the sequent to be false. Note also that though the example involves assigning an object to the extension of an atomic predicate at a world where the object does not exist—that is, 'F' is treated as existence neutral—this is not intrinsic to the contrast with nonmodal logic. For we can also show ~Fa $\nVDash_{S5}$ ($\exists$x)~Fx without putting objects in $w^*(F)$ or $u(F)$ that do not exist in the respective worlds (this is an exercise).

The interplay of quantifiers and modal operators becomes more complex when we consider formulae with many-place predicates. Looking at the simple dyadic formula-scheme ⌜(Qx)(Qy)Rxy⌝ where Q is a fixed quantifier, we see that there are three different places where a single modal operator M could appear, giving rise to seven different modal formulae based on ⌜(Qx)(Qy)Rxy⌝ and non-consecutive occurrences of '□', plus another seven using '◊', plus the formulae containing a mixture. There are few logical equivalences in this collection. For example:

Example 5: Show ($\forall$x)□($\forall$y)Rxy $\nVDash_{S5}$ ($\forall$x)($\forall$y)□Rxy.

Interpretation: $W = \{w^*, u\}$, $w^*(D) = \{\alpha\}$, $u(D) = \varnothing$, $w^*[R] = \{\langle\alpha,\alpha\rangle\}$, $u[R] = \varnothing$.

$$
\begin{array}{cc}
\bullet & \bullet \\
w^* & u \\
\{\alpha\} & \varnothing \\
\text{R: }\{\langle\alpha,\alpha\rangle\} & \text{R: }\varnothing
\end{array}
$$

Explanation: $u[Raa] = \bot$ because $\langle\alpha,\alpha\rangle \notin u[R]$, hence $w^*[\Box Raa] = \bot$ by (E□). But '□Raa' is an instance of '($\forall$y)□Ray' formed with an individual constant, 'a', whose reference α is in $w^*(D)$. So it follows that $w^*[(\forall y)\Box Ray] = \bot$, and hence by (E$\forall$) again, $w^*[(\forall x)(\forall y)\Box Rxy]$ is false as well. However, it is straightforward to check that '(∀x)□(∀y)Rxy' is true at w^*. For $w^*[Raa] = \top$, hence $w^*[(∀y)Ray] = \top$; and since $u(D)$ is empty, $u[(∀y)Ray] = \top$ as well; hence $w^*[□(∀y)Ray] = \top$. Since $w^*(D) = \{α\}$, this means that $w^*[(∀x)□(∀y)Rxy] = \top$.

Our last example involves identity, which exhibits interesting behavior in first-order modal logic (there will be further discussion of this in the next section).

Example 6: Show a ≠ b, □(∀x)□(Fx ∨ Gx), □(∀x)[(◇Fx & ◇~Fx) & (◇Gx & ◇~Gx)] ⊭$_{S5}$ ◇~(Fa ↔ Fb) ∨ ◇~(Ga ↔ Gb).

Interpretation: $W = \{w^*, u\}$, $w^*(D) = \{α, β\}$, $u(D) = \{α, β\}$, $w^*[F] = \{α, β\}$, $w^*[G] = \varnothing$, $u[F] = \varnothing$, $u[G] = \{α, β\}$:

$$
\begin{array}{cc}
\bullet & \bullet \\
w^* & u \\
\{α,β\} & \{α,β\} \\
F: \{α,β\} & F: \varnothing \\
G: \varnothing & G: \{α,β\}
\end{array}
$$

Explanation: Since $w^*[\sim(Fa ↔ Fb)] = \bot$ and $u[\sim(Fa ↔ Fb)] = \bot$, and $w^*[\sim(Ga ↔ Gb)]$ and $u[\sim(Ga ↔ Gb)] = \bot$, then $w^*[◇\sim(Fa ↔ Fb)] = \bot$ and $w^*[◇\sim(Ga ↔ Gb)] = \bot$, so the conclusion is false. And since $α$ and $β$ are different objects, the first premise is true. The effect of '□(∀x)' is to make a claim about all possible objects, since it means 'for each world, for every object in that world', and every possible object exists in at least one world. So the second premise says, of each possible object x, that at each world w, x satisfies either 'F' or 'G'. In our interpretation, both possible objects satisfy 'F' at w^* and both satisfy 'G' at u, hence premise 2 is true. Finally, premise 3 says that for each possible object, there is a world where it is in the extension of 'F' and a world where it is not, and similarly with respect to 'G'; that is, the properties *being F* and *being G* are contingent for every possible object. Since $α$ and $β$ are in the extension of 'F' at w^* and not at u, and $α$ and $β$ are in the extension of 'G' at u and not at w^*, premise 3 is true, completing the solution.

The gist of Example 6 is that distinct objects do not have to be distinguishable, since the conclusion formula, which is false in the interpretation, says that either there is a world where one of $α$ and $β$ satisfies 'F' and the other does not or there is a world where one satisfies 'G' and the other does not. When distinct objects have exactly the same properties at exactly the same worlds they are said to be *indiscernible*. So in the above interpretation, $α$ and $β$ are indiscernibles. Philosophers have sometimes held that objects which cannot be distinguished cannot be distinct. If this means 'cannot be distinguished by contingent monadic properties' then *prima facie* it seems that we have no trouble understanding how this view could be false. We consider other properties of identity in §6.

❏ Exercises

Show the following, with explanations:

(1) ◇(∃x)Fx ⊭$_{S5}$ (∃x)◇Fx
(2) (∃x)◇Fx ⊭$_{S5}$ ◇(∃x)Fx
(3) (∀x)□(Fx → Gx) ⊭$_{S5}$ (∀x)(Fx → □Gx)
(4) □(∀x)(Fx → Gx) ⊭$_{S5}$ □(∀x)(Fx → □Gx)
*(5) (∀x)(□Fx ↔ □Gx) ⊭$_{S5}$ (∀x)□(Fx ↔ Gx)
(6) ~Fa ⊭$_{S5}$ (∃x)~Fx
(7) (∀x)Fx ⊭$_{S5}$ Fa
(8) □(a = b) ⊭$_{S5}$ □(∃x)(x = b)
(9) (∀x)(Mx → □(Ex → Mx)) ⊭$_{S5}$ (∀x)□(Ex → Mx)
*(10) (∃x)◇(∃y)Rxy ⊭$_{S5}$ ◇(∃x)(∃y)Rxy
(11) □(∀x)(∀y)Rxy ⊭$_{S5}$ □(∀x)□(∀y)Rxy
(12) □(∀x)□(∀y)Rxy ⊭$_{S5}$ □(∀x)□(∀y)□Rxy
(13) ◇(∃x)□(∀y)Rxy ⊭$_{S5}$ ◇(∃x)(∀y)□Rxy
(14) (∀x)□(∃y)◇Rxy ⊭$_{S5}$ (∀x)◇(∃y)□Rxy
(15) □(∀x)(∀y)(∀z)((Oxy & Oxz) → □(Ex → (Oxy ∨ Oxz))) ⊭$_{S5}$
       □(∀x)(∀y)(∀z)((Oxy & Oxz) → □(Ex → (Oxy & Oxz)))
*(16) (∀x)□Ex ⊭$_{S5}$ ◇(∃x)Fx → (∃x)◇Fx
(17) (∀x)□Ex ⊭$_{S5}$ □[(∃x)◇Fx → ◇(∃x)Fx]
(18) □[□(∀x)Fx → (∀x)□Fx] ⊭$_{S5}$ (∀x)□Ex
(19) A & ◇(∀x)(Fx → □(A → ~Fx)) ⊭$_{S5}$ (∃x)◇~Fx
(20) ◇(∀x)(Fx → □(A → ~Fx)), (∃x)□(B → Fx) ⊭$_{S5}$ ~◇(A & B)

6 First-order modal logic: natural deduction

We would now like to formulate a set of first-order inference rules which are sound and complete with respect to the S5 semantics of the previous section. The most obvious way to proceed is to add the quantifier and identity rules from NK to the system S5 of §4 or, equivalently, to add the four modal operator rules of S5 to first-order NK. However, this will not achieve the desired effect: any system of inference which contains the quantifier rules of NK will be unsound with respect to ⊨$_{S5}$, for as we have just seen, Fa ⊭$_{S5}$ (∃x)Fx, and we also have ~Fa ⊭$_{S5}$ (∃x)~Fx (Exercise 5.6) and (∀x)Fx ⊭$_{S5}$ Fa (Exercise 5.7). A sound system of inference for S5 must match these failures of semantic consequence with corresponding failures of deducibility, and so the NK rules are inappropriate, since Fa ⊢$_{NK}$ (∃x)Fx, ~Fa ⊢$_{NK}$ (∃x)~Fx and (∀x)Fx ⊢$_{NK}$ Fa.

Inspection of the clauses (E∀) and (E∃) on page 316 reveals where the problem lies: at each world w, the quantifiers range over only the things which exist at w, while the individual constants of the language may refer at w to things which do not exist at w. Thus, for instance, ~Fa ⊭$_{S5}$ (∃x)~Fx, because even if w^*[~Fa] = ⊤, it does not follow that there *exists* an x which fails to satisfy 'F', since 'a' may stand for something which does not exist at w^*; that is, we may

$w^*[a] \notin w^*(D)$. However, the example suggests a condition under which an existential quantifier *may* be introduced; though it is not enough that we have established ϕt for some individual constant t, it *would* be enough if we could establish the conjunction $\ulcorner Et \ \& \ \phi t \urcorner$. In this formula, 'E_' is the existence predicate which abbreviates $\ulcorner (\exists v)(v = \_) \urcorner$ (recall the discussion on page 316). On the basis of $\ulcorner Et \ \& \ \phi t \urcorner$ we may infer $\ulcorner (\exists v)\phi v \urcorner$, and, conversely, for the purposes of $\exists E$ we say that an instance of $\ulcorner (\exists v)\phi v \urcorner$ is a formula of the form $\ulcorner Et \ \& \ \phi t \urcorner$.

Similarly, $(\forall x)Fx \nvDash_{S5} Fa$ because even if everything which exists satisfies 'F', 'Fa' is false if 'a' denotes something which both does not exist and does not satisfy 'F'. But though we cannot infer 'Fa' from '$(\forall x)Fx$', we *can* infer '*if a exists then Fa*', that is, 'Ea → Fa'. Conversely, if we can prove by the same steps every conditional $\ulcorner Et \rightarrow Ft \urcorner$, regardless of which possible object t denotes, that is, if we can prove it by reasoning that is generalizable with respect to goal formula, then we may infer '$(\forall x)Fx$'. So the quantifier rules we need will be like the quantifier rules of NK except that our notion of an *instance* of a quantified sentence will be altered to include a subformula containing 'E_'.

The variant of first-order NK which we are envisaging is known as *free logic*, since individual constants in it are free of existential commitment. We shall refer to our version as NKF, and we will approach first-order S5 in two steps, starting with the extension of sentential NK to NKF and then continuing with the extension of NKF, which is nonmodal, to first-order S5. To begin with, then, NKF is defined as the system which consists in (a) the rules of NK for the sentential connectives, (b) the rules of NK for identity, and (c) the following four 'free' rules for the quantifiers, given in schematic form:

∀-*Introduction:*

$$
\begin{array}{lll}
a_1,\ldots,a_n & (j) & Et \rightarrow \phi t \\
& \vdots & \\
a_1,\ldots,a_n & (k) & (\forall v)\phi v \qquad j \ \forall I
\end{array}
$$

where t is not in any of the formulae on lines $a_1,\ldots,a_n$. v, a variable not already in ϕt, replaces every occurrence of t in ϕt.

∀-*Elimination:*

$$
\begin{array}{lll}
a_1,\ldots,a_n & (j) & (\forall v)\phi v \\
& \vdots & \\
a_1,\ldots,a_n & (k) & Et \rightarrow \phi t \qquad j \ \forall E
\end{array}
$$

where t replaces every occurrence of v in ϕv.

∃-*Introduction:*

$$
\begin{array}{lll}
a_1,\ldots,a_n & (j) & Et \ \& \ \phi t \\
& \vdots & \\
a_1,\ldots,a_n & (k) & (\exists v)\phi v \qquad j \ \exists I
\end{array}
$$

where v, a new variable, replaces at least one occurrence of t in ϕt.

∃-Elimination:

$$
\begin{array}{llll}
a_1,...,a_n & \text{(i)} & (\exists v)\phi v & \\
& \vdots & & \\
j & \text{(j)} & Et \,\&\, \phi t & \text{Assumption} \\
& \vdots & & \\
b_1,...,b_u & \text{(k)} & \psi & \\
& \vdots & & \\
X & \text{(m)} & \psi & \text{i,j,k } \exists E
\end{array}
$$

where t is not in ψ, $\ulcorner(\exists v)\phi v\urcorner$ or any of the formulae at lines $b_1,...,b_u$ other than j. $X = \{a_1,...,a_n\} \cup \{b_1,...,b_u\}/j$; t replaces every occurrence of v in ϕv.

Though we have new quantifier rules, it should be noted that the identity rules remain the same. In particular, there is no need to change Identity Introduction, since the S5 evaluation clause (E=) for the identity symbol relies on the same features of identity as does the semantics for LFOL. It would be a different matter if we had required that, say, $\ulcorner t = t'\urcorner$ is true at w only if $w[t] \in w(D)$. But this sense of identity, in which identity entails existence (sometimes called 'strong' identity) is not the one we adopted.

Nonmodal free logic is worth studying in its own right, but we will immediately move on to its extension to modal logic. First-order S5, 'S5' for short, is defined to be the system which results from adding five rules governing modal operators to NKF. Four of these rules are the operator rules from sentential S5 described in §4 and repeated here as schemata, while the fifth is a new rule discussed below:

□-Introduction:

$$
\begin{array}{llll}
a_1,...,a_n & \text{(j)} & p & \\
& \vdots & & \\
a_1,...,a_n & \text{(k)} & \Box p & \text{j } \Box I
\end{array}
$$

where the formulae at lines $a_1,...,a_n$ are fully modalized.

□-Elimination:

$$
\begin{array}{llll}
a_1,...,a_n & \text{(j)} & \Box p & \\
& \vdots & & \\
a_1,...,a_n & \text{(k)} & p & \text{j } \Box E
\end{array}
$$

◇-Introduction:

$$
\begin{array}{llll}
a_1,...,a_n & \text{(j)} & p & \\
& \vdots & & \\
a_1,...,a_n & \text{(k)} & \Diamond p & \text{j } \Diamond I
\end{array}
$$

◇-Elimination:

$a_1,...,a$	(i)	$\diamond p$	
	⋮		
j	(j)	p	Assumption
	⋮		
$b_1,...,b_u$	(k)	q	
	⋮		
X	(m)	q	i,j,k ◇E

where all the formulae at lines $b_1,...,b_u$ other than j are fully modalized and q is fully modalized. $X = \{a_1,...,a_n\} \cup \{b_1,...,b_u\}/j$.

Before giving the final rule, something must be said about the notion of full modalization. In sentential S5, a formula is fully modalized if and only if every sentence-letter in it is within the scope of some modal operator. The simplest generalization of this to first-order S5 would be to say that a formula is fully modalized if and only if every atomic formula in it is within the scope of a modal operator. But this will not do. According to such a definition, '$(\forall x)\Box Mx$' is fully modalized, since its only atomic subformula, 'Mx', is within the scope of a '$\Box$', and so the following would be a correct proof of the sequent $(\forall x)\Box Mx$ $\vdash_{S5} \Box(\forall x)Mx$:

1	(1)	$(\forall x)\Box Mx$	Premise
2	(2)	Ea	Assumption
1	(3)	Ea → $\Box$Ma	1 ∀E
1,2	(4)	$\Box$Ma	3,2 →E
1,2	(5)	Ma	4 $\Box$E
1	(6)	Ea → Ma	2,5 →I
1	(7)	$(\forall x)$Mx	6 ∀I
1	(8)	$\Box(\forall x)$Mx	7 $\Box$I ◆

the last step being legitimate since (1) is fully modalized on the proposed account. So we have shown that $(\forall x)\Box Mx \vdash_{S5} \Box(\forall x)Mx$. However, by Example 1 of the previous section (page 316) we know that $(\forall x)\Box Mx \nvDash_{S5} \Box(\forall x)Mx$, and so the deductive consequence relation for S5 as we are currently defining it would be unsound with respect to our S5 semantics: an invalid sequent would have a proof.

The problem is with the definition of full modalization. Intuitively, the requirement of full modalization is supposed to pick out those premises which express conditions that hold no matter which of the ways things could have been is the way things are, since it is only when a statement follows from such conditions that we are entitled to conclude that it is necessary. But because the collection of things which exist can differ from world to world, the formula '$(\forall x)\Box Mx$' can be true without being such a condition; hence the criterion of full modalization must be strengthened to exclude such formulae. We adopt the following definition:

A formula of LFOML is *fully modalized* if and only if each atomic formula in it, and each quantifier in it, is within the scope of some modal operator.

On this account of full modalization the previous proof is incorrect; in particular, the last line is wrong, since '(∀x)Fx' at (7) depends on a premise that is not fully modalized, the '∀' in (1) not being within the scope of any modal operator.

Though it is now sound, our collection of rules is not yet complete. Free logic is so called because the use of a name is free of commitment to the existence of its referent, but in S5, every name denotes something which exists in at least one possible world. Thus use of a name brings with it a commitment to the *possible* existence of its referent. Commitment to the existence of the referent in standard first-order logic is reflected in the fact that every instance of ⌜(∃x)(x = t)⌝, where t is any individual constant, is a theorem of NK, proved by =I and ∃I. But because identity is not existence-entailing in S5, there are no corresponding derivations of instances of ⌜◇(∃x)(x = t)⌝, that is, ⌜◇Et⌝. Instead, we need a primitive rule to capture commitment to possible existence:

> **Rule of ◇∃**: At any line j in a proof, any instance of the formula ⌜◇Et⌝ may be written down, the line being labeled 'j, ◇∃', and depending on no premises or assumptions.

We call this rule Possible Existence. This rule and the other four modal operator rules, when added to NKF, produces a system of proof which is both sound and complete for the S5 semantics. Here are some examples of proofs:

Example 1: Show □(∃x)◇(Fx ∨ Gx) ⊢_S5 (∃x)(◇Fx ∨ ◇Gx).

1	(1)	□(∃x)◇(Fx ∨ Gx)	Premise
1	(2)	(∃x)◇(Fx ∨ Gx)	1 □E
3	(3)	Ea & ◇(Fa ∨ Ga)	Assumption
3	(4)	◇(Fa ∨ Ga)	3 &E
5	(5)	Fa ∨ Ga	Assumption
6	(6)	Fa	Assumption
6	(7)	◇Fa	6 ◇I
6	(8)	◇Fa ∨ ◇Ga	7 ∨I
9	(9)	Ga	Assumption
9	(10)	◇Ga	9 ◇I
9	(11)	◇Fa ∨ ◇Ga	10 ∨I
5	(12)	◇Fa ∨ ◇Ga	5,6,8,9,11 ∨E
3	(13)	◇Fa ∨ ◇Ga	4,5,12 ◇E
3	(14)	Ea	3 &E
3	(15)	Ea & (◇Fa ∨ ◇Ga)	14,13 &I
3	(16)	(∃x)(◇Fx ∨ ◇Gx)	15 ∃I
1	(17)	(∃x)(◇Fx ∨ ◇Gx)	2,3,16 ∃E ◆

The main point to notice about this proof is that it would not be correct to apply ∃I within ∨E so that the formula obtained by ∨E is '(∃x)(◇Fx ∨ ◇Gx)', for then one has to apply ◇E, which requires that the target formula be fully modalized, and '(∃x)(◇Fx ∨ ◇Gx)' is not fully modalized because its quantifier is not within the scope of a modal operator. Our use of ◇E at (13) is legitimate since '◇Fa ∨ ◇Ga' is fully modalized.

Apart from this point, there is a sense in which Example 1 is not very interesting: the crucial manipulations are those of sentential logic, with the quantifiers and modal operators being essentially decorative. More interesting are sequents which involve interplay between operators and quantifiers of the kind investigated semantically in the previous section. For example, we have seen that □(∀x)Mx ⊮_s5 (∀x)□Mx. However, a counterexample crucially involves a world where some actual object (element of w*(D)) does not exist. So if we were to add a premise ruling out the possible nonexistence of actual objects, we ought to be able to derive '(∀x)□Mx' from '□(∀x)Mx'. A formula which says that actual objects exist necessarily is '(∀x)□Ex', and so we should be able to give a proof of the sequent (∀x)□Ex, □(∀x)Mx ⊢_s5 (∀x)□Mx, or equivalently, the following problem:

Example 2: Show (∀x)□Ex ⊢_s5 □(∀x)Mx → (∀x)□Mx.

However, it is typical of proofs of such sequents that they run into the same kind of difficulty with the full-modalization requirements on □I and ◇E as did our proof on page 311 of Example 4.6, A ⊢_s5 □◇A. The technique of using conditionals as maximum formulae illustrated by that proof is useful here as well (recall that an occurrence of ⌜p → q⌝ as a line in a proof is a maximum formula if it is inferred by →I and then q is derived from it by →E using p at some other line). To prove Example 2 we will assume '□(∀x)Mx' and try to derive '(∀x)□Mx', presumably by ∀I. In that case we have to derive, say, 'Ea → □Ma', and our strategy will be to prove '□Ea → □Ma' and use it as a maximum formula. This yields the following proof:

1	(1)	(∀x)□Ex	Premise
2	(2)	□(∀x)Mx	Assumption
3	(3)	Ea	Assumption
1	(4)	Ea → □Ea	1 ∀E
1,3	(5)	□Ea	4,3 →E
6	(6)	□Ea	Assumption
6	(7)	Ea	6 □E
2	(8)	(∀x)Mx	2 □E
2	(9)	Ea → Ma	8 ∀E
2,6	(10)	Ma	9,7 →E
2,6	(11)	□Ma	10 □I
2	(12)	□Ea → □Ma	6,11 →I
1,2,3	(13)	□Ma	12,5 →E
1,2	(14)	Ea → □Ma	2,13 →I
1,2	(15)	(∀x)□Mx	14 ∀I
1	(16)	□(∀x)Mx → (∀x)□Mx	2,15 →I ◆

The crucial step here is the application of □I at line 11, requiring that (10) rest on premises and assumptions all of which are fully modalized. That is why we cannot use line 3 with line 9 in →E, since then 'Ma' would depend on an assumption that is not fully modalized. By deriving 'Ea' as (7) from the fully modalized (6), we circumvent this difficulty.

In cases where two conditionals $\ulcorner p \to q \urcorner$ and $\ulcorner q \to r \urcorner$ are involved, we can obtain shorter proofs if we use SI with the NK-sequent A → B, B → C ⊢ A → C. Since deriving a conditional $\ulcorner p \to r \urcorner$ from two others of the forms $\ulcorner p \to q \urcorner$ and $\ulcorner q \to r \urcorner$ is often called 'chaining', we will refer to this sequent as *Chain*. We can use Chain to get a 13-line proof of Example 2, applying it to (4) and (12) and avoiding the need for (3). This is perfectly legitimate but has the effect of 'hiding' the occurrence of the maximum formula, since the uses of →I and →E in the above proof would be absorbed by Chain.

The premise '(∀x)□Ex' of Example 2 requires that everything in the domain of the actual world be in the domain of every other world. A stronger condition is that the domains of all the worlds be the same. This is equivalent to the condition that anything which exists in some world exists in all worlds, and is expressed by the formula '□(∀x)□Ex', from which we can derive, for any possible object x, that x exists necessarily. The proof illustrates the need for Possible Existence, since to say that the domains of all worlds are the same is to say that *if* x exists in one world it exists in all, and so to conclude that x exists in all we need the antecedent that it exists in at least one.

Example 3: Show □(∀x)□Ex ⊢$_{S5}$ □Ea.

1	(1)	□(∀x)□Ex	Premise
	(2)	◇Ea	◇∃
3	(3)	Ea	Assumption
1	(4)	(∀x)□Ex	1 □E
1	(5)	Ea → □Ea	4 ∀E
1,3	(6)	□Ea	5,3 →E
1	(7)	□Ea	2,3,6 ◇E ◆

Line 7 is legitimate since (6) is fully modalized, and apart from (3), the only other premise or assumption on which (6) depends is (1), which is also fully modalized.

Example 3 is useful in the proof of a sequent on a similar theme to Example 2. We have established that ◇(∃x)Fx ⊬$_{S5}$ (∃x)◇Fx (Exercise 5.1), but any counterexample essentially involves the presence in a nonactual world of something which does not actually exist. So given a premise which rules out such objects, we would expect to be able to derive '(∃x)◇Fx' from '◇(∃x)Fx'. There is in fact no formula of LFOML which precisely expresses the condition that the domains of nonactual worlds contain only actual objects; in ordinary English the condition is 'there could not be things which do not actually exist' and we have no 'actually'. But this condition is implied by the stronger condition that the domains of all worlds are the same, for which we have '□(∀x)□Ex'. So we should be able to solve this problem:

Example 4: Show □(∀x)□Ex ⊢ₛₛ ◇(∃x)Fx → (∃x)◇Fx.

The restrictions on □I, ◇E and ∃E combine to make the proof unobvious. It turns out that once we have assumed '◇(∃x)Fx' we must aim for the stronger, fully modalized, formula '□(∃x)◇Fx' to use with ◇E, and then apply □E (compare the shortest proof of ◇□A ⊢ₛₛ A). To get '□(∃x)◇Fx' depending on the right premises and assumptions, we use the maximum formula '◇Fa → □(∃x)◇Fx'.

1	(1)	□(∀x)□Ex	Premise
2	(2)	◇(∃x)Fx	Assumption
3	(3)	(∃x)Fx	Assumption
4	(4)	Ea & Fa	Assumption
4	(5)	Fa	4 &E
4	(6)	◇Fa	5 ◇I
7	(7)	◇Fa	Assumption
1	(8)	□Ea	1 SI (Example 3)
1	(9)	Ea	8 □E
1,7	(10)	Ea & ◇Fa	9,7 &I
1,7	(11)	(∃x)◇Fx	10 ∃I
1,7	(12)	□(∃x)◇Fx	11 □I
1	(13)	◇Fa → □(∃x)◇Fx	7,12 →I
1,4	(14)	□(∃x)◇Fx	13,6 →E
1,3	(15)	□(∃x)◇Fx	3,4,14 ∃E
1,2	(16)	□(∃x)◇Fx	2,3,15 ◇E
1,2	(17)	(∃x)◇Fx	16 □E
1	(18)	◇(∃x)Fx → (∃x)◇Fx	2,17 →I ◆

The point of this strategy is that we can apply □I at line 12 because (11) depends on fully modalized assumptions. If we had tried to dispense with (7) and use (6) instead, '4' would have appeared on the left where we have '7', and (4) is not fully modalized, so the use of □I would be incorrect.

Another example, our first involving a binary predicate, gives a modal analogue of the implication from ∃∀ to ∀∃.

Example 5: Show ◇(∃x)□(Ex & (∀y)Rxy) ⊢ₛₛ □(∀y)(∃x)Rxy.

This time the maximum formula is '□(Ea & (∀y)Ray) → □(∀y)(∃x)Rxy'.

1	(1)	◇(∃x)□(Ex & (∀y)Rxy)	Premise
2	(2)	(∃x)□(Ex & (∀y)Rxy)	Assumption
3	(3)	Ea & □(Ea & (∀y)Ray)	Assumption
3	(4)	□(Ea & (∀y)Ray)	3 &E
5	(5)	□(Ea & (∀y)Ray)	Assumption
5	(6)	Ea & (∀y)Ray	5 □E
5	(7)	Ea	6 &E
5	(8)	(∀y)Ray	6 &E
5	(9)	Eb → Rab	8 ∀E

10	(10)	Eb	Assumption
5,10	(11)	Rab	9,10 →E
5,10	(12)	Ea & Rab	7,11 &I
5,10	(13)	(∃x)Rxb	12 ∃I
5	(14)	Eb → (∃x)Rxb	10,13 →I
5	(15)	(∀y)(∃x)Rxy	14 ∀I
5	(16)	□(∀y)(∃x)Rxy	15 □I
	(17)	□(Ea & (∀y)Ray) → □(∀y)(∃x)Rxy	5,16 →I
3	(18)	□(∀y)(∃x)Rxy	17,4 →E
2	(19)	□(∀y)(∃x)Rxy	2,3,18 ∃E
1	(20)	□(∀y)(∃x)Rxy	1,2,19 ◇E ◆

Here the main point is that □I at (16) requires that (15) depend only on assumptions and premises which are fully modalized. If we did not make the special assumption at line 5, we would have to use (4), which depends on (3), which is not fully modalized. This illustrates how the free elimination rule for '∃' creates the need for use of maximum formulae in proofs.

Our penultimate example embodies another way in which first-order S5 differs from nonmodal logic. In nonmodal logic, whenever p and q are LFOL formulae such that $p \vdash_{NK} q$ (equivalently, $p \vDash q$), then there is an LFOL formula r such that $p \vdash_{NK} r$ and $r \vdash_{NK} q$, and apart from logical constants and punctuation, any symbol which appears in r appears in both p and q; that is, r is constructed from the nonlogical vocabulary which is *common* to p and q. The formula r is called an *interpolant* for p and q. First-order S5 lacks the interpolation property, for there are p and q such that $p \vdash_{S5} q$ (equivalently, $p \vDash_{S5} q$) but no interpolant r exists. The following example was discovered by K. Fine (the two premises can be considered a single conjunction).

Example 6: Show A, □(∀x)□(A → Fx) ⊢_{S5} B → □(∀x)◇(B & Fx)

An interpolant would have to be built up from logical constants and the predicate 'F', but Fine has shown that there is no such formula. The problem arises from intrinsic limitations on the expressive power of LFOML. The premises of Example 6 say, respectively, that the actual world is an A-world and that every possible object is F in any A-world. The conclusion says that if B, then each possible object is F in some B-world. There is a condition which these premises entail and which entails the conclusion, that every possible object is F in the actual world; the conclusion follows from this because if not-B, it holds anyway, and if B, then the actual world is a world where each possible object is F. The catch is that the condition that every possible object is F in the actual world cannot be formulated in LFOML; in particular, 'is F in the actual world' is inexpressible when governed by 'every possible object'. Nevertheless, we can prove the sequent using '◇(A & B) → □(∀x)◇(B & Fx)' as a maximum formula.

1	(1)	A	Premise
2	(2)	□(∀x)□(A → Fx)	Premise
3	(3)	B	Assumption

	(4)	◇Ea	◇∃
5	(5)	Ea	Assumption
2	(6)	(∀x)□(A → Fx)	2 □E
2	(7)	Ea → □(A → Fa)	6 ∀E
2,5	(8)	□(A → Fa)	7,5 →E
2	(9)	□(A → Fa)	4,5,8 ◇E
10	(10)	◇(A & B)	Assumption
11	(11)	A & B	Assumption
2	(12)	A → Fa	9 □E
11	(13)	A	11 &E
2,11	(14)	Fa	12,13 →E
11	(15)	B	11 &E
2,11	(16)	B & Fa	15,14 &I
2,11	(17)	◇(B & Fa)	16 ◇I
2,10	(18)	◇(B & Fa)	10,11,17 ◇E
2,10	(19)	Ea → ◇(B & Fa)	18 SI (PMI)
2,10	(20)	(∀x)◇(B & Fx)	19 ∀I
2,10	(21)	□(∀x)◇(B & Fx)	20 □I
2	(22)	◇(A & B) → □(∀x)◇(B & Fx)	10,21 →I
1,3	(23)	A & B	1,3 &I
1,3	(24)	◇(A & B)	23 ◇I
1,2,3	(25)	□(∀x)◇(B & Fx)	22,24 →E
1,2	(26)	B → □(∀x)◇(B & Fx)	3,25 →I ◆

It is easy to be misled by the need to obtain line 20 in this proof, since this suggests proving (19), which in turn suggests assuming 'Ea' and discharging it later with →I. However, use of the not fully modalized 'Ea' as an assumption will interfere with uses of □I and ◇E which we wish to make. So we obtain (19) without assuming its antecedent. The same problem arises if we try to do without the assumption '◇(A & B)' and derive this formula instead by &I and ◇I (as (24) is derived). For then we will be unable to use □I as it is used at line 21, since '◇(A & B)' would be used in obtaining '(∀x)◇(B & Fx)', and so the latter would depend on (1) and (3) as well, and (1) and (3) are not fully modalized. The solution is to assume '◇(A & B)' and subsequently discharge the assumption.

Finally, as promised, we consider the logic of identity in first-order S5. Suppose an identity statement 'b = d' is true at the actual world. Then by (E=), $w*[b] = w*[d]$, and according to (RD), for every world u, $u[b] = u[d]$. Consequently, 'b = d' will be true at every world, so '□(b = d)' will be true at the actual world. That is, b = d ⊨$_{S5}$ □(b = d), so that given the completeness of our rules of inference, we must also have b = d ⊢$_{S5}$ □(b = d).

Example 7: Show b = d ⊢$_{S5}$ □(b = d).

1	(1)	b = d	Premise
	(2)	b = b	=I
	(3)	□(b = b)	2 □I
1	(4)	□(b = d)	1,3 =E ◆

The proof is quite straightforward, but the sequent itself may seem rather surprising in view of the following kind of consideration. 'Marilyn Monroe is Norma Jean Baker' is a true identity sentence but it hardly seems *necessary* that Marilyn Monroe is Norma Jean Baker. Could not Marilyn Monroe have been someone other than Norma Jean? However, the inclination we might feel to respond to affirmatively is dissipated when we recall the distinction from §1 between the *broadly logical* or *metaphysical* sense of 'possible' and 'necessary' and the *epistemic* sense. In the epistemic sense of 'could', it is true that Marilyn could have been someone other than Norma Jean. That Marilyn is Norma Jean is not implied by anything most people believe, so it is certainly epistemically possible for them that she is someone else. And if a private detective tracing Marilyn's past discovers the identity, it may still be true for him that Marilyn could have been someone else, in the sense that this was an epistemically possible outcome of his inquiry before it was initiated. But this is beside the point, since Example 7 says only that true identities are necessary in the broadly logical or metaphysical sense. *Granted* that Marilyn is Norma Jean, is there a way things could have gone in which she is someone else? Since Marilyn and Norma Jean are the same person, it seems that this would require a situation in which that person is someone else, and it is hard to make any sense of this. Example 7 is therefore more intuitively plausible than may be apparent at first sight.

❑ Exercises

Show the following:

(1) $(\exists x)\diamond\diamond Fx \vdash_{S5} (\exists x)\diamond Fx$

(2) $\square(\exists x)(\diamond Fx \vee \diamond Gx) \vdash_{S5} \square(\exists x)\diamond(Fx \vee Gx)$

(3) $(\exists x)\square(Ex \mathbin{\&} Fx) \vdash_{S5} \square(\exists x)Fx$

(4) $(\forall x)\square Fx, (\exists y)\diamond Gy \vdash_{S5} (\exists x)\diamond(Fx \mathbin{\&} Gx)$

*(5) $(\forall x)\square Ex \vdash_{S5} (\exists x)\diamond Fx \rightarrow \diamond(\exists x)Fx$

(6) $\square(\forall x)\square(Fx \rightarrow Ex) \vdash_{S5} (\exists x)\diamond Fx \rightarrow \diamond(\exists x)Fx$

(7) $\square(\forall x)\square(Fx \rightarrow {\sim}Ex) \dashv\vdash_{S5} {\sim}\diamond(\exists x)Fx$

(8) $\square(\forall x)Fx \rightarrow (\forall x)\square Fx \dashv\vdash_{S5} (\exists x)\diamond{\sim}Fx \rightarrow \diamond(\exists x){\sim}Fx$

(9) $(\forall x)\square Fx \rightarrow \square(\forall x)Fx \dashv\vdash_{S5} \diamond(\exists x){\sim}Fx \rightarrow (\exists x)\diamond{\sim}Fx$

(10) $\square(\forall x)\square Ex \vdash_{S5} \square(\forall x)Fx \leftrightarrow (\forall x)\square Fx$

(11) $\square(\forall x)\square Ex \vdash_{S5} \diamond(\exists x)Fx \leftrightarrow (\exists x)\diamond Fx$

*(12) $\diamond(\forall x)\square Ex \vdash_{S5} \diamond[\square(\forall x)Fx \rightarrow (\forall x)\square Fx]$

(13) $(\forall x)\square(\exists y)(Fx \rightarrow Gy), (\forall x)\square(\exists y)({\sim}Fx \rightarrow Gy), (\exists x)Hx \vdash_{S5} \square(\exists x)Gx$

(14) $\vdash_{S5} \square(\forall x)\square(\forall y)\square(x = y \rightarrow \square(x = y))$

(15) $a \neq b \vdash_{S5} \square(a \neq b)$

(16) $\square(\forall x)(Fx \rightarrow \square(\exists y)Rxy), \square(\forall x)\square(\forall y)\square(\forall z)\square((\diamond Rxy \mathbin{\&} \diamond Rxz) \rightarrow y = z)$
$\vdash_{S5} \square(\forall x)(Fx \rightarrow (\exists y)\square Rxy)$

(17) $\diamond(\exists x)\square(A \rightarrow {\sim}Ex) \vdash_{S5} \diamond A \rightarrow {\sim}\square(\forall x)\square Ex$

(18) $\diamond(\exists x)\diamond(\exists y)(x \neq y), A \mathbin{\&} \square(\forall x)\square(A \rightarrow Ex) \vdash_{S5} (\exists x)(\exists y)(x \neq y)$

*(19) $\square(\forall x)\square Ex, A, \diamond(\forall x)(Fx \rightarrow \square(A \rightarrow {\sim}Fx)) \vdash_{S5} \square(\exists x)\diamond{\sim}Fx$

(20) $\diamond(\forall x)(Fx \rightarrow \square(A \rightarrow {\sim}Fx)), \square(\exists x)(Fx \mathbin{\&} \square(B \rightarrow Fx)) \vdash_{S5} {\sim}\diamond(A \mathbin{\&} B)$

7 Summary

- Modal logic is the logic of the non-truth-functional operators 'possibly' or 'it is possible that', symbolized '◇', and 'necessarily' or 'it is necessary that', symbolized '□'.
- The modal system S5 is generated by treating '◇' and '□' as quantifiers over possible worlds.
- Sentential NK can be extended by adding introduction and elimination rules for '◇' and '□', resulting in a deductive system which is sound and complete for the S5 semantics.
- In order to accommodate contingent existence in first-order S5, the rules for quantifiers have to be changed from those of first-order NK to those of free logic.

10 Intuitionistic Logic

1 The motivation for intuitionistic logic

Classical logic is based on the Principle of Bivalence: every statement is either true or false and not both. One way of generating alternatives to classical logic is to reject bivalence, and of such alternatives, intuitionistic logic is the best known. Intuitionistic logic is the outgrowth of a philosophy of mathematics developed by L. J. Brouwer (1881–1966) known as intuitionism, and though intuitionistic logic can be motivated in many ways (see Dummett), a certain motivation from the philosophy of mathematics is the most accessible.

The central question in the philosophy of mathematics is, What, if anything, is pure mathematics *about?* The historically dominant philosophy of mathematics, known as *Platonism* or *realism*, answers this question by saying that pure mathematics is about a domain of mathematical entities—numbers, functions, sets and so on—a domain which exists in as robust a sense as the ordinary physical world exists. Pure mathematics is thus an enterprise of discovery, not invention, and every statement of mathematics has a truth-value, either true or false, regardless of whether anyone knows what its truth-value is. For instance, Goldbach's Conjecture, that every even number greater than 2 is the sum of two prime numbers, is either true or false according to the Platonist, though (in mid-1993) no one knows which.

The Platonist philosophy of mathematics is often contested on the grounds that the idea of an independently existing realm of mathematical entities is a myth. An alternative view which has been advanced is that mathematical entities are simply the product of human mathematical thought, and the nature of mathematical entities, that is, the nature of mathematical fact, is *determined* by the production of proofs about their properties. Consequently, if the question whether every even number greater than 2 is the sum of two primes is still open, we should not think that there is some mathematical reality which has already settled the matter, and which we have just not been very successful in discerning. For there is no more to mathematical reality than what has been, or what we can recognize could mechanically have been, proved about mathematical objects. As far as Goldbach's Conjecture is concerned, therefore, it is not merely the conjecture that is open, in the sense that it is yet to be resolved one way or the other, but mathematical reality is itself open.

From this perspective, truth and provability coincide. However, the degree of coincidence should not be overstated. It would be too strict to identify what is true with what mathematicians have in fact proved. For example, there are many values of x, y and z, in particular large ones, for which $x + y = z$ has not been proved or refuted, simply because no one has bothered to make the computation. But we can allow that for all values of x, y and z there is a mathematical fact about whether or not $x + y = z$, because although there are some values of x, y and z for which the matter has not been decided by us, we know we are in possession of methods which would decide it for those values, were we to apply one of them: any of the usual algorithms for addition is such a method.[1] By contrast, we are not in possession of any procedure of which we can now recognize that it would settle Goldbach's Conjecture if we were to apply it. Consequently, the conjecture is not settled by mathematical reality either, since that is a creature of our own articulation by proof.

If truth and provability coincide in the way indicated, we cannot explain the meanings of the logical constants in terms of the classical truth-tables. There are statements, such as Goldbach's Conjecture, which are neither provable nor refutable from our current perspective: we have not yet proved them, and not yet refuted them, and we possess no method that we know would lead to a proof or refutation if it were applied. A statement has a determinate truth-value only if it concerns a part of reality that is determinate, so statements like Goldbach's Conjecture lack determinate truth-value, an option the classical apparatus does not permit. What then should replace the classical explanations of the connectives? A truth-table explains the meaning of '&' by saying under what conditions a conjunction is true (that both conjuncts are true). Hence if truth is restricted to the realm of the provable, the idea that suggests itself is that we should explain the meaning of '&' by saying under what conditions a conjunction is *provable*. In intuitionistic jargon, the conditions under which something is provable are called its *warranted assertability conditions*.

Intuitionistic logic is the logic obtained from explanations of the meanings of the logical constants in terms of warranted assertability conditions rather than classical truth-conditions. As our previous discussion indicates, warranted assertability is time-relativized in a way that truth is not: the Platonist's mathematical reality is unchanging, but the intuitionist's mathematical reality is constantly developing as new proofs and refutations are constructed. What is warrantedly assertable at any time depends on our information at that time and does not change until that information changes. So more precisely, we will explain the meanings of connectives relative to *states of information*. We say that a state of information σ' is a *development* of a state of information σ if everything warrantedly assertable in σ is warrantedly assertable in σ'; σ' is a *proper* development of σ if in addition there are statements warrantedly assertable in σ' but not in σ.

---

[1] If x, y and z are very large, none of our methods for settling whether or not $x + y = z$ may be practical: we may not have computers fast enough, robust enough and with enough storage space, to complete the computation. However, all that intuitionists require is methods which we can recognize would *in principle* settle the matter. For further discussion of the notion of decidability in principle, see §6 of Chapter 6.

Conjunction and disjunction are the most straightforward logical constants to explain. We say:

(*Int*-&): A sentence of the form $\ulcorner p \& q \urcorner$ is warrantedly assertable in a state of information σ if and only if p is warrantedly assertable in σ and q is warrantedly assertable in σ.

(*Int*-∨): A sentence of the form $\ulcorner p \vee q \urcorner$ is warrantedly assertable in a state of information σ if and only if either p is warrantedly assertable in σ or q is warrantedly assertable in σ, or we are in possession of a method which we can recognize would extend σ to a state of information σ' in which one or the other of p and q would be warrantedly assertable.

In these clauses we use connectives such as 'if and only if' in the metalanguage, and there are two different ways of taking this. One possibility is that the connectives are classical, so that we are explaining an intuitionistic object language in a classical metalanguage. Another possibility is that these metalanguage connectives are themselves intuitionistic. If we take the latter view, we must be careful not to try to reason about the object language in an essentially classical way. Taking the metalanguage to be intuitionistic might be objected to on the grounds that we are not then *explaining* the meanings of the intuitionistic logical constants. But in the classical case there was a *prima facie* comparable circularity, since the explanations of the connectives of LSL were given in a metalanguage whose own connectives were classical.

The clause (*Int*-∨) has a certain complexity, but it would not do to say merely that a disjunction is warrantedly assertable if and only if at least one disjunct is. For each number n, $\ulcorner \mathbf{n}$ is prime or $\mathbf{n}$ is composite$\urcorner$ is warrantedly assertable by intuitionistic lights,[2] since a computer can be programmed to test any number for primality, no matter how large the number is, and it will eventually get the right answer if it has enough storage space, time, power and robustness. However, if the computation has not been carried out for a particular number n, then $\ulcorner \mathbf{n}$ is prime$\urcorner$ is not warrantedly assertable (perhaps the computer would discover that n is composite) and $\ulcorner \mathbf{n}$ is composite$\urcorner$ is not warrantedly assertable (perhaps the computer would discover that n is prime).

Since the biconditional will be defined as usual, we can complete our informal account of intuitionistic sentential logic by explaining negation and the conditional. From the classical point of view, what the intuitionist means by $\ulcorner p \rightarrow q \urcorner$ is something like $\ulcorner$we know that if p is provable then q is provable$\urcorner$. When might such a statement be warrantedly assertable? To assert it with warrant, we would have to have in our possession some method which we can see would transform any warrant for asserting p which some development of our current state of information might provide, into a warrant for asserting q. In the mathematical case, this would be a method for transforming proofs of p into proofs of q. For instance, if q is derivable from p as hypothesis, then the

---

[2] We use boldface 'n' to abbreviate 'the standard numeral for the number n'. Thus "$\ulcorner \mathbf{n}$ is prime or $\mathbf{n}$ is composite$\urcorner$" abbreviates "the result of writing the standard numeral for the number n followed by 'is prime or' followed by the standard numeral for the number n followed by 'is composite'".

method would simply be to attach the derivation of q to the end of any proof of p we might originate. These remarks motivate the following account of '→':

(*Int*-→): A sentence of the form ⌜$p \to q$⌝ is warrantedly assertable in a state of information σ if and only if there is in σ a recognizable guarantee that if σ develops into a state of information σ' warranting the assertion of p, then σ' can be extended to a state of information σ'' in which the assertion of q is warranted.

Since development need not be proper, (*Int*-→) allows for the case where p is already warrantedly assertable in σ.

Finally, what of negation? It will not do to say that ⌜$\sim p$⌝ is warrantedly assertable in σ if p is not warrantedly assertable in σ. Goldbach's Conjecture is not warrantedly assertable, but neither is its negation. Such a simple clause would also violate the idea of a created reality that becomes further articulated as investigation proceeds. In this process, the already determinate parts of reality do not alter their nature, and whatever is warrantedly assertable corresponds to a particular determination of reality. However, a statement p may not be warrantedly assertable in σ but may become warrantedly assertable in a later development σ' of σ. On the current proposal about negation, this would mean that ⌜$\sim p$⌝ would go from being warrantedly assertable to not being warrantedly assertable, violating the 'once determinate, fixed forever' feature of the 'creationist' conception of reality.

This shows that we have to bestow on negation a meaning such that if ⌜$\sim p$⌝ is ever warrantedly assertable in a state of information σ, no subsequent development of σ warrants p. The surest way to guarantee that this condition holds is to *use* it to define '∼'. So we obtain:

(*Int*-∼): A sentence of the form ⌜$\sim p$⌝ is warrantedly assertable in a state of information σ if and only if σ is recognizably such that no future development of it will warrant p.

In the mathematical case it is clear how a state of information could provide such a *carte blanche* warrant about the future, that is, if we have a proof of ⌜$p \to \wedge$⌝. But in empirical cases it is not obvious how to characterize warrant so that its requirements are neither too strong, rendering it impossible for us ever to be in a position to assert a negation with warrant, nor too weak, warranting ⌜$\sim p$⌝ when it is merely some contingent obstacle which is preventing us from discovering what is in fact the case, namely, that p.

How do NK's rules of inference or their equivalents fare in the light of these informal explanations of the meanings of the sentential connectives? It is apparent on reflection that the rules DN, LEM, Classical Reductio and Nonconstructive Dilemma have to be rejected.[3] This is most obvious in the case of LEM: if p is Goldbach's Conjecture, then p is not warrantedly assertable in the current state of information, nor is its negation, nor do we know of any way of

---

[3] For discussion of these rules, see §10 of Chapter 4.

developing the current state of information into one which would warrant either p or its negation. As for DN, if we can assert $\ulcorner \sim\sim p \urcorner$ with warrant, this means that we know that no development of our current state of information will warrant $\ulcorner \sim p \urcorner$; in other words, applying (*Int-*$\sim$) again, we know that no development of our current state of information will warrant the claim that no development of that state of information will warrant p. But there is a clear gap between this and being in a position to assert p with warrant. For example, though $\ulcorner (p \lor \sim p) \urcorner$ may not be warrantedly assertable for certain sentences p, $\ulcorner \sim\sim(p \lor \sim p) \urcorner$ is warrantedly assertable for every sentence p, as the following argument shows. To be in a position to assert $\ulcorner \sim(p \lor \sim p) \urcorner$ we would have to know that no development of our current state of information will warrant $\ulcorner (p \lor \sim p) \urcorner$, which means that we have to know that no development will warrant p and none will warrant $\ulcorner \sim p \urcorner$. But that is impossible, since if we know that no development will warrant p then by (*Int-*$\sim$) that is already sufficient warrant for $\ulcorner \sim p \urcorner$. This argument shows us that we will never be in a position warrantedly to assert $\ulcorner \sim(p \lor \sim p) \urcorner$. Again by (*Int-*$\sim$), therefore, we are always warranted in asserting $\ulcorner \sim\sim(p \lor \sim p) \urcorner$. But if p is a statement like Goldbach's Conjecture, then $\ulcorner (p \lor \sim p) \urcorner$ is not warranted. Hence DN is intuitionistically incorrect.

However, the other rules of NK are all intuitionistically acceptable. For example, if q has been derived from p together with other premises and assumptions Σ, we may assert $\ulcorner p \rightarrow q \urcorner$ depending just on Σ, since given any derivation d of p in an expansion of a state of information including Σ, we can derive q by appending to d the derivation of q from p and Σ. Similarly, if 'λ' has been derived from p together with other premises and assumptions Σ, we may infer $\ulcorner \sim p \urcorner$, depending just on Σ. For we know we will never have warrant for 'λ', and consequently, if we are in a state of information where every member of Σ is warranted, then we know we will never have warrant for p, otherwise we *would* have warrant for λ. Hence by (*Int-*$\sim$) we have warrant for $\ulcorner \sim p \urcorner$. Thus $\rightarrow$I and $\sim$I are accepted by intuitionists.

Finally, the rule EFQ, to infer any sentence whatsoever from premises which entail 'λ' (see page 135), is also intuitionistically acceptable: there is no danger of having warrant to assert everything, since if $\Sigma \vdash \lambda$, no state of information will ever warrant all the members of Σ. Consequently, all ten I and E rules for the five connectives are intuitionistically acceptable, and in addition EFQ is acceptable. These eleven rules, plus the Rule of Assumptions, constitute the subsystem of NK known as NJ, and, as we have just shown, NJ is an acceptable logic for intuitionism. However, there are systems of logic strictly between NK and NJ,[4] so it is a substantial question whether every sequent that is intuitionistically acceptable is provable in NJ, or whether NJ is too weak to capture all such sequents, so that we need instead one of the systems lying between NJ and NK (it is even conceivable that *no* system of natural deduction captures

---

[4] This means that there are systems S such that every NJ-provable sequent is S-provable, but not conversely, and every S-provable sequent is NK-provable, but not conversely. One example is the system LC which consists in NJ together with the rule D (for 'Dummett') that at any line in a proof, any instance of $\ulcorner (p \rightarrow q) \lor (q \rightarrow p) \urcorner$ may be written down, depending on no assumptions. The effect of this is gained in NK by TI, since $\ulcorner (p \rightarrow q) \lor (q \rightarrow p) \urcorner$ is a theorem of NK, but the effect of DN cannot be gained in LC, since the corresponding sequent is not LC-provable.

exactly the intuitionistically acceptable sequents). That is, it is a substantial question whether NJ is *complete* for the intuitionistic semantics we have sketched. However, before we can tackle this question, we have to be more precise about exactly what that semantics is.

2 Semantic consequence in intuitionistic sentential logic

There is more than one way of making the informal semantics for intuitionism given in the previous section more rigorous, but the easiest to work was formulated by Kripke (see Van Dalen). In intuitionistic logic as in classical, an argument-form is valid, and its conclusion is a semantic consequence of its premises, if and only if there is no interpretation on which the premises are, as we might neutrally put it, *correct*, and the conclusion incorrect. However, an interpretation in intuitionistic logic is a very different sort of thing from its classical namesake. The major difference arises from the need to relativize warranted assertability to states of information and to impose some order upon the latter to determine which states of information are developments of which so that we can evaluate negations and conditionals.

The kind of ordering which is required is called a *tree-ordering*, since the picture we want is that of a structure with a root node (the current state of information) which branches upward from there through a collection of nodes representing possible future developments of the current state of information. We define this order using the concepts introduced in §5 of Chapter 8 (they also apply to the trees we have met in other contexts earlier in this book).

Let X be a set and let R be a binary relation on X, that is, a set of pairs of elements of X. R is said to be a tree-ordering of X if and only if

(i) R is reflexive, antisymmetric and transitive;
(ii) there is an R-minimal element α in X, that is, an element α such that for all x in X, $R\alpha x$;
(iii) every element of X other than α has an immediate R-predecessor, that is, for every x ($\neq \alpha$) in X there is y in X such that: (a) Ryx; and (b) there is no z such that Ryz and Rzx;
(iv) for every x in X there is a unique R-path back from x to α, that is, for every x, y and z, if Ryx and Rzx then Ryz or Rzy.

Condition (iv) is the one that makes the difference between tree-orderings and other kinds of ordering, for example, a lattice ordering. If X = $\{\alpha,\beta,\gamma,\delta\}$ then R = $\{\langle\alpha,\beta\rangle,\langle\beta,\gamma\rangle,\langle\alpha,\gamma\rangle,\langle\alpha,\delta\rangle\}$ is a tree-ordering of X whose minimal element is α (the presence in R of all pairs $\langle x,x\rangle$ is understood). As a picture, X ordered by R is represented on the left at the top of the next page. But if we add $\langle\delta,\gamma\rangle$ to R to obtain R', we get the lattice on its right. In the lattice, clause (iv) of the definition of tree-ordering is violated, since we have $R\beta\gamma$ and $R\delta\gamma$ but neither $R\beta\delta$ nor $R\delta\beta$. As a result, there are *two* paths from γ back to α, one through β, the other through δ.

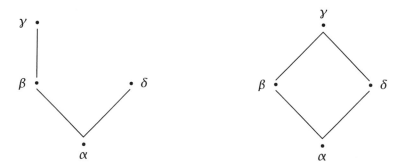

We now define an *intuitionistic interpretation* $\mathcal{K}$ for an LSL argument-form to be a quadruple $\langle S, \leqslant, \alpha, warr \rangle$ where

(a) S is a set of states of information;
(b) $\leqslant$ is a tree-ordering of S;
(c) α is the $\leqslant$-minimal element of S;
(d) *warr* is a function which associates each σ in S with some subset, possibly empty, of sentence-letters in the argument-form other than '$\wedge$'. *warr* is subject to the constraint that whenever $\sigma \leqslant \sigma'$, $warr(\sigma)$ is a subset, not necessarily proper, of $warr(\sigma')$. Intuitively, *warr* determines which atomic sentences are warrantedly assertable in which states of information, and is subject to the constraint that when an atomic sentence is warrantedly assertable in a state of information, it is warrantedly assertable in all developments of that state.

What does it mean that there may be two or more states of information which are immediate successors of a given state? Such branching represents different courses inquiry might take, according to what is investigated and what is not. For instance, we might develop our current state of information either by establishing Goldbach's Conjecture while paying no attention to the Prime Pairs Problem, or by solving the Prime Pairs Problem, paying no attention to Goldbach's Conjecture.[5]

We have described what an intuitionist interpretation of LSL is. It remains to say when such an interpretation makes a sentence correct. The *correct* sentences are the ones which are warranted at the minimal state α, but our evaluation rules have to be stated for an arbitrary member σ of S, since evaluation of conditionals and negations at α will lead us to evaluate their subformulae at developments of α. So we define the general notion 'state of information σ war-

---

[5] The Prime Pairs Problem is to prove that there are infinitely many pairs of prime numbers $\langle p, p' \rangle$ such that $p' = p + 2$. The standard example in the literature on intuitionism of an arithmetical hypothesis whose truth-value is unknown is Fermat's Last Theorem (the conjecture that $x^n + y^n = z^n$ has no solution in the positive integers for $n \geqslant 3$). Pierre de Fermat announced this theorem, but did not provide a proof, in 1637. However, on June 28th 1993, Andrew Wiles gave the first public presentation of a proof of the result. So the literature on intuitionism needs a new standard example.

rants sentence p', abbreviated '$\sigma \Vdash p$', as follows. For any sentences p and q of LSL:

(1) if p is a sentence-letter of LSL, then $\sigma \Vdash p$ if and only if p belongs to *warr*(σ);
(2) $\sigma \Vdash p$ & q if and only if $\sigma \Vdash p$ and $\sigma \Vdash q$;
(3) $\sigma \Vdash p \vee q$ if and only if $\sigma \Vdash p$ or $\sigma \Vdash q$;
(4) $\sigma \Vdash p \rightarrow q$ if and only if, for every σ' such that $\sigma \leqslant \sigma'$, if $\sigma' \Vdash p$ then $\sigma' \Vdash q$;
(5) $\sigma \Vdash \sim p$ if and only if, for every σ' such that $\sigma \leqslant \sigma'$, $\sigma' \nVdash p$.

Note that by these clauses and constraint (d) on *warr* in the definition of intuitionist interpretation, we find that in any interpretation, for any σ, $\sigma \nVdash \wedge$. Also, clause (3) is a simplified version of (*Int*-$\vee$) in the previous section. We could present the semantics in a way that is more faithful to (*Int*-$\vee$), but it turns out that the result is equivalent to the current definitions, though we will not prove this here.

We complete the story with these definitions:

(6) If $\mathcal{K} = \langle S, \leqslant, \alpha, warr \rangle$ is an intuitionistic interpretation and p is a sentence of LSL then $\mathcal{K}$ warrants p if and only if $\alpha \Vdash p$.
(7) $p_1,...,p_n \vDash_I q$ if and only if there is no intuitionistic interpretation $\mathcal{K}$ which warrants $p_1,...,p_n$ but does not warrant q. In other words, for no $\mathcal{K}$ do we have $\alpha \Vdash p_i$, $1 \leqslant i \leqslant n$, but $\alpha \nVdash q$.

To illustrate the semantics, we establish that $\sim\sim A \nvDash_I A$ and that $\sim(A \& B) \nvDash_I \sim A \vee \sim B$, by contrast with classical semantic consequence. To shorten the description of interpretations, in the specification of $\leqslant$ it is to be understood that all pairs $\langle \sigma, \sigma \rangle$ for σ in S are included.

Example 1: Show $\sim\sim A \nvDash_I A$.

Interpretation: $S = \{\alpha, \beta\}$, $\leqslant = \{\langle \alpha, \beta \rangle\}$, *warr*($\alpha$) = $\varnothing$, *warr*(β) = $\{'A'\}$, or as pictured below:

Explanation: $\alpha \nVdash A$ since 'A' $\notin$ *warr*(α). But $\alpha \Vdash \sim\sim A$ by (5), since $\alpha \nVdash \sim A$ and $\beta \nVdash \sim A$.

Note that if we had left *warr*(β) empty, we would have $\beta \Vdash \sim A$, since $\beta \nVdash A$, and trivially no other development of β warrants 'A' since there are no other developments of β. More generally, in the interpretation $S = \{\alpha\}$, *warr*(α) = $\varnothing$, we have $\alpha \Vdash \sim p$ for every sentence-letter p.

Example 2: Show ~(A & B) $\nVdash_I$ ~A ∨ ~B.

Interpretation: $S = \{\alpha,\beta,\gamma\}$, ⩽ $= \{\langle\alpha,\beta\rangle,\langle\alpha,\gamma\rangle\}$, *warr*($\alpha$) = ∅, *warr*($\beta$) = {'A'}, *warr*($\gamma$) = {'B'}, as pictured below:

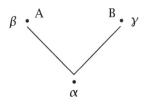

Explanation: No state in S warrants 'A & B', so α ⊩ ~(A & B) (as do β and γ). But α ⊮ ~A since we have both (i) α ⩽ β and (ii) β ⊩ A; similarly, α ⊮ ~B, since α ⩽ γ and γ ⊩ B. Consequently, α ⊮ ~A ∨ ~B.

Many other familiar validities of classical logic are intuitionistically invalid; some of them are given as exercises below. However, one relationship between $\vDash_I$ and $\vDash_C$ is worth remarking on. If Γ is a set of LSL sentences, let (~~)Γ be the set which results from prefixing '~~' to every member of Γ. Then we have the following 'embedding' of classical sentential logic in intuitionistic logic:

E: $\Gamma \vDash_C q$ if and only if (~~)$\Gamma \vDash_I$ ~~q.

The right-to-left half of E is established in two steps:

(a) If (~~)$\Gamma \vDash_I$ ~~q then (~~)$\Gamma \vDash_C$ ~~q

and

(b) If (~~)$\Gamma \vDash_C$ ~~q then $\Gamma \vDash_C q$.

(a) holds because for every classical interpretation $\mathcal{I}$ there is an equivalent intuitionistic interpretation $\mathcal{K} = \langle S,⩽,\alpha,warr\rangle$ in which $S = \{\alpha\}$, and for each sentence-letter p, $p \in warr(\alpha)$ if and only if $\mathcal{I}$ assigns ⊤ to p. Since there are no other states in S, we get as a consequence that α ⊩ ~p for every sentence-letter to which $\mathcal{I}$ assigns ⊥. More generally, since α is the only state of information, the intuitionistic clauses agree with the classical ones for every formula. Consequently, $\mathcal{K}$ warrants exactly the same sentences as $\mathcal{I}$ makes true. So if a classical interpretation $\mathcal{I}$ makes every member of (~~)Γ true and ⌜~~q⌝ false, the corresponding $\mathcal{K}$ warrants every member of (~~)Γ and does not warrant ⌜~~q⌝. This proves (a) via its contrapositive. (b) is an immediate consequence of the fact that a sentence and its double negation are logically equivalent on classical semantics. Thus we have established the right-to-left direction of E.

The left-to-right direction of E is trickier. We prove it in its contrapositive formulation: if (~~)$\Gamma \nvDash_I$ ~~q then $\Gamma \nvDash_C q$. Suppose (~~)$\Gamma \nvDash_I$ ~~q. Then we have an intuitionistic interpretation which warrants every member of $(\sim\sim)\Gamma$ but not $\ulcorner\sim\sim q\urcorner$. The problem is to define a classical interpretation in which every member of $(\sim\sim)\Gamma$ is true and $\ulcorner\sim\sim q\urcorner$ false. To do this we need the following result about intuitionistic sentential logic, which we will not prove (for a proof, see Van Dalen):

> F: If there is any intuitionistic interpretation which warrants every member of $(\sim\sim)\Gamma$ but not $\ulcorner\sim\sim q\urcorner$, then there is an intuitionistic interpretation which does this in which S is finite.

So suppose $\mathcal{K}$ is a finite intuitionistic interpretation $\langle S,\leqslant,\alpha,warr\rangle$ such that α warrants every member of $(\sim\sim)\Gamma$ and α does not warrant $\ulcorner\sim q\urcorner$. It follows that there must be some state β such that $\alpha \leqslant \beta$, β warrants every member of $(\sim\sim)\Gamma$, and $\beta \Vdash \sim\sim q$. Say that a state γ is an *endstate* if and only if it has no successors: $\sim(\exists\sigma)(\gamma < \sigma)$. Since $\mathcal{K}$ is finite, there must be some endstate γ in $\mathcal{K}$ such that $\beta \leqslant \gamma$, and such a γ will also warrant every member of $(\sim\sim)\Gamma$ and $\ulcorner\sim q\urcorner$.[6] Choose an endstate γ in $\mathcal{K}$ above β and define the classical interpretation $\mathcal{I}$ by: for every sentence-letter p, $\mathcal{I}$ assigns $\top$ to p if $p \in warr(\gamma)$, $\mathcal{I}$ assigns $\bot$ to p if $p \notin warr(\gamma)$. We can then prove:

> H: For every sentence p of LSL, $\mathcal{I}$ verifies p if and only if σ warrants p.

H guarantees that $\mathcal{I}$ is the classical interpretation we want: $\mathcal{I}$ establishes that $(\sim\sim)\Gamma \nvDash_C \sim\sim q$ and therefore that $\Gamma \nvDash_C q$, thus completing the proof of (the contrapositive of) the left-to-right direction of E. H is itself proved by a technique known as induction on the structure of formulae, which is beyond the scope of this book.

Finally, we return to the question which prompted the construction of a formal account of intuitionistic semantic consequence, as to whether or not there is a conception of intuitionistic validity according to which sentential NJ (= NK except that EFQ replaces DN) is a sound and complete system of inference. The answer is that Kripke's semantics, which we have just outlined, defines such a conception for sentential logic. In other words, we can prove: for any set of LSL sentences Γ and LSL sentence q,

(Snd_I): If $\Gamma \vdash_{NJ} q$ then $\Gamma \vDash_I q$

and

(Comp_I): If $\Gamma \vDash_I q$ then $\Gamma \vdash_{NJ} q$.

As in the classical case, (Snd_I) is fairly obviously true: the proof essentially reca-

---

[6] Why? Here we are relying on another feature of the semantics, that it is *persistent*. It is built into the semantics that atomic sentences which are warranted at a state σ are warranted at every development of σ. It is possible to prove from this that every sentence whatsoever which is warranted at a state σ is warranted at every development of σ. This captures the idea of the unrevisability of the ideal realm which is created by mathematical thought.

pitulates the informal justifications of the rules which we gave in the previous section. Also as in the classical case, (Comp$_I$) is a substantial claim, since we have no easy way of telling if all the rules needed to prove the validities of Kripke's semantics are present in NJ. In fact, NJ has all the necessary rules. However, there is a contrast with the classical case over the hypothetical question of what would have to be done if the rules had turned out to be incomplete with respect to the semantics. The correctness of the classical semantics has a kind of self-evidence, granted the fundamental assumption of bivalence: there is really nothing controversial about the method of truth-tables or the definition of ⊨$_C$. So if NK were to prove incomplete, that would simply mean that some further rule of inference is needed; we would not think that the problem lies with the semantics. But if NJ were to prove incomplete with respect to Kripke's semantics, it would be just as plausible to regard the problem as lying in the semantics as in the system of rules; for even given the intuitive intuitionistic explanations of the meanings of the logical constants, the semantics is not *self-evidently* correct. Why, for instance, insist on ordering states of information as *trees?* So if NJ were incomplete with respect to Kripke's semantics, and if there were no intuitionistically acceptable way of making a nonconservative extension of NJ, the conclusion to draw would be that some aspect of Kripke's semantics outruns the content of the intuitive explanations of the logical constants. Given the direct intuitive justification of the rules of NJ, these principles of inference would represent the fundamental way of making the meanings of the connectives precise, and any semantics would have to answer to them.

Even though sentential NJ is in fact complete for Kripke's semantics, the foregoing comments are not thereby rendered beside the point, for it is still conceivable that there is an aspect of the semantics which is intuitionistically dubious, so that it is to some extent accidental that NJ is complete with respect to them. Some intuitionists hold that this situation obtains with respect to the extension of Kripke semantics to first-order logic, to which we now turn. For them, therefore, the semantics is really no more than a formal device for establishing unprovability in NJ.

❏ Exercises

I   Use intuitionistic interpretations of LSL to establish the following (all are classically valid):

(1) ⊬$_I$ A ∨ ~A
(2) ⊬$_I$ ~A ∨ ~~A
(3) ⊬$_I$ (A → B) ∨ (B → A)
(4) ~~A → A ⊬$_I$ A ∨ ~A
*(5) ~B → ~A ⊬$_I$ A → B
(6) (A → B) → B ⊬$_I$ A ∨ B
(7) (A → B) → A ⊬$_I$ A
(8) (A → B) ⊬$_I$ ~A ∨ B
(9) ~(A → B) ⊬$_I$ A & ~B

(10) ~(~A ∨ ~B) ⊬$_I$ A & B
(11) ~(~A & B), ~(~A & ~B) ⊬$_I$ A
*(12) ⊬$_I$ A ∨ (A → B)
(13) A → (B ∨ C) ⊬$_I$ (A → B) ∨ (A → C)
(14) (A ∨ B) → (A ∨ C) ⊬$_I$ A ∨ (B → C)
(15) A ↔ B ⊬$_I$ (A & B) ∨ (~A & ~B)

II   Recall that NJ is the system obtained from NK by replacing DN with EFQ (see page 135 for a statement of EFQ). Show the following. You may use SI only if you have already proved the sequent to which you appeal *in NJ:*

(1) ~A ⊢$_{NJ}$ A → B
(2) A ∨ B, ~A ⊢$_{NJ}$ B
(3) ~A ∨ B ⊢$_{NJ}$ A → B
(4) A ∨ ~A ⊢$_{NJ}$ ~~A → A
(5) ⊢$_{NJ}$ ~~(~A ∨ ~~A)
(6) ~~⋏ ⊢$_{NJ}$ ⋏

3 Intuitionistic monadic predicate logic

In classical mathematics there are two ways of establishing an existential sentence $\ulcorner(\exists v)\phi v\urcorner$, where ϕv is some mathematical formula. The *constructive* method is to derive a specific instance ϕt, or to provide an effective means for finding such an instance, while the *nonconstructive* method is, essentially, to show that $\ulcorner\sim(\exists v)\phi v\urcorner$ leads to a contradiction. Since we need DN to infer $\ulcorner(\exists v)\phi v\urcorner$ from this, intuitionists allow only constructive methods in mathematical proofs, and the constructivity requirement is treated as part of the meaning of '∃':

(Int-∃): A sentence of the form $\ulcorner(\exists v)\phi v\urcorner$ is warrantedly assertable in a state of information σ if and only if we have a method of effectively producing in σ an object x such that we are warranted in asserting ϕt, where t is a term referring to x.

In seeking a companion clause for '∀', a complication arises from the fact that objects are conceived of as coming into existence as the products of mathematical thinking. This means that one way in which a state σ' can be a development of σ is that more objects may exist in σ' than do in σ. Therefore, it is not sufficient for the warranted assertability of a universal sentence in σ that all its instances merely in σ be warrantedly assertable. For this condition is consistent with a counterexample appearing among objects which come into existence at a later state of information, and since a correct assertion determines reality unrevisably, we cannot have a universal sentence warranted today but not tomorrow. So a clause for '∀' must refer to future states in the same way the clauses for '~' and '→' did:

(Int-∀): A sentence of the form $\ulcorner(\forall v)\phi v\urcorner$ is warrantedly assertable in a state of information σ if and only if we have a method of which we can recognize that its application in σ or any subsequent state of information will transform any warrant for asserting that x exists into warrant for asserting ϕt, where t is a term referring to x.

As a result of (Int-∃) and (Int-∀), quantifiers in intuitionistic first-order logic are no longer interdefinable, since '~(∀x)Fx' does not entail '(∃x)~Fx'. A state of information warrants '~(∀x)Fx' if we can recognize in it that neither it nor any subsequent development will provide an effective procedure of the kind described in (Int-∀). But recognizing that we are in a position to assert some sentence $\ulcorner$~Ft$\urcorner$ with warrant is only one way of recognizing that there will never be such a procedure; perhaps all we are in a position to assert with warrant is the conditional '(∀x)Fx → ⋏'. Consequently, '~(∀x)Fx' is intuitionistically much weaker than '(∃x)~Fx'.

To give a formal definition of '⊨$_I$' for intuitionistic logic incorporating (Int-∀) and (Int-∃), we simply add some apparatus to the Kripke semantics for LSL (we consider only monadic logic, since the main issues arise in its framework). Each state of information σ must be provided with its own domain of discourse, subject to the constraint that if σ' develops from σ, then anything which exists in σ must also exist in σ' (once mathematical thought has introduced, say, complex numbers, there is no going back on it). In addition, to each σ, *warr* assigns a set whose members are either atomic sentence-letters (never '⋏') or else are pairs whose first element is a monadic predicate symbol and whose second member is an element of $dom(\sigma)$. Intuitively, if the pair $\langle\phi,j\rangle$ is in $warr(\sigma)$, this means that the predicate ϕ is warrantedly predicable of j in σ. There is a concomitant constraint that if a predicate is warrantedly predicable of an object in σ, it remains so in all σ's developments.

With these explanations we define an intuitionistic interpretation for an argument-form in LMPL to be a sextuple

$$\mathcal{K} = \langle S,\leqslant,\alpha,dom,ref,warr\rangle$$

in which S, $\leqslant$ and α are as before. *dom* is a function which assigns a nonempty set to each $\sigma \in S$ such that if $\sigma \leqslant \sigma'$, then $dom(\sigma)$ is a subset of $dom(\sigma')$. *ref* is a function which assigns each individual constant in the argument-form a reference. $ref(t)$ is always an element of $dom(\alpha)$, since in the object language we can talk only about things which currently exist: α is the current state, and we must ascend to the metalanguage for the 'God's-eye view' on which the possible future developments of α are laid out. Finally, *warr* is a function which assigns each $\sigma \in S$ a set whose members are sentence-letters or predicate/object pairs, as described earlier; again, if $\sigma \leqslant \sigma'$, then $warr(\sigma)$ is a subset of $warr(\sigma')$.

Since we are already using Greek letters in the metalanguage for states of information, we will use the dingbats '❶', '❷' and so on to stand for elements of the domains of states. But in forming instances of quantified sentences there is the complication that we cannot use object-language names for things that do not exist in $dom(\alpha)$. So by contrast with the classical semantics for monadic predicate logic, we will say that an object-language existential sentence is true if and only if it has a true instance formed from the object-language predicate and either an object-language *or* a metalanguage individual constant. For purposes of evaluating quantified sentences, therefore, we assume that the function *ref* of $\mathcal{K}$ is also defined for the metalanguage terms used in specifying the domains of the states in S. The *ref*-clauses for these metalanguage terms are 'trivial', in that we give the reference of a metalanguage name in the metalanguage by using that very name. Thus our assumption amounts to *ref* having the implicit clauses $ref(\text{'}\mathbf{0}\text{'}) = \mathbf{0}$, $ref(\text{'}\mathbf{2}\text{'}) = \mathbf{2}$ and so on.

The complete scheme of evaluation clauses for intuitionistic monadic predicate logic is as follows:

(1a) $\sigma \Vdash p$ if and only if $p \in warr(\sigma)$, for any atomic sentence p of LSL;

(1b) $\sigma \Vdash \phi t$ if and only if $\langle \phi, ref(t) \rangle \in warr(\sigma)$, for any atomic predicate ϕ of LMPL and individual constant t of LMPL or the metalanguage

(2) $\sigma \Vdash p \,\&\, q$ if and only if $\sigma \Vdash p$ and $\sigma \Vdash q$;

(3) $\sigma \Vdash p \lor q$ if and only if $\sigma \Vdash p$ or $\sigma \Vdash q$;

(4) $\sigma \Vdash p \to q$ if and only if, for every σ' such that $\sigma \leqslant \sigma'$, if $\sigma' \Vdash p$ then $\sigma' \Vdash q$;

(5) $\sigma \Vdash {\sim}p$ if and only if, for every σ' such that $\sigma \leqslant \sigma'$, $\sigma' \nVdash p$;

(6) $\sigma \Vdash (\exists v)\phi v$ if and only if there is some object x in $dom(\sigma)$ such that $\sigma \Vdash \phi t$ and $ref(t) = x$;

(7) $\sigma \Vdash (\forall v)\phi v$ if and only if for every σ', if $\sigma \leqslant \sigma'$ then for every $x \in dom(\sigma')$, $\sigma' \Vdash \phi t$, where $ref(t) = x$.

As before, an intuitionistic interpretation $\mathcal{K}$ of LMPL warrants a sentence if that sentence holds at the minimal element α of $\mathcal{K}$, while a sequent is intuitionistically valid if and only if no intuitionistic interpretation of it warrants its premises but not its conclusion.

To illustrate the semantics, here is a demonstration of the intuitionistic invalidity of a classical validity:

Example 1: Show $(\forall x)(Fx \lor A) \nvDash_I (\forall x)Fx \lor A$

Interpretation: $S = \{\alpha,\beta\}$, $\leqslant = \{\langle\alpha,\beta\rangle\}$, $dom(\alpha) = \{\mathbf{0}\}$, $dom(\beta) = \{\mathbf{0},\mathbf{2}\}$, $warr(\alpha) = \{\langle\text{'F'},\mathbf{0}\rangle\}$, $warr(\beta) = \{\langle\text{'F'},\mathbf{0}\rangle, \text{'A'}\}$. This is pictured below, with the domains of the states displayed inside them:

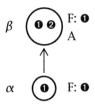

Explanation: $\alpha \Vdash F\mathbf{0} \lor A$, since $\alpha \Vdash F\mathbf{0}$, while $\beta \Vdash F\mathbf{0} \lor A$ and $\beta \Vdash F\mathbf{2} \lor A$ (since

$\beta \Vdash A$). By clause (7), therefore, we have $\alpha \Vdash (\forall x)(Fx \vee A)$. However, $\alpha \nVdash A$, since 'A' $\notin warr(\alpha)$, and $\alpha \nVdash (\forall x)Fx$, since $\beta \nVdash F\mathbf{2}$. Thus $\alpha \nVdash (\forall x)Fx \vee A$.

A number of classical validities can be shown to be intuitionistically invalid by comparably simple interpretations. However, unlike classical monadic predicate logic, intuitionistic monadic predicate logic is undecidable, so some sequents must have only infinite counterexamples (otherwise the procedure outlined in §7 of Chapter 8 would be a decision procedure, granted the completeness of NJ with respect to the current semantics). Here is an example of such a sequent:

Example 2: Show $(\forall x)\sim\sim Fx \nvDash_I \sim\sim(\forall x)Fx$.

Interpretation: $S = \{\sigma_1, \sigma_2, \sigma_3,...\}$, $\alpha = \sigma_1$, $\beta = \sigma_2$, $\gamma = \sigma_3$ and so on, $\leqslant = \{\langle\sigma_i,\sigma_j\rangle:$ $i \leqslant j\}$, $dom(\sigma_i) =$ the first i elements in the sequence $\mathbf{1}\ \mathbf{2}\ \mathbf{3}\ \mathbf{4}\ \mathbf{5}...$, $warr(\sigma_1) = \emptyset$, $warr(\sigma_2) = warr(\sigma_1) \cup \{\langle 'F',\mathbf{1}\rangle\}$, $warr(\sigma_3) = warr(\sigma_2) \cup \{\langle 'F',\mathbf{2}\rangle\}$ and so on, as pictured below:

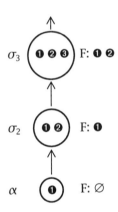

Explanation: $\alpha \Vdash (\forall x)\sim\sim Fx$, since for every σ_i, $\sigma_i \Vdash \sim\sim Fn$ for all $n \in dom(\sigma_i)$, because given any σ_i, there is no $\sigma_j \geqslant \sigma_i$ such that $\sigma_j\Vdash \sim Fn$. Rather, if $n < j$, then $\sigma_j \Vdash Fn$, and if $n \geqslant j$, then $\sigma_{n+1} \Vdash Fn$, and $\sigma_i \leqslant \sigma_{i+1}$. On the other hand, $\alpha \nVdash \sim\sim(\forall x)Fx$, since for all i, $\sigma_i \Vdash \sim(\forall x)Fx$. This is because for all $j \geqslant i$, $\sigma_j \nVdash (\forall x)Fx$, since in each σ_j there is an instance $\ulcorner Ft \urcorner$ of '$(\forall x)Fx$' with $ref(t) \in dom(\sigma_j)$, such that $\sigma_j \nVdash Ft$.

It is also easy to see that no counterexample with finite S is possible. For if $\alpha \Vdash (\forall x)\sim\sim Fx$ then every state above α warrants '$(\forall x)\sim\sim Fx$', including the end-states that must exist if S is finite. Let σ be an endstate. Then $\sigma \Vdash (\forall x)\sim\sim Fx$, so for every n in $dom(\sigma)$, $\sigma \Vdash \sim\sim Fn$. Hence $\sigma \nVdash \sim Fn$, so $\sigma' \Vdash Fn$ for some $\sigma' \geqslant \sigma$. But σ is the only such σ', so $\sigma \Vdash Fn$. Thus $\sigma \Vdash (\forall x)Fx$, and so $\alpha \Vdash \sim\sim(\forall x)Fx$.

This completes our characterization of $\vDash_I$ for LMPL. As a final note, we remark that if we add our familiar I and E rules for '$\forall$' and '$\exists$' to sentential NJ, we get a system of rules that is sound and complete with respect to these semantics. So no new issues are raised by the derivation relation, which is why we have focused on the semantics.

❑ Exercises

Use intuitionistic interpretations of LMPL to show the following:

   (1) ~(∀x)Fx ⊭$_I$ (∃x)~Fx
   (2) ~(∃x)~Fx ⊭$_I$ (∀x)Fx
   (3) ~~(∃x)Fx ⊭$_I$ (∃x)Fx
   (4) ~(∃x)(Fx & ~Gx) ⊭$_I$ (∀x)(Fx → Gx)
   (5) A → (∃x)Fx ⊭$_I$ (∃x)(A → Fx)
 *(6) (∃x)Fx → (∃x)Gx ⊭$_I$ (∃x)(Fx → Gx)
   (7) ⊭$_I$ ~~[(∀x)Fx ∨ (∃x)~Fx]
   (8) ⊭$_I$ ~~[(∀x)(Fx ∨ ~Fx)]
 *(9) ⊭$_I$ (∃x)(∀y)(Fx → Fy)
 (10) ⊭$_I$ (∃x)[(∃y)Fy → Fx]
 (11) ⊭$_I$ (∃x)(Fx → (∀y)Fy)
 (12) ⊭$_I$ (∀x)((Fx → Gx) ∨ (Gx → Fx))
 (13) (∀x)(∃y)(Fx → Gy), (∀x)(∃y)(~Fx → Gy) ⊭$_I$ (∃z)Gz
 (14) (∀x)Fx → (∀x)Gx ⊭$_I$ (∃x)(∀y)(Fx → Gy)
 (15) (∀x)(∃y)(Fx → Gy) ⊭$_I$ (∃x)(∀y)(Fy → Gx)

11 Fuzzy Logic

1 Sorites paradoxes

Imagine a thousand giraffes $g_1,...,g_{1000}$ crossing the Serengeti plain in single file, with the tallest giraffe g_{1000} leading the way. Suppose that careful measurement has established that g_1 is exactly 2.005 meters tall and that each other giraffe is five millimeters taller than the giraffe immediately behind it, so that g_{1000} is 7 meters tall. There is therefore a very perceptible difference in height between g_1 and g_{1000}, but the height difference between any two successive giraffes is entirely imperceptible to the naked eye.

g_{250} g_{500} g_{750} g_{1000}

The following three statements appear to be true in this situation:

(A) g_1 is short.
(B) For every i, $1 \leqslant i \leqslant 1000$, if g_i is short then g_{i+1} is short.
(C) g_{1000} is not short.

Understanding 'short' as 'short for a giraffe', (A) and (C) are hardly open to discussion. (B) might be queried, but it is difficult to see how it could be *false*. Someone who claims that (B) is false has to hold that there is a counterexample to it: there are successive giraffes g_k and g_{k+1}, g_k immediately behind g_{k+1}, such that g_k is short and g_{k+1} is not short. But no matter how hard anyone stares at any pair of successive giraffes, no difference in height will be discernible. So how could it be that one of them is short and the other not? Of course, there is in fact a difference in height between adjacent giraffes, but it is so slight as not to be noticeable, and such a difference is too small to constitute grounds for applying 'short' to one giraffe and withholding it from the other. The predicate 'short' is one which is applied on the basis of how things ordinarily look, and we cannot impose an exact cut-off point on its application without abandoning this central point of having it in the language.

Yet as the reader has probably already seen, (A), (B) and (C) form an NK-inconsistent set of sentences (recall that Γ is NK-inconsistent if and only if $\Gamma \vdash_{NK} \curlywedge$). We can demonstrate this in sentential logic, since if (B) is true then there are 999 true conditionals, one for each giraffe, to the effect that if it is short so is the giraffe in front of it. Consequently, we can construct the following argument:

$$
\begin{array}{lll}
\text{S} & (1) & g_1 \text{ is short} \\
& (2) & g_1 \text{ is short} \rightarrow g_2 \text{ is short} \\
& (3) & g_2 \text{ is short} \rightarrow g_3 \text{ is short} \\
& \vdots & \vdots \\
& \vdots & \vdots \\
& (1000) & g_{999} \text{ is short} \rightarrow g_{1000} \text{ is short} \\
& (1001) & \therefore g_{1000} \text{ is short.}
\end{array}
$$

S is provable in NK, by 999 applications of $\rightarrow$E, and its conclusion contradicts (C), hence the derivability of '$\curlywedge$'.

The argument S is one of a family called *Sorites paradoxes*, or *paradoxes of vagueness*, in which we use apparently incontestable reasoning to carry us from truths to falsehoods. Other examples include: (i) the paradox of the heap—100,000 grains of sand make a heap, removing one grain from a heap leaves a heap, consequently one grain of sand makes a heap; (ii) the paradox of the bald man—if x is not bald, then if y has one hair less on his head than x then y is not bald, therefore no-one is bald; (iii) the black-is-white paradox—if an object x is black, then any colorwise indistinguishable object is black, so a white object, arrived at through a sequence of shades of gray in which the percentage of white increases smoothly and in which each shade is visually indistinguishable from the previous one, is black; (iv) the voting age paradox—if a person is not sufficiently mature to merit the right to vote at a time t then that person is not sufficiently mature to merit the right to vote at time $t + \delta$, where δ is as small a magnitude as you please, hence since we are all at some point not sufficiently mature to merit the right to vote, none of us are ever sufficiently mature to merit the right; finally, on a topical note, (v) the paradox of human existence—a newly-born baby is biologically a human being, anything which is biologically a human being at a time t was biologically a human being at time $t - \delta$, where δ is as small a magnitude as you please, therefore the fertilized egg from which the baby develops is biologically a human being.

All these arguments exploit the same phenomenon. There is some predicate—'short', 'black', 'mature', 'human being'—which is applied to objects on
the basis of their possessing certain features—height, color, stage of psychological or biological development—and these features have a range of variation
within which they can alter in smaller increments than are marked by predicates for those features. Each predicate continues to apply if the change in the
relevant feature is not too great but gives way to a different predicate for that
sort of feature when the change is significant. However, the paradoxical reasoning forces us to apply the same predicate to two cases which do differ significantly in the relevant respects by approaching one case from the other through
a sequence of intermediate cases where each change is insignificant. The problem is that many insignificant changes add up to a significant one.

Clearly, something has to give: granted that the first premise of a Sorites
paradox is true and its conclusion false, then either some conditional premise
must be false, or there must be a problem with the semantic framework within
which we are working. In the next section, we will consider an approach to the
paradoxes based on the idea that the classical semantic framework is the
source of the trouble.

2 Degrees of truth

There are different approaches to the treatment of 'unsharp' predicates like
'short', 'white', 'mature', and so on, but the one we present here is intrinsically
very plausible, and in the computer-science community at least, has won out
over its rivals. In virtue of the Principle of Bivalence, classical logic offers us
only two possibilities for a predicate ϕ and an object x: either ϕ applies to x or
ϕ does not apply to x. But when we consider the examples of the previous section, it is natural to ask for more options: as the features of a thing change by
small increments in a particular direction, one wants to say that the predicate
in question becomes *less and less* applicable to the object, until it is no longer
applicable. Suppose we have some way of quantifying how well a predicate
applies to an object; then we can call this value the *degree* to which the predicate applies to the object. So in our line of progressively taller giraffes, we start
out with a giraffe to which the predicate 'short' applies to the maximum degree.
But as we move along the line, the degree to which the predicate applies begins
to drop, while the degree of applicability of the predicate 'medium height'
begins to increase, until somewhere toward the middle we reach a point where
'short' is completely inapplicable while 'medium height' applies to maximum
degree.

Of course, any particular assignment of degrees of shortness will have arti-
ficialities; for instance, there will be a last giraffe which is short to the maximum degree, while nothing special distinguishes that giraffe from those close
to it. But degrees of shortness are not assigned on the basis of how things look
unless the degrees are very rough: 'fairly short', 'quite short' and so on. If we
are thinking of degrees based on exact measurement, then although there is no

one assignment of degrees which is the right one, there are many which are equally reasonable, and the question to ask is how the Sorites paradoxes stand up in a framework in which the correct logical principles are those that hold for any reasonable assignment of degrees.

If there are degrees to which a predicate applies, then there must also be degrees of truth. For example, if g_{250} is fairly short, then the statement 'g_{250} is short' can be said to be fairly true; in general, a statement to the effect that x is ϕ will be true to the degree that x satisfies ϕ. In ordinary language we speak of statements being wholly true, partly true, almost wholly false and so on. These phrases can be understood as standing for *fuzzy* degrees of truth: degrees which do not have precise boundaries. But we can also assign precise degrees of truth. Since we are concerned with features for which the range of variability may permit arbitrarily small changes, we should choose a continuous set to represent degrees of truth. The segment of the real number line from 0 to 1 inclusive, [0,1], is the simplest model, and we shall work with it. So we are abandoning the Principle of Bivalence: far from there being merely two truth-values, there is a continuum of truth-values ranging from wholly true, 1, to wholly false, 0. The classical truth-values $\top$ and $\bot$ are simply the endpoints of this continuum. *Fuzzy logic* is that logic which is generated by replacing the classical evaluation clauses for the connectives by counterparts which take account of the whole range of truth-values. We now present one particular scheme which is appropriate for understanding Sorites paradoxes. The principles are the same as those that regulate the fuzzy controllers of many contemporary electronic appliances.

3 Fuzzy semantics for LFOL

A classical interpretation for the language of first-order logic is a domain D and an assignment of truth-values to atomic sentences, that is, sentence-letters and structured atomic sentences containing names of elements of the domain (we exploit the equivalence between a sequence of objects being in the extension of a predicate-letter and a corresponding atomic sentence being true). The set of truth-values is now [0,1] instead of $\{\bot,\top\}$, so we define a *fuzzy interpretation* to be a triple $\mathcal{F} = \langle D,deg,ref \rangle$, where (i) D is a domain of discourse; (ii) *deg* is an assignment of degrees of truth in [0,1] to atomic sentences containing names of elements of the domain; and (iii) *ref* specifies the reference of each individual constant in the atomic sentences for which *deg* is defined. A classical interpretation is therefore simply a special case of a fuzzy interpretation, one where the only values in the interval [0,1] which are used are 0 and 1.

We now have to change the evaluation clauses for the logical constants, since they only tell us how the truth-value of a compound sentence is derived from its constituents when the truth-values are 0 and 1. However, we have some guidelines in arriving at new clauses, since new clauses should agree with the classical ones when 0 and 1 are the only degrees of truth in question, that is, they should agree with classical ones on classical interpretations. They should also be in reasonable accord with our intuitions about such sentences as make up the premises and conclusion of Sorites paradoxes. Writing '$\mathcal{F}[\phi] = n$' to mean that the degree of truth of formula ϕ on interpretation $\mathcal{F}$ is n, we begin with the stipulation

(*F-at*): If p is an atomic sentence of LFOL, $\mathcal{F}[p] = deg(p)$; $\mathcal{F}[\curlywedge] = 0$

where '*F*' indicates 'fuzzy', not 'formation'.

Negation, conjunction and disjunction are straightforward. A statement which is wholly true is not false to any degree, and vice versa, and a statement which is only partly true is false to a *concomitant* degree; for instance, something which is almost wholly true should be hardly false at all. This suggests the following. For any formula p of LFOL,

(*F-~*): $\mathcal{F}[\ulcorner \sim p \urcorner] = 1 - \mathcal{F}[p]$.

So if the degree of truth of 'A' is 0.75, then the degree of truth of '~A' is 0.25.

In the classical case, a conjunction takes the worse of the two values of its conjuncts and a disjunction the better. We can use exactly the same condition for degrees of truth. We write '$max\{...\}$' for the largest element of the set $\{...\}$ and '$min\{...\}$' for the smallest.

(*F-&*): $\mathcal{F}[\ulcorner p \ \& \ q \urcorner] = min\{\mathcal{F}[p], \mathcal{F}[q]\}$.
(*F-∨*): $\mathcal{F}[\ulcorner p \lor q \urcorner] = max\{\mathcal{F}[p], \mathcal{F}[q]\}$.

These clauses are not arbitrary, since there are constraints applicable to any semantics for '~', '&' and '∨' and it turns out that the three functions $1 - x$, *min* and *max* are the *only* functions which satisfy them. Let $h_\&$ and $h_\lor$ be binary functions into [0,1] defined for all pairs of real numbers x, y with $x, y \in [0,1]$ and which satisfy the six conditions below; then it can be proved, though we will not go into the details, that $h_\&$ must be *min* and $h_\lor$ must be *max*. The six conditions are:

(i) $\mathcal{F}[\ulcorner p \ \& \ q \urcorner] = h_\&(\mathcal{F}[p], \mathcal{F}[q])$ and $\mathcal{F}[\ulcorner p \lor q \urcorner] = h_\lor(\mathcal{F}[[p], \mathcal{F}[q])$;
(ii) $h_\&$ and $h_\lor$ are commutative, associative and mutually distributive;
(iii) $h_\&$ and $h_\lor$ are continuous and non-decreasing with respect to each input;
(iv) $h_\&(\mathcal{F}[p], \mathcal{F}[p])$ and $h_\lor(\mathcal{F}[p], \mathcal{F}[p])$ are strictly increasing;
(v) $h_\&(\mathcal{F}[p], \mathcal{F}[q]) \leqslant min\{\mathcal{F}[p], \mathcal{F}[q]\}$ and $h_\lor(\mathcal{F}[p], \mathcal{F}[q]) \leqslant max\{\mathcal{F}[p], \mathcal{F}[q]\}$;
(vi) $h_\&(1,1) = 1$ and $h_\lor(0,0) = 0$.

Here (i) requires that the value of a conjunction be determined only by the function $h_\&$ applied to the values of the conjuncts, *mutatis mutandis* for disjunction. (ii) is required for standard properties of conjunction and disjunction, as expressed in the NK sequents Com (page 123), Example 4.5.6 (page 115) and Dist (page 123). (iii) says that if the degree of truth of p increases or decreases through a range of values without jumps, so the degree of truth of ⌜p & q⌝ changes, if at all, without jumps, and does not decrease unless the degree of truth of p decreases (similarly for q). The other conditions are self-explanatory.

Similarly, $1 - x$ is the only function h which satisfies the following conditions:

(i) $\mathcal{F}[⌜{\sim}p⌝] = h(p)$;
(ii) $h(1) = 0$, $h(0) = 1$;
(iii) h is continuous and strictly monotonically decreasing;
(iv) $h(h(\mathcal{F}[p])) = \mathcal{F}[p]$;
(v) if $\mathcal{F}[p] + \mathcal{F}[q] = 1$ then $h(p) + h(q) = 1$.

Condition (iii) requires that the degree of truth of a negation be a continuous function in the sense familiar from topology or calculus, and that if q has a higher degree of truth than p then ⌜${\sim}q$⌝ has a lower degree of truth than ⌜${\sim}p$⌝. Condition (iv) ensures that negation behaves classically by guaranteeing the correctness of DN. Condition (v) involves the quantity $\mathcal{F}[p] + \mathcal{F}[q]$, which does not correspond to any intuitive semantic property; however, this condition does not by itself exclude any natural alternative to $1 - x$ (see Kandel).

The remaining sentential connective is the conditional (as usual, we treat '↔' as defined). Classically, a conditional is wholly true when its antecedent is no more true than its consequent, otherwise it is wholly false. How should we generalize this to the fuzzy case? It seems reasonable to leave the first part as it stands: if the antecedent is no more true than the consequent, the conditional is wholly true. But if we say that otherwise it is wholly false, then we do not discriminate between a conditional 'A → B' where 'A' is wholly true and 'B' is wholly false and a conditional 'A → B' where the degree of truth of 'B' is only slightly lower than that of 'A'. But the latter case is much closer to one where we have already said the result is 'wholly true' than it is to the classical case of ⊤ → ⊥. What makes it 'closer' is that if 'B' is only slightly less true than 'A', then the amount of truth that is lost in going from antecedent to consequent is small, and a small loss of truth is closer to no loss than it is to maximum loss. The idea, then, is to give a clause for the conditional on which the degree of truth of a conditional reflects *how much* truth is lost in the passage from antecedent to consequent: the bigger the loss, the less true the conditional. The simplest clause with this effect is the following:

$$(F\text{-}{\to}){:}\quad \mathcal{F}[⌜p \to q⌝] = 1 - (\mathcal{F}[p] - \mathcal{F}[q]) \text{ if } \mathcal{F}[p] > \mathcal{F}[q],$$
$$= 1 \text{ otherwise.}$$

The reader can confirm that this agrees with the classical truth-table when only 0 and 1 are considered, and assigns a high degree of truth to conditionals where the antecedent is only a little more true than the consequent (e.g., $0.2 \to 0.1 = 0.9$) and a low degree of truth when there is a large gap ($0.9 \to 0.1 = 0.2$).

These clauses for the sentential connectives determine clauses for the quantifiers, since in the framework of fuzzy logic there is no reason to abandon the analogy between '∃' and disjunction and '∀' and conjunction. So an existential sentence will have the degree of truth of its *best* instance and a universal sentence the degree of truth of its *worst:*

(F-∀): $\mathcal{F}[\ulcorner(\forall v)\phi v)\urcorner] = min\{\phi t:\ ref(t) \in D\}$
(F-∃): $\mathcal{F}[\ulcorner(\exists v)\phi v)\urcorner] = max\{\phi t:\ ref(t) \in D\}$

(of course, we are assuming that for every $x \in D$ there is a t such that $ref(t) = x$).[1]

It remains to complete the development of fuzzy logic by defining an appropriate notion of semantic consequence. We shall consider only finite sets of premises, so if Γ is a finite set of LFOL sentences, a useful piece of notation is to write $\mathcal{F}[\Gamma]$ to mean the minimum of the degrees of truth of the members of Γ on the interpretation $\mathcal{F}$ (if we allowed infinite Γ we would have to accommodate the case where there is no such minimum). How, then, should we generalize the classical notion of consequence? One possibility is not to generalize at all. This gives us:

(Df1): $\Gamma \vDash_F p$ if and only if there is no $\mathcal{F}$ such that $\mathcal{F}[\Gamma] = 1$ and $\mathcal{F}[p] = 0$.

But this is unsatisfactory, since it would allow premises to entail a conclusion even if the conclusion is less true than any premise. Yet a valid argument should surely not allow loss of truth in going from premises to conclusion. We can avoid this feature of (Df1) by changing it to

(Df2): $\Gamma \vDash_F p$ if and only if there is no $\mathcal{F}$ such that $\mathcal{F}[\Gamma] = 1$ and $\mathcal{F}[p] < 1$.

This proves to be a reasonable definition but still leaves open something which we may wish to exclude, namely, that an argument-form can be valid even if it has interpretations in which the conclusion is less true than the least true premise(s). (Df2) allows this since we might have an interpretation $\mathcal{F}$ on which there is a sentence in Γ which is not wholly true, $\mathcal{F}[\Gamma] > \mathcal{F}[p]$, but there is no interpretation $\mathcal{F}$ such that $\mathcal{F}[\Gamma] = 1$ and $\mathcal{F}[p] < 1$, as required for failure of semantic consequence by (Df2). To rule this out, the definition we need is:

(Df3): $\Gamma \vDash_F p$ if and only if there is no $\mathcal{F}$ such that $\mathcal{F}[p] < \mathcal{F}[\Gamma]$.

In the discussion below, we refer to the notion of semantic consequence defined in (Df2) as the *weak* notion ($\vDash_{F2}$) and the notion defined in (Df3) as the *strong* notion ($\vDash_{F3}$). It is obvious that if p is a consequence of Γ in the strong sense then it is so in the weak sense; but the converse does not hold, and we will exhibit a specific counterexample to it.

---

[1] When D is infinite, there may be no such thing as the maximum or the minimum of the degrees of truth of the instances of a universal or existential sentence. In such a case, in place of the maximum of the degrees of truth of the instances, we use the least degree greater than the degree of every instance, and in place of the minimum, we use the greatest degree less than the degree of every instance. The structure of the real numbers guarantees that such degrees always exist.

4 Resolution of the paradoxes

How does the new semantic framework help us diagnose Sorites paradoxes? The essential point is that we no longer need say that the premises of a Sorites paradox are all true, that is, all wholly true. For if the degree to which g_j is short is greater than the degree to which g_{j+1} is short, then 'g_j is short $\to$ g_{j+1} is short' is not wholly true: there is a loss of truth in going from antecedent to consequent and so the conditional, by ($F\to$), is less then wholly true, even if only slightly. Consequently, the Sorites paradoxes are not *sound* arguments at all, if we define a sound argument to be a valid argument whose premises are *wholly* true. And it is only sound arguments whose conclusions are guaranteed to be true. Similarly, the universally quantified statement 'any giraffe indiscernibly taller than a short giraffe is itself short' is at best almost wholly true, since it has instances which fall short of complete truth.

It seems, therefore, that the force of the paradoxes lies in their being presented in a context in which the Principle of Bivalence is assumed to hold. If our only options are 'true' and 'false' then statements which are almost wholly true are going to be classified as true, since this is the least unsatisfactory choice available. But the dilemma is avoidable, and when we make explicit allowances for predicates which apply in various degrees to different objects, the difficulty vanishes.

Not merely is a Sorites argument unsound, but on our strong sense of validity it is also invalid, for an almost wholly true conditional and its antecedent can have higher degrees of truth than its consequent. For example, if $\mathcal{F}[A] = 0.5$ and $\mathcal{F}[B] = 0.4$, then $min\{\mathcal{F}[A], \mathcal{F}[A \to B]\} = 0.5$. This shows that A, A $\to$ B $\nvDash_{F3}$ B. However, A, A $\to$ B $\vDash_{F2}$ B, since if 'A' and 'A $\to$ B' are wholly true, 'B' must be wholly true as well. This is our promised example of a weakly valid sequent which is not strongly valid. Correspondingly, $\to$E is a weakly sound but strongly unsound rule. Of course, this can be taken two ways. If the correctness of $\to$E seems undeniable, then $\vDash_{F3}$ must be rejected. But if we insist that validity should be inconsistent with any loss of truth, no matter the amount at the starting point, then $\to$E must be rejected in favor of $\vDash_{F3}$. However, $\vDash_{F3}$ does have some odd properties. For example, the classical relationship between consequence and tautologous conditionals breaks down: though A, A$\to$ B $\nvDash_{F3}$ B, we do have $\vDash_{F3}$ A $\to$ ((A $\to$ B) $\to$ B), as we can see using $1 - (x - y) = (1 - x) + y$. If $\mathcal{F}[A] = a$, $\mathcal{F}[B] = b$, then

$$
\begin{aligned}
\mathcal{F}[(A \to B) \to B] &= 1 - ([1 - (a - b)] - b) \\
&= 1 - ([(1 - a) + b] - b) \\
&= 1 - (1 - a) \\
&= (1 - 1) + a \\
&= a \\
&= \mathcal{F}[A].
\end{aligned}
$$

How do the other rules of NK fare according to the two definitions of fuzzy consequence? If $\mathcal{F}[A] = 0.5$, then $\mathcal{F}[A \lor \sim A] = \mathcal{F}[\sim(A \,\&\, \sim A)] = 0.5$, and therefore $\nvDash_{F2,3}$ A $\lor$ $\sim$A, $\nvDash_{F2,3}$ $\sim$(A $\&$ $\sim$A). The NK-proofs of LEM and the Law of Noncontradiction use various rules which are bo⁺h weakly and strongly sound: &E, DN and so on. But they also use ~E and ~I, and these rules are objectionable. ~E turns out to be weakly sound, but ~I is weakly unsound. For example: B, A & ~A ⊨$_{F2}$ ⋏, since for no $\mathcal{F}$ do we have $\mathcal{F}$[A & ~A] = 1, but B ⊭$_{F2}$ ~(A & ~A)—consider the interpretation $\mathcal{F}$[B] = 1, $\mathcal{F}$[A] = 0.5. As this example suggests, however, there are restricted versions of ~I which are weakly sound. For instance, if 'A' is a sentence-letter which does not occur in Γ, then from Γ,A ⊨$_{F2}$ ⋏ we can infer that Γ ⊨$_{F2}$ ~A. For if 'A' is a sentence-letter not in Γ, then for every degree of truth d for which there is an interpretation $\mathcal{F}$ with $\mathcal{F}[\Gamma]$ = d, there is an interpretation $\mathcal{F}'$ such that $\mathcal{F}'[\Gamma]$ = d, $\mathcal{F}'$[A] = 1. Therefore, since Γ,A ⊨$_{F2}$ ⋏, there is no interpretation $\mathcal{F}$ such that $\mathcal{F}[\Gamma]$ = 1; so Γ ⊨$_{F2}$ ~A. However, the general problem of giving a sound and complete set of NK-like rules for fuzzy logic is an open one.

❏ Exercises

I   Construct three graphs in which the points on the X-axis represent all possible heights from one meter to two meters (or 3½ to 6½ feet) either for a male or for a female adult from a fixed population, and the Y-axis represents the real interval [0,1] of degrees of truth. In the first graph, draw a path which displays what seems to you to be a reasonable assignment of degrees of shortness to the heights on the X-axis (thus the path should begin at (0,1) and end at (1,0)). Do the same in the second graph for degrees of medium heightedness and in the third graph for degrees of tallness. In each graph, indicate by shading around the path you draw the minimum area within which you think all reasonable assignments of degrees for the property in question lie.

II   Show that ~E is strongly unsound. Explain why →I is strongly sound.

III   Show the following:

(1)   A → B ⊭$_{F3}$ ~A ∨ B
(2)   A ↔ B ⊭$_{F3}$ (A & B) ∨ (~A & ~B)
(3)   B → ~B ⊭$_{F3}$ ~B
(4)   A → B, A → ~B ⊭$_{F3}$ ~A

IV   In systems where '⋏' is a primitive, it is sometimes proposed to treat negation as a defined connective like '↔' (see Exercise 4.10.5, page 139, on the rule Df~). The proposed definition is:

$$\ulcorner{\sim}p\urcorner \overset{\text{def}}{=} \ulcorner p \to \curlywedge\urcorner.$$

Is this definition acceptable in fuzzy logic? Consider both weak and strong semantic consequence.

Appendix: Using *MacLogic*

As I remarked in the Preface, I wrote *Modern Logic* because I wanted to have a textbook that I could use in conjunction with the Apple Macintosh® program *MacLogic*. *MacLogic* is the award-winning creation of the MALT (machine-assisted logic teaching) team at St Andrews University, led by Stephen Read and Roy Dyckhoff.[1] In this appendix I shall briefly describe how *MacLogic* works.

The program has two modes of operation, proof checking and proof constructing, and in each mode it can handle a variety of systems of logic: minimal, intuitionistic, classical and modal (S4, not covered in this book, and S5). In proof-checking mode, the user enters a proof line by line in a line-entry command window, typing premise and assumption numbers, formulae and rule-application line numbers in their own sub-windows, and choosing a rule to apply by checking a radio-button in a palette of rule-buttons. If a mistake is made a brief explanatory message appears and the user can fix the problem by making appropriate changes in the line-entry command window. The other mode of operation of the program is proof construction, in which a problem is broken down into simpler problems and at each stage the user chooses a rule to reduce the current problem to something simpler (faulty choices are flagged). The program comes with a library of problems, to which more can be added, and there is a dialog box which allows entry of the user's own problems. If an invalid sequent is entered, then so long as the problem does not involve the identity or modal rules, *MacLogic*'s built-in theorem prover can catch the mistake.

So far as the compatibility of *Modern Logic* with *MacLogic* is concerned, there are three areas where there are differences. First, *MacLogic* refers to both premises and assumptions as assumptions, and reserves the word 'premise' for use in the phrase 'premise of a rule application'. Thus if a proof has '5,6 &I' on its right, lines 5 and 6 are said to be the *premises for* that application of &I.

---

[1] An evaluation copy of *MacLogic* may be obtained over the Internet. For details send e-mail to Dyckhoff at rd@dcs.st-andrews.ac.uk or consult the web page http://www-theory.dcs.st-and.ac.uk:80/~rd; alternatively, send Roy Dyckhoff a floppy disk and return postage costs at the Computer Science Division, The University, St Andrews, Scotland KY16 9SS. (Mention that you are using *Modern Logic* in case there is a customized version of *MacLogic* available.) There is a generous site-licensing arrangement which permits a university unlimited copying for academic purposes (students may not be charged for copies beyond the cost of media and duplication). In 1989 *MacLogic* won the Philosophy Software Contest arranged by the Philosophy Documentation Center at Bowling Green State University.

Second, QS is not available for SI in predicate logic. However, any sentential logic sequent can be used for SI. There is an SI rule which allows the user to choose a sequent from a list of those most commonly employed and to specify what formulae are to be substituted for its sentence-letters. There is also a rule of Tautology which allows SI in connection with sequents not in the program's own list. Third, the quantifier rules of *MacLogic* allow open sentences to be inferred from quantified sentences, whereas in this book I have insisted that every line of a proof be a closed sentence. However, *MacLogic* can be made more like *Modern Logic* in this respect by an instructor who knows how to use ResEdit™, Apple's resource editor for the Macintosh: replace the contents of STR resource 31150 with 'constant' or 'individual constant'. *MacLogic* will then move individual constants to the head of the list of terms that the user chooses to substitute for bound variables, and the program will use individual constants when it makes substitutions itself.

Most students should find that learning *MacLogic* leads to a significant improvement in their mastery of natural deduction. However, there are some pitfalls an instructor should beware of. Every student is capable of using the program in checking mode, though in my experience most are reluctant to do so, presumably because of the extra effort required on top of working out a proof in the first place. Unless directed otherwise, students tend to try constructing mode before they have a firm grasp of how to go about constructing proofs. Lacking an overall strategy, their rule-choices are rejected for reasons they do not understand, so they become discouraged and lose interest in the program. The cure for this syndrome is to insist on using only checking mode for the first few sessions or until students have grasped the idea of working out a proof from the bottom up. It is this latter procedure in which *MacLogic* most effectively develops skill.

Solutions to Selected Exercises

(A complete solutions manual is available for instructors. Requests should be made on official letterhead directly to the author at the Department of Philosophy, Tulane University, New Orleans, LA 70118, USA. Please state your course number and/or title, and if possible supply an e-mail address.)

Chapter 2, §2

(6)   I: Inflation is falling
        G: The government can guide the economy wisely
        R: The government can regain its popularity

      I & ~(G & R)

["can't _ and at the same time _" is the denial of a conjunction: not both are true. So we should have '~(_ & _)'.]

(9)   S: Smith is a lawyer   B: Brown is a lawyer
        R: Robinson is a lawyer   X: Smith is honest
        Y: Brown is honest   Z: Robinson is honest

      (S & (B & R)) & ((X & Y) ∨ ((Y & Z) ∨ (X & Z)))

['Still' is another way of expressing conjunction. 'At least two of them are honest' should be expanded into full sentential form as a claim explicitly about Smith, Brown and Robinson. If at least two are honest, then of the three, either the first two are honest ('X & Y'), or the second two are ('Y & Z'), or the first and third are ('X & Z').]

Chapter 2, §3

(5)   P: Grades may be posted
        R: Students request grades to be posted
        S: Students are identified by name

      (R & ~S) → P

(To say that a condition p will hold provided q holds is to say that q is sufficient for p, as in 'you'll be notified provided you're registered'—this implies that there is nothing else you must do. But it leaves it open that there are ways an unregistered person may be notified, say by some special arrangement.)

(8)   R: One has the right to run for president
        C: One is an American citizen

        R ↔ C

(12)   E: One makes an effort
        P: One passes the exam

        E → P

(15)   F: There is an increase in government funding for education
        Q: There is an improvement in the quality of education

        (Q → F) & ~(F → Q)

Chapter 2, §4

(4)   H: Homer existed
        C: The Odyssey was written by a committee
        B: Butler was right
        W: The Odyssey was written by a woman

        ~H → (C ∨ (B → W)), ~W ∴ C

(9)   P: Parking is prohibited in the center of town
        B: People will buy bicycles
        O: The general inconvenience is offset
        U: City shops pick up some business

        P → B, B → (O & U), P → ~U ∴ ~B

['In which case' means 'in the case that people buy bicycles'. The best we can do for 'the general inconvenience is offset by the fact that...' is to say that the general inconvenience is offset *and* shops pick up some business.]

Chapter 2, §5

(I.e) In (2) the '&' is within the scope of the '~', since the scope of the '~', which by definition is the formula at the node in the parse tree for (2) immediately beneath the node where the '~' is introduced, is '(P & Q)', which contains the '&' in question. In (3) the '&' is again within the scope of the '~', since in the parse tree for (3), the '&' appears in the formula which is the scope of the '~', the formula at the node where the '~' is introduced (the root node).

(II.4) The outer parentheses in (4) are incorrect, since in its structure the last step (reading bottom to top) is to apply the formation rule (f-~), and this rule (page 36) does not put outer parentheses around the formula which results from its application.

Chapter 2, §6

(II.1) (a) If p = 'snow is' and q = 'white', then $\ulcorner p \; q \urcorner$ (note the space) is 'snow is white'. (b) The result of writing p followed by 'q' is 'snow is$q$' or, with optional space, 'snow is q'.

(IV.1) This is incorrect—what is meant is that 'Rome' is the name of a city. Rome is a city, not a city's name.

(IV.4) This is grammatically correct as it stands. It is also true: if p is an indicative sentence of English, so is the result of prefixing 'it is not the case that' to it.

(IV.11) The reading on which the statement is syntactically incorrect is the one on which what is meant is that if we prefix 'it is obvious to every English speaker that' to any syntactically correct indicative sentence we obtain a syntactically correct English sentence. This claim, which is in fact true, requires corners around 'it is obvious to every English speaker that p'. On the other reading (can you see what it is?) the claim is syntactically correct but false.

Chapter 3, §1

(4) If A is a Knight, then since that means what he says is true, B is a Knave and so speaks falsely. Thus A and C are of *different* types, and therefore C is a Knave. This is a possible solution, but for it to be *determined* what C is, there must be no solution on which C is a Knight. So what happens if A is a Knave? Then B is a Knight since A speaks falsely, so A and C are of the same type, since B speaks truly, being a Knight, so again C is a Knave. Thus being a Knave is the only possibility for C.

Chapter 3, §2

(I.4)

A B	(A ↔ B) & (A & ~B)
⊤ ⊤	⊤ ⊥ ⊥
⊤ ⊥	⊥ ⊥ ⊤
⊥ ⊤	⊥ ⊥ ⊥
⊥ ⊥	⊤ ⊥ ⊥

Since the final column of the table for formula (1d) contains only ⊥s, (1d) is a contradiction.

Chapter 3, §4

(I.4) There are two cases, (1) 'A' is true, 'C' is false, and (2) the converse. In case 1, premise 1 false, so only case 2 need be considered. This requires 'D' to be true for premise 3 to be true and hence 'B' must be true for premise 2 to be true. This interpretation makes all premises true and the conclusion false, so the argument-form is invalid, as demonstrated by the interpretation assigning ⊥ to 'A' and ⊤ to 'B', 'C' and 'D'.

(II.4)   F: Yossarian flies his missions
               D: Yossarian puts himself in danger
               R: Yossarian is rational
               A: Yossarian asks to be grounded

               (F → D) & (D → ~R), (R → A) & (~F → A), (~F → ~R) & (A → R)
               ∴ (R ∨ ~R) → F

['Only irrational people are grounded' particularized to Yossarian means 'Yossarian will be grounded (i.e. doesn't fly his missions) only if Yossarian is not rational'; 'a request to be grounded is proof of rationality' particularized to Yossarian means 'if Yossarian asks to be grounded then Yossarian is rational'.]

    To test for validity, we note that 'F' must be false for the conclusion to be false; hence 'A' must be true for premise 2 to be true, and so 'R' must be true for the second conjunct of premise 3 to be true; but 'R' has to be false for the first conjunct to be true. Thus the argument-form is valid.

Chapter 3, §5

(I.4) A → (B & C), D → (B ∨ A), C → D ⊨ A ↔ C

The solution is displayed on the next page. In this inverted tree there are ten paths, eight of which are closed. The two open paths, the fifth and ninth (reading along the leaves from the left), determine the same interpretation, the one assigning ⊥ to 'A' and ⊤ to 'B', 'C' and 'D'. So this result agrees with the one obtained by the method of constructing an interpretation.

Chapter 3, §6

(3) If $p \vDash (q \& r)$, then on no assignment do we have p true, $(q \& r)$ false; so none of (a), p is true, q is true, r is false, (b) p is true, q is false, r is true, (c) p is true, q is false, r is false, is possible. If $(p \leftrightarrow q) \nvDash (p \leftrightarrow r)$, then there are assignments of truth-values to sentence-letters in p, q and r on which $(p \leftrightarrow q)$ is true, $(p \leftrightarrow r)$ is false. This is possible, since, comparing (a), (b) and (c), we see that we have not ruled out an assignment on which e.g. p is false, q is false, r is true. So it does not follow that $(p \leftrightarrow q) \vDash (p \leftrightarrow r)$.

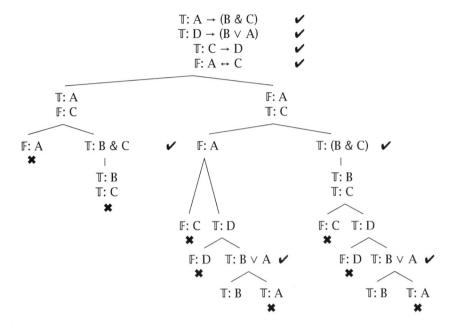

In this inverted tree there are ten paths, eight of which are closed. The two open paths, the fifth and ninth reading along the leaves from the left, determine the same interpretation, the one assigning ⊥ to 'A' and ⊤ to 'B', 'C' and 'D'. So this result agrees with the one obtained by the method of constructing an interpretation.

Chapter 3, §6 (continued)

(8) Semantic consequence holds when no interpretation makes the premises true and the conclusion false. Consequently, if no interpretation at all makes the conclusion false, then no interpretation makes the premises true and the conclusion false, and so semantic consequence holds no matter what the premises are. Thus every semantic sequent with a tautology (such as '(A ∨ ~A)') as its conclusion is correct.

Chapter 3, §7

(I) The formula in {~,&,∨} for table 2 is:

(A & ~B & C) ∨ (A & ~B & ~C) ∨ (~A & ~B & C) ∨ (~A & ~B & ~C).

To eliminate all occurrences of '∨' we have to apply the substitution-rule given on page 78 three times. Eliminating the first occurrence of '∨' yields:

~[~(A & ~B & C) & ~[(A & ~B & ~C) ∨ (~A & ~B & C) ∨ (~A & ~B & ~C)]].

Here we put '(A & ~B & C)' for *r* and '(A & ~B & ~C) ∨ (~A & ~B & C) ∨ (~A & ~B & ~C)' for *s* in the substitution rule. Eliminating the first occurrence of '∨' in the latter formula yields

~[~(A & ~B & C) & ~{~[(A & ~B & ~C) & ~[(~A & ~B & C) ∨ (~A & ~B & ~C)]]}].

Finally, we eliminate the remaining '∨' to obtain

~[~(A & ~B & C) & ~{~[(A & ~B & ~C) & ~[~{~(~A & ~B & C) & ~(~A & ~B & ~C)}]]}].

In this last step, the subformula ⌜*r* ∨ *s*⌝ for which substitution has been made is '(~A & ~B & C) ∨ (~A & ~B & ~C)'.

(II.3) ⌜*p* ← *q*⌝ has the table ⊤ ⊤ ⊥ ⊤. Since {~,&,∨} is functionally complete, then given any truth-table, we can find a formula *p* in {~,&,∨} which has that table. We need two substitution rules, one of which allows us to replace the occurrences of '&' in *p* (if any), the other of which allows us to replace the occurrences of '∨' (if any). Since the rules will replace formulae with logically equivalent formulae, the result will be a formula in {~,←} which also has the given truth-table. The following rules are correct:

(1)   Replace every subformula of *p* of the form ⌜*r* & *s*⌝ with ⌜~(~*r* ← *s*)⌝.
(2)   Replace every subformula of *p* of the form ⌜*r* ∨ *s*⌝ with ⌜*r* ← ~*s*⌝.

(3) The simplest solution is to show that, using just '~' and '↔', there is no way of expressing a two-place truth-function which over the four possible inputs has exactly one or exactly three ⊤s in its output. In other words, no formula in '~' and '↔' with two sentence-letters *p* and *q* has an odd number of ⊤s in its four-row truth-table. First we establish the **Minor Lemma**, that if any formula has an even number of ⊤s in its table, so does its negation. *Proof:* The number of rows in any truth-table is even, so a formula with an even number of ⊤s also has an even number of ⊥s, producing an even number of ⊤s for the formula's negation.

Next, we prove the **Major Lemma**, that if two columns L and R in a four-row truth-table each have an even number of ⊤s, so does the result of applying the biconditional truth-function to these columns. *Proof:* Assume that L and R each have an even number of ⊤s. Then there are three possibilities, (1) no ⊤s in L, (2) two ⊤s in L, (3) four ⊤s in L. Each of cases (1), (2) and (3) subdivides into three further cases (a), (b) and (c). *Case 1:* (a) No ⊤s in R. Then each column has four ⊥s, so the resulting table has four ⊤s; (b) Two ⊤s in R. For Case 1 this means two rows have the same values and two rows have different values; hence the resulting table has two ⊤s; (c) Four ⊤s in R. Then the resulting table has no ⊤s. *Case 2:* (a) No ⊤s in R. This is essentially the same as Case 1b; (b) Two ⊤s in R. If these ⊤s face the two ⊤s in L then there are four ⊤s in the table. If neither faces a ⊤ in L then there are no ⊤s in the table. If just one faces a ⊤

in L then two rows have the same value and two rows have different values, so there are two $\top$s in the table; *Case 3:* (a) No $\top$s in R. Essentially the same as Case 1c; (b) Two $\top$s in R. See Case 2c; (c) Four $\top$s in R. Then all rows have $\top$s on them and so the table has four $\top$s as well. This proves the Major Lemma. (The argument is a special case of a more general argument that can be given for any number 2^n of rows.)

Now let B be an arbitrary formula in p, q, '$\sim$' and '$\leftrightarrow$'. We show that B's 4-row truth-table contains an even number of $\top$s (and hence of $\bot$s) in its final column. (a) Suppose B contains no occurrences of '$\leftrightarrow$'. Then B is either a sentence-letter, in which case its column contains an even number of $\top$s, or a sentence-letter prefixed by one or more '$\sim$'s, in which case, by repeated applications of the Minor Lemma, it still contains an even number of $\top$s. (b) Suppose B contains one occurrence of '$\leftrightarrow$'. Then in case (b1), B is a biconditional with two sides, each of which contains no occurrence of '$\leftrightarrow$', and so by (a), each side of B has an even number of $\top$s in its table. Hence by the Major Lemma, B has an even number of $\top$s in its table. In case (b2), B consists in a biconditional C prefixed with one or more occurrences of '$\sim$'; by (b1), C has an even number of $\top$s in its table, so by repeated applications of the Minor Lemma, B has an even number of $\top$s in its table. (c) Suppose B contains two occurrences of '$\leftrightarrow$'. Then in case (c1), B is a biconditional with two sides, each of which contains one or no occurrence of '$\leftrightarrow$', and so by (a) and (b), each side of B has an even number of $\top$s in its table. Hence by the Major Lemma, B has an even number of $\top$s in its table. In case (c2), B consists in a biconditional C prefixed with one or more occurrences of '$\sim$'; by (c1), C has an even number of $\top$s in its table, so by repeated applications of the Minor Lemma, B has an even number of $\top$s in its table.

Continuing in this way we can show that for any n, if B is a formula in two sentence-letters p, q, '$\sim$' and '$\leftrightarrow$' with n occurrences of '$\leftrightarrow$', B has an even number of $\top$s in its table (the reader who is familiar with strong mathematical induction should think about how we could go about making 'continuing in this way' rigorous). Hence $\{\sim, \leftrightarrow\}$ is expressively incomplete; for example, '$\rightarrow$' is undefinable.

Chapter 3, §8

(3) (i) Let 'A' mean 'the speed of light does not vary with the motion of its source'. Then 'A' is true and also 'It is surprising that A' is true. (ii) Let 'A' mean 'there are more robberies when the police are on strike'. Then 'A' is true but 'It is surprising that A' is false. Thus no entry can be made in the top row of a purported function-table for 'It is surprising that'.

(5) (i) Let 'A' mean 'lead sinks in water' and 'B' mean 'lead is denser than water'. Then 'A' and 'B' are both true, and in addition, 'A, which means that B' is true. (ii) Let 'A' mean 'lead sinks in water' and 'B' = 'Moses wrote the Pentateuch'. Then 'A' and 'B' are true, but 'A, which means that B' is false. So no entry can be made in the top row of a purported function-table for '..., which means that...'.

Chapter 4, §2

(2)	1	(1)	A → B	Premise
	2	(2)	A → C	Premise
	3	(3)	A	Assumption
	1,3	(4)	B	1,3 →E
	2,3	(5)	C	2,3 →E
	1,2,3	(6)	B & C	4,5 &I
	1,2	(7)	A → (B & C)	3,6 →I ◆

(5)	1	(1)	(A & B) → C	Premise
	2	(2)	A	Assumption
	3	(3)	B	Assumption
	2,3	(4)	A & B	2,3 &I
	1,2,3	(5)	C	1,4 →E
	1,2	(6)	B → C	3,5 →I
	1	(7)	A → (B → C)	2,6 →I ◆

(16)	1	(1)	A → B	Premise
	2	(2)	A	Assumption
	3	(3)	C	Assumption
	1,2	(4)	B	1,2 →E
	1,2	(5)	C → B	3,4 →I
	1	(6)	A → (C → B)	2,5 →I ◆

Chapter 4, §3

(4)	1	(1)	A → (B & C)	Assumption
	2	(2)	A	Assumption
	1,2	(3)	B & C	1,2 →E
	1,2	(4)	B	3 &E
	1	(5)	A → B	2,4 →I
	1,2	(6)	C	3 &E
	1	(7)	A → C	2,6 →I
	1	(8)	(A → B) & (A → C)	5,7 &I
		(9)	[A → (B & C)] → [(A → B) & (A → C)]	1,8 →I ◆

Chapter 4, §4

(1)	1	(1)	A → ~B	Premise
	2	(2)	B	Assumption
	3	(3)	A	Assumption
	1,3	(4)	~B	1,3 →E
	1,2,3	(5)	⋏	4,2 ~E
	1,2	(6)	~A	3,5 ~I
	1	(7)	B → ~A	2,6 →I ◆

(8)	1	(1)	A & ~A	Assumption
	2	(2)	~B	Assumption
	1	(3)	A	1 &E
	1	(4)	~A	1 &E
	1	(5)	人	3,4 ~E
	1	(6)	~~B	2,5 ~I
	1	(7)	B	6 DN
		(8)	(A & ~A) → B	1,7 →I ◆

(13)	1	(1)	A → B	Premise
	2	(2)	B → ~A	Premise
	3	(3)	A	Assumption
	1,3	(4)	B	1,3 →E
	1,2,3	(5)	~A	2,4 →E
	1,2,3	(6)	人	5,3 ~E
	1,2	(7)	~A	3,6 ~I ◆

(18)	1	(1)	(A & ~B) → ~A	Premise
	2	(2)	A	Assumption
	3	(3)	~B	Assumption
	2,3	(4)	A & ~B	2,3 &I
	1,2,3	(5)	~A	1,4 →E
	1,2,3	(6)	人	5,2 ~E
	1,2	(7)	~~B	3,6 ~I
	1,2	(8)	B	7 DN
	1	(9)	A → B	2,8 →I ◆

Chapter 4, §5

(4)	1	(1)	A ∨ ~~B	Premise
	2	(2)	A	Assumption
	2	(3)	A ∨ B	2 ∨I
	4	(4)	~~B	Assumption
	4	(5)	B	4 DN
	4	(6)	A ∨ B	5 ∨I
	1	(7)	A ∨ B	1,2,3,4,6 ∨E ◆

(10)	1	(1)	A ∨ B	Premise
	2	(2)	A → B	Assumption
	3	(3)	A	Assumption
	2,3	(4)	B	2,3 →E
	5	(5)	B	Assumption
	1,2	(6)	B	1,3,4,5,5 ∨E
	1	(7)	(A → B) → B	2,6 →I ◆

(13) | 1 | (1) | A ∨ B | Premise
| 2 | (2) | ~A & ~B | Assumption
| 3 | (3) | A | Assumption
| 2 | (4) | ~A | 2 &E
| 2,3 | (5) | ⋏ | 4,3 ~E
| 6 | (6) | B | Assumption
| 2 | (7) | ~B | 2 &E
| 2,6 | (8) | ⋏ | 7,6 ~E
| 1,2 | (9) | ⋏ | 1,3,5,6,8 ∨E
| 1 | (10) | ~(~A & ~B) | 2,9 ~I   ◆

Chapter 4, §6

(4) | 1 | (1) | (A ∨ B) ↔ A | Premise
| 2 | (2) | B | Assumption
| 1 | (3) | ((A ∨ B) → A) & (A → (A ∨ B)) | 1 Df
| 1 | (4) | (A ∨ B) → A | 3 &E
| 2 | (5) | A ∨ B | 2 ∨I
| 1,2 | (6) | A | 4,5 →E
| 1 | (7) | B → A | 2,6 →I   ◆

(10) | 1 | (1) | (A ∨ B) ∨ C | Premise
| 2 | (2) | B ↔ C | Premise
| 2 | (3) | (B → C) & (C → B) | Df
| 4 | (4) | A ∨ B | Assumption
| 5 | (5) | A | Assumption
| 5 | (6) | C ∨ A | 5 ∨I
| 7 | (7) | B | Assumption
| 2 | (8) | B → C | 3 &E
| 2,7 | (9) | C | 8,7 →E
| 2,7 | (10) | C ∨ A | 9 ∨I
| 2,4 | (11) | C ∨ A | 4,5,6,7,10 ∨E
| 12 | (12) | C | Assumption
| 12 | (13) | C ∨ A | 12 ∨I
| 1,2 | (14) | C ∨ A | 1,4,11,12,13 ∨E ◆

Chapter 4, §8

(I.5) '~(M ∨ N) ∨ (W & U) ⊢_NK (M ∨ N) → (W & U)' is a substitution-instance of (Imp): this sequent may be obtained from Imp by putting 'M ∨ N' for 'A' and 'W ∨ U' for 'B'.

(II.iii) This sequent is a substitution-instance of (a). To obtain (iii) from (a), put '~~(R & S)' for 'A', '~T & S' for 'B', '~~W' for 'C' and '~T' for 'D'.

(III.4)   1   (1)   ~B → A   Premise
          2   (2)   B → A   Assumption
              (3)   B ∨ ~B   TI (LEM)
          4   (4)   B   Assumption
        2,4   (5)   A   2,4 E
          6   (6)   ~B   Assumption
        1,6   (7)   A   1,6 →E
        1,2   (8)   A   3,4,5,6,7 ∨E
          1   (9)   (B → A) → A   2,8 →I   ◆

(III.8)   1   (1)   (A ∨ B) → (A ∨ C)   Premise
          2   (2)   ~[A ∨ (B → C)]   Assumption
          2   (3)   ~A & ~(B → C)   2 SI (DeM)
          2   (4)   ~A   3 &E
          2   (5)   ~(B → C)   3 &E
          6   (6)   B   Assumption
          6   (7)   A ∨ B   6 ∨I
        1,6   (8)   A ∨ C   1,7 →E
      1,2,6   (9)   C   8,4 SI (DS)
        1,2 (10)   B → C   6,9 →I
        1,2 (11)   ⅄   5,10 ~E
          1 (12)   ~~[A ∨ (B → C)]   2,11 ~I
          1 (13)   A ∨ (B → C)   12 DN   ◆

(III.13) 1   (1)   (A ↔ B) ↔ (C ↔ D)   Premise
          1   (2)   A ↔ [B ↔ (C ↔ D)]   1 SI ((15), page 119)
          1   (3)   {A → [B ↔ (C ↔ D)]} & {[B ↔ (C ↔ D)] → A}   2 Df
          1   (4)   A → [B ↔ (C ↔ D)]   3 &E
          1   (5)   [B ↔ (C ↔ D)] → A   3 &E
          6   (6)   A   Assumption
        1,6   (7)   B ↔ (C ↔ D)   4,6 →E
        1,6   (8)   (B ↔ C) ↔ D   7 SI ((15), page 119)
        1,6   (9)   [(B ↔ C) → D] & [D → (B ↔ C)]   8 Df
        1,6 (10)   (B ↔ C) → D   9 &E
        1,6 (11)   D → (B ↔ C)   9 &E
         12 (12)   D   Assumption
     1,6,12 (13)   B ↔ C   11,12 →E
     1,6,12 (14)   C ↔ B   13 SI ((2), page 119)
        1,6 (15)   D → (C ↔ B)   12,14 →I
         16 (16)   C ↔ B   Assumption
         16 (17)   B ↔ C   16 SI ((2), page 119)
     1,6,16 (18)   D   10,17 →E
        1,6 (19)   (C ↔ B) → D   16,18 →I
        1,6 (20)   [(C ↔ B) → D] & [D → (C ↔ B)]   19,15 &I
        1,6 (21)   (C ↔ B) ↔ D   20 Df
        1,6 (22)   C ↔ (B ↔ D)   21 SI ((15), page 119)
          1 (23)   A → [C ↔ (B ↔ D)]   6,22 →I

24	(24)	C ↔ (B ↔ D)	Assumption
24	(25)	(C ↔ B) ↔ D	24 SI ((15), page 119)
26	(26)	B ↔ C	Assumption
26	(27)	C ↔ B	26 SI ((2), page 119)
24	(28)	[(C ↔ B) → D] & [D → (C ↔ B)]	25 Df
24	(29)	(C ↔ B) → D	28 &E
24	(30)	D → (C ↔ B)	28 &E
24,26	(31)	D	29,27 →E
24	(32)	(B ↔ C) → D	26,31 →I
33	(33)	D	Assumption
24,33	(34)	C ↔ B	30,33 →E
24,33	(35)	B ↔ C	34 SI ((2), page 119)
24	(36)	D → (B ↔ C)	33,35 →I
24	(37)	[(B ↔ C) → D] & [D → (B ↔ C)]	32,36 &I
24	(38)	(B ↔ C) ↔ D	37 Df
24	(39)	B ↔ (C ↔ D)	38 SI ((15), page 119)
1,24	(40)	A	5,39 →E
1	(41)	[C ↔ (B ↔ D)] → A	24,40 →I
1	(42)	{A → [C ↔ (B ↔ D)]} & {[C ↔ (B ↔ D)] → A} 23,41 &I	
1	(43)	A ↔ [C ↔ (B ↔ D)]	42 &E
1	(44)	(A ↔ C) ↔ (B ↔ D)	43 SI (15, page 119) ◆

Chapter 4, §9

(1) In the same style as the tree-format schema for ∨E on page 130, the schemata for the remaining rules are as follows:

In the rules →I and ~I, the notation indicates that we discharge every occurrence of p as leaf node on a path that contains the nodes labelled q and $\curlywedge$ respectively. Also, the rules permit but do not require the making and discharging of the assumption p. Thus the two-line proof A//B → A is a correct proof by →I in tree format of the sequent A ⊢$_{NK}$ B → A. One reason for this divergence from Lemmon format, in which the assumption 'B' would be displayed, is that in tree format there is no sensible place to put the assumption 'B', as some experimentation will reveal.

(2) A tree-format proof of Example 5.6, A ∨ (B ∨ C) ⊢$_{NK}$ (A ∨ B) ∨ C:

$$
\begin{array}{c}
\dfrac{
\dfrac{
\dfrac{\overline{\rule{1.5em}{0.4pt}}\ (3)}{A}\ \vI
}{A \vee B}\ \vI
}{A \vee (B \vee C)}\quad
\dfrac{
\dfrac{
\dfrac{A \vee B}{(A \vee B) \vee C}\ \vI
}{}
}{}
\end{array}
$$

(proof tree — see image)

Proof tree structure:

- From $\overline{}$ (3), A, by ∨I: A ∨ B, by ∨I: (A ∨ B) ∨ C
- From $\overline{}$ (1), B, by ∨I: A ∨ B, by ∨I: (A ∨ B) ∨ C
- From $\overline{}$ (2), C, by ∨I: (A ∨ B) ∨ C
- (A ∨ B) ∨ C by ∨E (1,2)
- $\overline{}$ (4), B ∨ C
- (A ∨ B) ∨ C by ∨E (3,4)
- A ∨ (B ∨ C), (A ∨ B) ∨ C
- (A ∨ B) ∨ C

Chapter 4, §10

(3) First we show that every S-sequent is an NK-sequent. Suppose an S-proof contains a step of CR at line (m), and that this is the first application of CR. At line j we have the assumption ⌜~p⌝ and at line k we have '$\curlywedge$', so in NK we may repeat the S-proof to line k and then by →I add the extra line ⌜~p → $\curlywedge$⌝, depending on {a$_1$,...,a$_n$}/j. Then if we also have the sequent ~A → $\curlywedge$ ⊢$_{NK}$ A, we can apply SI to infer p in NK. To show ~A → $\curlywedge$ ⊢$_{NK}$ A:

1	(1)	~A → $\curlywedge$	Premise
2	(2)	~A	Assumption
1,2	(3)	$\curlywedge$	1,2 →E
1	(4)	~~A	2,3 ~I
1	(5)	A	4 DN ◆

Every application of CR in an S-proof can be dealt with by the same combination of →I and SI; consequently, every sequent provable in S is also provable in NK. Conversely, we show that every sequent provable in NK is provable in S. NK proofs may use ~I and DN, which S lacks. But we can use SI in S to get the effect of an NK-application of DN, since we have ~~A ⊢$_S$ A.

1	(1)	~~A	Premise
2	(2)	~A	Assumption
1,2	(3)	$\curlywedge$	1,2 ~E
1	(4)	A	2,3 CR ◆

To get the effect of ~I in S we show A → λ ⊢$_S$ ~A. We already know ~~A ⊢$_S$ A, so we can use this sequent.

1	(1)	A → λ	Premise
2	(2)	~~A	Assumption
2	(3)	A	2, SI (~~A ⊢$_S$ A)
1,2	(4)	λ	1,3 →E
1	(5)	~A	2,4 CR ◆

(6) The difficult part is showing that every S-provable sequent is NK-provable, since Df~$_{Gen}$ can be applied to an arbitrary subformula of a formula on a line in an S-proof, and we need to provide a general explanation of why the same effect can always be obtained in NK. For example, in an S-proof we can move in one step from 'A ∨ (~(B & ~~C) → D)' to 'A ∨ (~(B & ~(C → λ)) → D)'. But what guarantee do we have in advance that there is an NK-derivation of 'A ∨ (~(B & ~(C → λ)) → D)' from 'A ∨ (~(B & ~~C) → D)'? The guarantee is provided by a procedure or algorithm $\mathcal{P}$ which can be applied to any formula ϕ to produce an NK-derivation of ψ from ϕ if ψ is S-derivable from ϕ by one application of Df~$_{Gen}$. The procedure is given as a collection of instructions, which may be applied to generate the next line of the derivation according to the main connective of the formula on the current line. If ψ is obtained from ϕ in S by applying Df~$_{Gen}$ to the subformula $\ulcorner$~$r$$\urcorner$, we call $\ulcorner$~$r$$\urcorner$ the *target negation* and $\ulcorner r \to \lambda\urcorner$ its *transform*. If ξ is a subformula of ϕ which contains the target negation $\ulcorner$~$r$$\urcorner$ we write ξ* for the result of replacing $\ulcorner$~$r$$\urcorner$ with $\ulcorner r \to \lambda\urcorner$ in ξ. If ξ is a subformula of ψ which contains the target negation's transform $\ulcorner r \to \lambda\urcorner$ we write ξ* for the result of replacing $\ulcorner r \to \lambda\urcorner$ with $\ulcorner$~$r$$\urcorner$ in ξ (because of the way →I and ~I work, we may find ourselves working with formulae in which substitution of $\ulcorner r \to \lambda\urcorner$ for $\ulcorner$~$r$$\urcorner$ has already been made). In understanding how the following procedure works the reader would be well-advised to apply it to 'A ∨ (~(B & ~~C) → D)' to see how it leads to a proof of 'A ∨ (~(B & ~(C → λ)) → D)'.

(&) If $\ulcorner p\ \&\ q\urcorner$ is the formula at the current stage and p is the conjunct containing the target negation or its transform, apply &E to obtain p and q and continue to execute $\mathcal{P}$ on p until p* is obtained; then apply &I to obtain $\ulcorner p*\ \&\ q\urcorner$. If q is the conjunct containing the target negation or its transform, apply &E to obtain p and q and continue to execute $\mathcal{P}$ on q until q* is obtained; then apply &I to obtain $\ulcorner p\ \&\ q*\urcorner$.

(∨) If $\ulcorner p \lor q\urcorner$ is the formula at the current stage and p is the disjunct containing the target negation or its transform, assume p and continue to execute $\mathcal{P}$ until p* is obtained, then infer $\ulcorner p* \lor q\urcorner$ by ∨I; then assume q and immediately infer $\ulcorner p* \lor q\urcorner$ by ∨I. Finally, infer $\ulcorner p* \lor q\urcorner$ by ∨E. If q contains the target negation or its transform, proceed analogously to obtain $\ulcorner p \lor q*\urcorner$ by ∨E.

(→) If $\ulcorner p \to q\urcorner$ is the formula at the current stage and the antecedent p contains the target negation or its transform (as a proper subformula), assume p* (note: *not p*) and continue to execute $\mathcal{P}$ until p is obtained. Then apply →E using $\ulcorner p \to$

q' to obtain q, then →I to obtain $\ulcorner p^* \rightarrow q \urcorner$. If $\ulcorner p \rightarrow q \urcorner$ is the formula at the current stage and the consequent q contains the target negation or its transform, assume p, derive q by →E, then continue with $\mathcal{P}$ until q^* is obtained, and then apply →I to obtain $\ulcorner p \rightarrow q^* \urcorner$.

(~) If $\ulcorner \sim p \urcorner$ is the formula at the current stage and p contains the target negation (as a proper subformula) or its transform, then assume p^* and continue to execute $\mathcal{P}$ until p is obtained; then apply ~E using $\ulcorner \sim p \urcorner$ followed by ~I, thereby rejecting p^* to obtain $\ulcorner \sim p^* \urcorner$.

(SI) If none of the previous instructions apply then the formula at the current stage is the target negation or its transform. In this case we apply SI using $\sim A \dashv \vdash_{NK} A \rightarrow \curlywedge$.

Since the same instructions serve to produce a derivation of ϕ from ψ, we have shown that every S-provable sequent is NK-provable.

Chapter 4, §11

(4) Given $\Sigma \vdash_{NK} A$, then $\Sigma, \sim A \vdash_{NK} A$ also, by definition of '$\vdash_{NK}$' ('$\Sigma \vdash_{NK} p$' requires only that p depend on a subset of Σ, so we get $\Sigma, \sim A \vdash_{NK} A$ by using the subset Σ of $\Sigma, \sim A$). Obviously, $\Sigma, \sim A \vdash_{NK} \sim A$, hence by ~E, $\Sigma, \sim A \vdash_{NK} \curlywedge$. Thus $\Sigma, \sim A$ is inconsistent. Conversely, if $\Sigma, \sim A \vdash_{NK} \curlywedge$, then by ~I, $\Sigma \vdash_{NK} \sim\sim A$, and hence by DN, $\Sigma \vdash_{NK} A$.

Chapter 5, §2

(3) 'It is not the case that there is at least one person x such that x is a mathematician and x is famous': '~(∃x)(Mx & Fx)'; where 'F_' is '_ is famous' and 'M_' is '_ is a mathematician'.

(12) 'There is at least one thing x such that x is polluted, x is a city and x is smoggy and there is at least one thing x such that x is a city and x is polluted and x is not smoggy': '(∃x)((Px & Cx) & Sx) & (∃x)((Px & Cx) & ~Sx)'; where 'C_' is '_ is a city', 'P_' is '_ is polluted' and 'S_' is '_ is smoggy'.

(15) 'If there is at least one person x such that x is wealthy and x is an economist, then there is at least one person x such that x is famous and x is a mathematician': '(∃x)(Wx & Ex) → (∃x)(Fx & Mx)'; symbols as in (3), plus 'W_' is '_ is wealthy' and 'E_' is '_ is an economist'.

Chapter 5, §3

(3) 'For all x, if x is an expensive university then x is private': '(∀x)((Ux & Ex) → Px)'. The italics in the English rule out the readings 'for all x, if x is private and

expensive, then x is a university', which says that the only expensive private things are universities ('only private *universities* are expensive'), and 'for all x, if x is expensive then x is a private university', which says that the only expensive things are private universities ('only *private universities* are expensive').

(7) 'It is not the case that for all x, if x glitters then x is gold'. Or: 'there is at least one thing x such that x glitters and x is not gold'. So: '~(∀x)(Gx → Ox)', or '(∃x)(Gx & ~Ox)'. Note that '(∀x)(Gx → ~Ox)' is wrong, since it says that everything which glitters is not gold, i.e., nothing which glitters is gold, and this is not what the English saying means.

(12) 'For all x, if x is an elected politician, then x is corrupt'; or, 'there does not exist an x such that x is an elected politician and x is incorrupt'. '(∀x)((Ex & Px) → Cx)' or '~(∃x)((Ex & Px) & ~Cx)'.

(16) 'For all x, if x is a wealthy logician then x is a textbook author'; '(∀x)((Wx & Lx) → Tx)'.

Chapter 5, §4

(I.5) This formula is not a wff since the formation rule *f*-~ does not introduce parentheses around the formula which is formed when this rule is applied. The formula should be: '(∃x)~(∃y)(Fx & ~Gy)'.

(II.3) '~' is within the scope of '&' since the '~' already occurs in the formula formed at the node where the '&' is introduced by *f*-& in the parse tree for (iii). In other words, the node where the '&' is introduced dominates the node where the '~' is introduced.

Chapter 6, §1

(3) '(∃x)(Fx & Gx)' is false because 'Fa & Ga', 'Fb & Gb' and 'Fc & Gc' are all false.

(8) '(∀x)(Hx → (∃y)(Jx & Iy))' has three instances, 'Ha → (∃y)(Ja & Iy)', 'Hb → (∃y)(Jb & Iy)' and 'Hc → (∃y)(Jc & Iy)'. The first instance is true since its antecedent is true and its consequent, '(∃y)(Ja & Iy)', is true (since 'Ja & Ic' is true). But the second instance is false, since 'Hb' is true but '(∃y)(Jb & Iy)' is false, because 'Jb & Ia', 'Jb & Ib' and 'Jb & Ic' are all false, 'Jb' being false in each case.

Chapter 6, §2

(I.4) D = {α}, Ext(F) = ∅, Ext(G) = ∅, Ext(H) = ∅. Then '(∀x)(Fx ∨ Gx) ∨ (∀x)(Fx ∨ Hx)' is false since '(∀x)(Fx ∨ Gx)' is false (because 'Fa ∨ Ga' is false) and '(∀x)(Fx ∨ Hx)' is false (because 'Fa ∨ Ha' is false). '(∀x)((Fx & Gx) → Hx)' is true because '(Fa & Ga) → Ha' is true, since its antecedent is false.

(I.13) D = {α,β}, Ext(F) = {α,β}, Ext(G) = {α}, Ext(H) = {α}, Ext(J) = ∅. '(∃x)(Fx & Gx) → (∀x)(Hx → Jx)' is false because '(∃x)(Fx & Gx)' is true ('Fa & Ga' is true) while '(∀x)(Hx → Jx)' is false ('Ha → Ja' is false). '(∀x)(Fx → Gx) → (∀x)(Hx → Jx)' is true since '(∀x)(Fx → Gx)' is false, because 'Fb → Gb' is false.

(I.20) D = {α,β}, Ext(F) = {α}, Ext(G) = ∅. '(∀x)(Fx → (∃y)Gy)' is false because 'Fa → (∃y)Gy' is false, since 'Fa' is true and '(∃y)Gy' is false ('Ga' and 'Gb' are both false). '(∀x)Fx → (∃y)Gy' is true because '(∀x)Fx' is false, since 'Fb' is false.

(I.27) D = {α,β}, Ext(F) = {α}, Ext(G) = {α}. '(∀x)[(∃y)Gy → Fx]' is false because '(∃y)Gy → Fb' is false ('(∃y)Gy' is true because 'Ga' is true, 'Fb' is false). '(∀x)(∃y)(Gy → Fx)' is true because '(∃y)(Gy → Fa)' and '(∃y)(Gy → Fb)' are both ⊤, respectively because 'Gb → Fa' and 'Gb → Fb' are ⊤.

(II.2) Suppose 𝐼 is an interpretation which verifies the sentence σ, that 𝐼's domain D contains n objects, and that $F_1...F_k$ are all the monadic predicates to which 𝐼 assigns an extension from D. Choose any element x in D. Then x has a *signature* in 𝐼, which we can take to be a k-membered sequence of pluses and minuses, where a plus in jth position in the signature indicates that x is in the extension of F_j and a minus that it is not in the extension of F_j. Two objects in D are said to be *indistinguishable* if they have the same signature (in terms of matrices, x and y have the same signature if they have the same pattern of pluses and minuses across their rows in the matrix). Though it can be proved rigorously, reflection on examples should be sufficient to convince the reader that adding an object to an interpretation and giving it the same signature as an object already in the interpretation does not affect the truth-value of any sentence in the interpretation. Hence, if 𝐼 verifies σ and 𝐼's domain contains n objects, expanding 𝐼 by adding an object with the same signature as one already in 𝐼 yields an interpretation with $n + 1$ objects on which σ is still true.

Chapter 6, §3

(I.3)

1	(1)	(∀x)(Fx → Gx)	Premise
2	(2)	(∀x)Fx	Assumption
1	(3)	Fa → Ga	1 ∀E
2	(4)	Fa	2 ∀E
1,2	(5)	Ga	3,4 →E
1,2	(6)	(∀x)Gx	5 ∀I
1	(7)	(∀x)Fx → (∀x)Gx	2,6 →I ◆

(I.7)

1	(1)	(∃x)Fx → Ga	Premise
2	(2)	Fa	Assumption
2	(3)	(∃x)Fx	2 ∃I
1,2	(4)	Ga	1,3 →E
1	(5)	Fa → Ga	2,4 →I
1	(6)	(∃x)(Fx → Gx)	5 ∃I ◆

(II.2)   All tigers are fierce: $(\forall x)(Tx \rightarrow Fx)$
       No antelope is fierce: $(\forall x)(Ax \rightarrow \sim Fx)$
       No antelope is a tiger: $(\forall x)(Ax \rightarrow \sim Tx)$

This symbolization is the most convenient. If you symbolize 'no antelope is fierce' as '$\sim(\exists x)(Ax \& Fx)$' you will probably have trouble finding a proof.

1	(1)	$(\forall x)(Tx \rightarrow Fx)$	Premise
2	(2)	$(\forall x)(Ax \rightarrow \sim Fx)$	Premise
3	(3)	Ac	Assumption
2	(4)	$Ac \rightarrow \sim Fc$	2 $\forall$E
2,3	(5)	$\sim Fc$	4,3 $\rightarrow$E
1	(6)	$Tc \rightarrow Fc$	1 $\forall$E
1,2,3	(7)	$\sim Tc$	6,5 SI (MT)
1,2	(8)	$Ac \rightarrow \sim Tc$	3,7 $\rightarrow$I
1,2	(9)	$(\forall x)(Ax \rightarrow \sim Tx)$	8 $\forall$I ◆

Chapter 6, §4

(I.2)

1	(1)	$(\exists x)Fx \vee (\exists x)Gx$	Premise
2	(2)	$(\exists x)Fx$	Assumption
3	(3)	Fa	Assumption
3	(4)	$Fa \vee Ga$	3 $\vee$I
3	(5)	$(\exists x)(Fx \vee Gx)$	4 $\exists$I
2	(6)	$(\exists x)(Fx \vee Gx)$	2,3,5 $\exists$E
7	(7)	$(\exists x)Gx$	Assumption
8	(8)	Ga	Assumption
8	(9)	$Fa \vee Ga$	8 $\vee$I
8	(10)	$(\exists x)(Fx \vee Gx)$	9 $\exists$I
7	(11)	$(\exists x)(Fx \vee Gx)$	7,8,10 $\exists$E
1	(12)	$(\exists x)(Fx \vee Gx)$	1,2,6,7,11 $\vee$E ◆

(I.8)

1	(1)	$(\exists x)(Fx \vee (Gx \& Hx))$	Premise
2	(2)	$(\forall x)(\sim Gx \vee \sim Hx)$	Premise
3	(3)	$Fa \vee (Ga \& Ha)$	Assumption
2	(4)	$\sim Ga \vee \sim Ha$	2 $\forall$E
2	(5)	$\sim(Ga \& Ha)$	4 SI (DeM)
2,3	(6)	Fa	3,5 SI (DS)
2,3	(7)	$(\exists x)Fx$	6 $\exists$I
1,2	(8)	$(\exists x)Fx$	1,3,7 $\exists$E ◆

(I.13)

1	(1)	$(\exists x)Gx$	Premise
2	(2)	Ga	Assumption
2	(3)	$Fb \rightarrow Ga$	2 SI (PMI)
2	(4)	$(\exists y)(Fb \rightarrow Gy)$	3 $\exists$I
2	(5)	$(\forall x)(\exists y)(Fx \rightarrow Gy)$	4 $\forall$I
1	(6)	$(\forall x)(\exists y)(Fx \rightarrow Gy)$	1,2,5 $\exists$E ◆

(I.17)	1	(1)	~(∃x)(∀y)(Fx → Fy)	Assumption
	2	(2)	~Fa	Assumption
	2	(3)	Fa → Fb	2 SI (PMI)
	2	(4)	(∀y)(Fa → Fy)	3 ∀I
	2	(5)	(∃x)(∀y)(Fx → Fy)	4 ∃I
	1,2	(6)	人	1,5 ~E
	1	(7)	~~Fa	2, 6 ~I
	1	(8)	Fa	7 DN
	1	(9)	Fb → Fa	8 SI (PMI)
	1	(10)	(∀y)(Fb → Fy)	9 ∀I
	1	(11)	(∃x)(∀y)(Fx → Fy)	12 ∃I
	1	(12)	人	1,11 ~E
		(13)	~~(∃x)(∀y)(Fx → Fy)	1,12 ~I
		(14)	(∃x)(∀y)(Fx → Fy)	13 DN ◆

(II.2 (→))	1	(1)	(∃x)(A ∨ Fx)	Premise
	2	(2)	A ∨ Fb	Assumption
	3	(3)	A	Assumption
	3	(4)	A ∨ (∃x)Fx	3 ∨I
	5	(5)	Fb	Assumption
	5	(6)	(∃x)Fx	5 ∃I
	5	(7)	A ∨ (∃x)Fx	6 ∨I
	2	(8)	A ∨ (∃x)Fx	2,3,4,5,7 ∨E
	1	(9)	A ∨ (∃x)Fx	1,2,8 ∃E ◆

(III.1)		(1)	(∃y)Fy ∨ ~(∃y)Fy	TI (LEM)
	2	(2)	(∃y)Fy	Assumption
	3	(3)	Fb	Assumption
	3	(4)	(∃y)Fy → Fb	3 SI (PMI)
	3	(5)	(∃x)[(∃y)Fy → Fx]	4 ∃I
	2	(6)	(∃x)[(∃y)Fy → Fx]	2,3,5 ∃E
	7	(7)	~(∃y)Fy	Assumption
	7	(8)	(∃y)Fy → Fb	7 SI (PMI)
	7	(9)	(∃x)[(∃y)Fy → Fx]	8 ∃I
		(10)	(∃x)[(∃y)Fy → Fx]	1,2,6,7,9 ∨E ◆

Chapter 6, §5

(8)	1	(1)	~(∃x)(∀y)(Fx → Fy)	Assumption
	1	(2)	(∀x)~(∀y)(Fx → Fy)	1 SI (QS)
	1	(3)	~(∀y)(Fa → Fy)	2 ∀E
	1	(4)	(∃y)~(Fa → Fy)	3 SI (QS)
	5	(5)	~(Fa → Fb)	Assumption
	5	(6)	Fa & ~Fb	5 SI (Neg-Imp)
	5	(7)	~Fb	6 &E
	5	(8)	Fb → Fc	7 SI (PMI)
	5	(9)	(∀y)(Fb → Fy)	8 ∀I

5 (10)	(∃x)(∀y)(Fx → Fy)	9 ∃I
1,5 (11)	𝆏	1,10 ~E
1 (12)	𝆏	4,5,11 ∃E
(13)	~~(∃x)(∀y)(Fx → Fy)	1,12 ~I
(14)	(∃x)(∀y)(Fx → Fy)	13 DN ◆

Chapter 6, §7

(19) Determine whether (∃x)(Fx ↔ Gx) ⊨ (∀x)Fx ↔ (∀x)Gx.

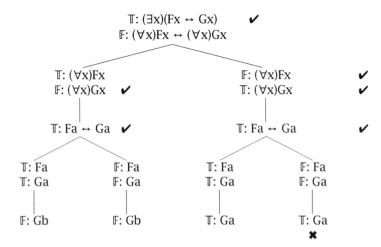

Since three branches do not close, (∃x)(Fx ↔ Gx) ⊭ (∀x)Fx ↔ (∀x)Gx.

Chapter 6, §8

(I.2) Show (∃x)Fx ∨ (∃x)Gx ⊢$_{NK}$ (∃x)(Fx ∨ Gx).

$$
\begin{array}{c}
\begin{array}{cc}
\quad & \begin{array}{c} \text{━━ (1)} \\ Fa \\ \overline{} \text{ ∨I} \\ Fa \lor Ga \end{array} \quad\quad \begin{array}{c} \text{━━ (2)} \\ Ga \\ \overline{} \text{ ∨I} \\ Fa \lor Ga \end{array}
\end{array} \\
\end{array}
$$

━━━ (3)
(∃x)Fx   (∃x)(Fx ∨ Gx)   ∃I   ━━━ (4)
(∃x)Gx   (∃x)(Fx ∨ Gx)   ∃I

(∃x)Fx ∨ (∃x)Gx   (∃x)(Fx ∨ Gx)   ∃E (1)   (∃x)(Fx ∨ Gx)   ∃E (2)

━━━ ∨E (3,4)

(∃x)(Fx ∨ Gx)

Chapter 7, §1

(I.3) '~(∃x) no one loves x'; 'no one loves x' = '~(∃y)Lyx'; so '~(∃x)~(∃y)Lyx' or '(∀x)(∃y)Lyx'.

(I.6) '(∀x)(if x loves x then x is a lover)'. 'x is a lover' = 'x loves someone' = '(∃y)Lxy'. So: '(∀x)(Lxx → (∃y)Lxy)'.

(III.4) '(∃x) x is a student and x does not read any books'; 'x does not read any books' = '(∀y) if y is a book then x does not read y'. So: '(∃x)(Sx & (∀y)(By → ~Rxy))'. Domain: things.

(III.12) '(∀x) if x is a play attributed to Shakespeare then Marlowe wrote x'; 'x is a play attributed to Shakespeare' = 'x is a play and x is attributed to Shakespeare'. So: '(∀x)((Px & Axa) → Wbx)'. Domain: things.

(III.17) '(∃x) x is a number and (x is prime if and only if all numbers ≥ x are composite)'. 'all numbers ≥ x are composite' = '(∀y) if y is at least as large as x then y is not prime'. So: '(∃x)(Nx & [Px ↔ (∀y)((Ny & Qyx) → ~Py)])'. [N.B.: The statement is true—let x be any composite number. Then because there are infinitely many primes, both sides of the biconditional are false.] Domain: things.

(III.22) '~(∃x) x is a person who trusts a politician who makes promises he can't keep' = '~(∃x)(Px & (∃y)(y is a politician who makes promises y can't keep & x trusts y))'. 'y is a politician who makes promises y can't keep' = 'Ly & (∃z) z is a promise & y makes z & y can't keep z'. So: '~(∃x)(∃y)(Px & (Ly & [(∃z)(Rz & Myz & ~Kyz)) & Txy]))'. Domain: things.

(III.27) '(∃x)(x is a composer and x is liked by anyone who likes any composer at all)'; 'x is liked by anyone who likes any composer at all' = '(∀y)(if y likes any composer at all then y likes x)'; 'y likes any composer at all' = 'y likes at least one composer'. So: '(∃x)(Cx & (∀y)[(∃z)(Cz & Lyz) → Lyx])'. Domain: people.

(III.30) '(∀x) if x pities those who pity themselves then x is lonely'. 'x pities those who pity themselves' = '(∀y)(y pities y → x pities y)'. So: '(∀x)[(∀y)(Pyy → Pxy) → Lx]'. Domain: people.

(IV.2) 'Some sequents have only finite counterexamples.'

Chapter 7, §2

(I.6) '~(∃x) x is wiser than someone else'; 'x is wiser than someone else' = '(∃y)(y ≠ x & Wxy)'. So: '~(∃x)(∃y)(y ≠ x & Wxy)'. (The English means that all people are equally wise.)

(I.13) '(∀x)(∀y)(∀z)(if (Cx & Cy & Cz) then (if x answers every question & y

answers every question & z answers every question then x = y ∨ y = z ∨ x = z))';
'x answers every question' = '(∀w)(Qw → Axw)', 'y answers every question' = '(∀w)(Qw → Ayw)', 'z answers every question' = '(∀w)(Qw → Azw)'. So:

$$(\forall x)(\forall y)(\forall z)\{(Cx \ \& \ Cy \ \& \ Cz) \rightarrow ([(\forall w)(Qw \rightarrow Axw) \ \& \ (\forall w)(Qw \rightarrow Ayw) \ \& \ (\forall w)(Qw \rightarrow Azw)] \rightarrow (x = y \lor y = z \lor x = z))\}.$$

We can use a single quantifier for the three "every"'s, which allows us to simplify the antecedent of the internal conditional to '(∀w)(Qw → (Axw & Ayw & Azw))'.

(18) There is exactly one composer liked by anyone who likes any composer at all and he is (identical to) Mozart.

'(∃x) (x is a composer liked by anyone who likes any composer at all and (∀y)(if y is a composer liked by anyone who likes any composer at all then y = x) and x = Mozart))'.

'x is a composer liked by anyone who likes any composer at all' = '(∀y)(if y likes any composer at all then y likes x)'; 'y likes any composer at all' = 'y likes at least one composer' = '(∃z)(Cz & Lyz)'. For this treatment of 'any', see the discussion of Examples 5.3.18–5.3.20 on page 162.

So: 'x is liked by anyone who likes any composer at all' = '(∀y)[(∃z)(Cz & Lyz) → Lyx]'. Thus 'y is a composer liked by anyone who likes any composer at all' = '(∀w)[(∃z)(Cz & Lwz) → Lwy]'. So for the whole formula:

$$(\exists x)[\{Cx \ \& \ (\forall y)([(\exists z)(Cz \ \& \ Lyz) \rightarrow Lyx]\} \ \& \ (\forall y)(\{Cy \ \& \ (\forall w)[(\exists z)(Cz \ \& \ Lwz) \rightarrow Lwy]\} \rightarrow y = x) \ \& \ x = m].$$

Chapter 7, §3

(I.3) The left parenthesis between the two quantifiers should not be there if the subformula following it was formed by an application of (f-q), since (f-q) does not put outer parentheses around the formulae formed using it. And either this parenthesis or the one immediately preceding 'Rxy' has no matching right parenthesis.

Chapter 7, §4

(I.4) The two readings are (a) that he would sell to no one, and (b) that he would not sell to just anyone, that is, there are certain people he would not sell to.

    (a) John would sell a picture to no one: (∀x)(Tx → ~(∃y)(Py & Sjxy))
    (b) John would not sell to just anyone: (∀x)(Tx → (∃y)~(Py & Sjxy)).

So the ambiguity lies in the relative scopes of '∃' and '~'.

(II.2) The three readings are: (a) only *private universities* are expensive (nothing else is expensive); (b) only *private* universities are expensive (no other kind of university is); (c) only private *universities* are expensive (no other private things are).

(a) $(\forall x)(Ex \rightarrow (Px \,\&\, Ux))$; (b) $(\forall x)((Ex \,\&\, Ux) \rightarrow Px)$; (c) $(\forall x)((Ex \,\&\, Px) \rightarrow Ux)$.

Chapter 8, §1

(I.5) '$(\forall x)(Rxx \rightarrow (\exists z)Sxz)$' is false because 'Rdd $\rightarrow (\exists z)Sdz$' is false, since 'Rdd' is true and '$(\exists z)Sdz$' is false. '$(\exists z)Sdz$' is false because 'Sda', 'Sdb', 'Sdc' and 'Sdd' are all false, since none of $\langle \delta,\alpha \rangle$, $\langle \delta,\beta \rangle$, $\langle \delta,\gamma \rangle$ or $\langle \delta,\delta \rangle$ belongs to Ext(S).

(II.4) '$(\forall x)(\forall y)(Rxy \leftrightarrow {\sim}Syx)$' is false because '$(\forall y)(Ray \leftrightarrow {\sim}Sya)$' is false, because 'Rab $\leftrightarrow {\sim}$Sba' is false, because 'Rab' is true and 'Sba' is true.

(III.4) '$(\exists x)(\forall y)(y \neq a \rightarrow (\exists z)Sxyz)$' is true because '$(\forall y)(y \neq a \rightarrow (\exists z)S1yz)$' is true, since for every element y from the domain other than 10, $z = 1 + y$ is a member of the domain.

(IV.3) '$(\forall x)(x \neq 1 \rightarrow (\exists y)\, y > x)$' is true since for any number $n \in [0,1]$, '$\mathbf{n} \neq 1 \rightarrow (\exists y)\, y > \mathbf{n}$' is true; when $n = 1$ the antecedent is false, and for every other n, '$(\exists y)\, y > \mathbf{n}$' is true since '$1 > \mathbf{n}$' is true. Here we use boldface '$\mathbf{n}$' to mean the standard numeral for the number n. (In fact, for any particular $n \neq 1$, there are infinitely many numbers y in [0,1] such that $y > n$, but 1 can be used in every case).

Chapter 8, §2

(I.4) D = $\{\alpha,\beta\}$, Ext(R) = $\{\langle \alpha,\beta \rangle, \langle \beta,\beta \rangle\}$. '$(\exists x){\sim}Rxb$' is false because '${\sim}Rab$' is false and '${\sim}Rbb$' is false. '$(\forall x)(Rxa \rightarrow {\sim}Rxb)$' is true because 'Raa $\rightarrow {\sim}$Rab' and 'Rba $\rightarrow {\sim}$Rbb' are both ⊤. '$(\forall x)(\exists y)Rxy$' is true because '$(\exists y)Ray$' and '$(\exists y)Rby$' are both ⊤, respectively because 'Rab' and 'Rbb' are both ⊤.

(I.12) D = $\{\alpha,\beta\}$, Ext(F) = Ext(G) = Ext(H) = $\{\alpha\}$, Ext(R) = $\{\langle \alpha,\alpha \rangle\}$. '$(\forall x)(\forall y)((Fx \,\&\, Hx) \rightarrow Rxy)$' is false because '(Fa & Ha) $\rightarrow$ Rab' is false. '$(\forall x)(Hx \rightarrow Gx)$' is true because 'Ha $\rightarrow$ Ga' and 'Hb $\rightarrow$ Gb' are true. '$(\forall x)(Fx \rightarrow (\forall y)(Gy \rightarrow Rxy))$' is true because 'Fb $\rightarrow (\forall y)(Gy \rightarrow$ Rby)' is true (false antecedent) and 'Fa $\rightarrow (\forall y)(Gy \rightarrow$ Ray)' is true because 'Fa' is true and '$(\forall y)(Gy \rightarrow$ Ray)' is true. '$(\forall y)(Gy \rightarrow$ Ray)' is true since 'Gb $\rightarrow$ Rab' is true (false antecedent) and 'Ga $\rightarrow$ Raa' is true (⊤ $\rightarrow$ ⊤).

(II.3) D = $\{\alpha\}$, Ext(F) = $\{\alpha\}$. Then 'Ga' is false, and 'Ga $\rightarrow$ Fa' is true so '$(\forall x)(Gx \rightarrow Fx)$' is true. α is the only thing which is F, so the first premise is also true.

(III.2) D = $\{\alpha,\beta\}$, Ext($\in$) = $\{\langle \alpha,\beta \rangle, \langle \beta,\alpha \rangle\}$, Ext(S) = $\{\alpha,\beta\}$. (E) is true since $\alpha \neq \beta$,

$\alpha \in \beta$, $\beta \notin \beta$. But '$(\exists x)(\forall y)y \notin x$' is false since '$(\forall y)y \notin a$' and '$(\forall y)y \notin b$' are both false.

Chapter 8, §3

(I.3)

1	(1)	$(\exists x)Fx \rightarrow (\forall x)(Fx \rightarrow Gxa)$	Premise
2	(2)	$(\exists x)Hx \rightarrow (\forall x)(Hx \rightarrow Jxa)$	Premise
3	(3)	$(\exists x)(Fx \;\&\; Hx)$	Assumption
4	(4)	$Fb \;\&\; Hb$	Assumption
4	(5)	Fb	4 &E
4	(6)	Hb	4 &E
4	(7)	$(\exists x)Fx$	5 ∃I
1,4	(8)	$(\forall x)(Fx \rightarrow Gxa)$	1,7 →E
4	(9)	$(\exists x)Hx$	6 ∃I
2,4	(10)	$(\forall x)(Hx \rightarrow Jxa)$	2,9 →E
1,4	(11)	$Fb \rightarrow Gba$	8 ∀E
1,4	(12)	Gba	11,5 →E
2,4	(13)	$Hb \rightarrow Jba$	10 ∀E
2,4	(14)	Jba	13,6 →E
1,2,4	(15)	$Gba \;\&\; Jba$	12,14 &I
1,2,4	(16)	$(\exists y)(Gby \;\&\; Jby)$	15 ∃I
1,2,4	(17)	$(\exists x)(\exists y)(Gxy \;\&\; Jxy)$	16 ∃I
1,2,3	(18)	$(\exists x)(\exists y)(Gxy \;\&\; Jxy)$	3,4,17 ∃E
1,2	(19)	$(\exists x)(Fx \;\&\; Hx) \rightarrow (\exists x)(\exists y)(Gxy \;\&\; Jxy)$	3,18 →I ◆

(I.7)

1	(1)	$(\forall x)(\forall y)(Exy \rightarrow Eyx)$	Premise
2	(2)	$(\forall x)(\forall y)(Exy \rightarrow Exx)$	Premise
3	(3)	$(\exists y)Eya$	Assumption
4	(4)	Eba	Assumption
1	(5)	$(\forall y)(Eby \rightarrow Eyb)$	1 ∀E
1	(6)	$Eba \rightarrow Eab$	5 ∀E
1,4	(7)	Eab	6,4 →E
2	(8)	$(\forall y)(Eay \rightarrow Eaa)$	2 ∀E
2	(9)	$Eab \rightarrow Eaa$	8 ∀E
1,2,4	(10)	Eaa	9,7 →E
1,2,3	(11)	Eaa	3,4,10 ∃E
1,2	(12)	$(\exists y)Eya \rightarrow Eaa$	3,11 →I
1,2	(13)	$(\forall x)[(\exists y)Eyx \rightarrow Exx]$	12 ∀I ◆

Note that ∃E at (11) is legal since the name used to form the instance of (3) is 'b', and 'b' does not occur in 3, 10, or 1 and 2.

(I.12)

1	(1)	$(\exists x)(\forall y)((Fx \lor Gy) \rightarrow (\forall z)(Hxy \rightarrow Hyz))$	Premise
2	(2)	$(\exists z)(\forall x)\sim Hxz$	Premise
3	(3)	$(\forall y)((Fa \lor Gy) \rightarrow (\forall z)(Hay \rightarrow Hyz))$	Assumption
4	(4)	$(\forall x)\sim Hxb$	Assumption

5	(5)	Fa	Assumption
3	(6)	(Fa ∨ Gc) → (∀z)(Hac → Hcz)	3 ∀E
5	(7)	Fa ∨ Gc	5 ∨I
3,5	(8)	(∀z)(Hac → Hcz)	6,7 →E
3,5	(9)	Hac → Hcb	8 ∀E
4	(10)	~Hcb	4 ∀E
3,4,5	(11)	~Hac	9,10 SI (MT)
3,4	(12)	Fa → ~Hac	5,11 →I
3,4	(13)	(∀x)(Fa → ~Hax)	12 ∀I
3,4	(14)	(∃y)(∀x)(Fy → ~Hyx)	13 ∃I
1,3	(15)	(∃y)(∀x)(Fy → ~Hyx)	2,4,14 ∃E
1,2	(16)	(∃y)(∀x)(Fy → ~Hyx)	1,3,15 ∃E ◆

(II.4) T_: _ is tall; A_: _ applied; L_,_: _ is taller than _; b: John; c: Mary

1	(1)	(∀x)(Ax → ~Tx)	Premise
2	(2)	Tc & Lbc	Premise
3	(3)	(∀x)[(∃y)(Ty & Lxy) → Tx]	Premise
4	(4)	Ab	Assumption
1	(5)	Ab → ~Tb	1 ∀E
1,4	(6)	~Tb	4,5 →E
3	(7)	(∃y)(Ty & Lby) → Tb	3 ∀E
2	(8)	(∃y)(Ty & Lby)	2 ∃I
2,3	(9)	Tb	7,8 →E
1,2,3,4	(10)	⋏	6,9 ~E
1,2,3	(11)	~Ab	4,10 ~I ◆

(II.7) S_: _ is a student; C_: _ cheated; P_: _ is a professor; B_,_: _ bribed _; A_: _ was accused

1	(1)	(∀x)((Sx & Cx) → (∃y)(Py & Bxy))	Premise
2	(2)	(∃x)[(Sx & Ax) & (∀y)((Py & Bxy) → Ay)]	Premise
3	(3)	(∀x)((Sx & Ax) → Cx)	Premise
4	(4)	(Sd & Ad) & (∀y)((Py & Bdy) → Ay)	Assumption
4	(5)	Sd & Ad	4 &E
3	(6)	(Sd & Ad) → Cd	3 ∀E
3,4	(7)	Cd	6,5 →E
4	(8)	Sd	5 &E
3,4	(9)	Sd & Cd	8,7 &I
1	(10)	(Sd & Cd) → (∃y)(Py & Bdy)	1 ∀E
1,3,4	(11)	(∃y)(Py & Bdy)	10,9 →E
12	(12)	Pe & Bde	Assumption
4	(13)	(∀y)((Py & Bdy) → Ay)	4 &E
4	(14)	(Pe & Bde) → Ae	13 ∀E
4,12	(15)	Ae	14,12 →E
3	(16)	(Se & Ae) → Ce	3 ∀E
17	(17)	Se	Assumption

4,12,17	(18)	Se & Ae	17,15 &I
3,4,12,17	(19)	Ce	16,18 →E
3,4,12	(20)	Se → Ce	17,19 →I
3,4,12	(21)	~Se ∨ Ce	20 SI (Imp)
12	(22)	Pe	12 &E
3,4,12	(23)	Pe & (~Se ∨ Ce)	22,21 &I
3,4,12	(24)	(∃x)(Px & (~Sx ∨ Cx))	23 ∃I
1,3,4	(25)	(∃x)(Px & (~Sx ∨ Cx))	11,12,24 ∃E
1,2,3	(26)	(∃x)(Px & (~Sx ∨ Cx))	2,4,25 ∃E ◆

Chapter 8, §4

(I.7)

1	(1)	(∃x)(∃y)Hxy	Premise
2	(2)	(∀y)(∀z)(Dyz ↔ Hzy)	Premise
3	(3)	(∀x)(∀y)(~Hxy ∨ x = y)	Premise
4	(4)	(∃y)Hay	Assumption
5	(5)	Hab	Assumption
2	(6)	Dba ↔ Hab	2 ∀E twice
2,5	(7)	Dba	5,6 SI
3	(8)	~Hab ∨ a = b	3 ∀E twice
3	(9)	Hab → a = b	8 SI (Imp)
3,5	(10)	a = b	9,5 →E
3,5	(11)	Hbb	10,5 =E
2,5	(12)	Dbb	10,7 =E
2,3,5	(13)	Hbb & Dbb	11,12 &I
2,3,5	(14)	(∃x)(Hxx & Dxx)	13 ∃I
2,3,4	(15)	(∃x)(Hxx & Dxx)	4,5,14 ∃E
1,2,3	(16)	(∃x)(Hxx & Dxx)	1,4,15 ∃E ◆

(III.2)

1	(1)	(∀x)(∀y)((Sx & Sy) → (x = y ↔ (∀z)(z ∈ x ↔ z ∈ y)))	Premise
2	(2)	(Sa & Sb) & [~(∃z)z ∈ a & ~(∃z)z ∈ b]	Assumption
1	(3)	(Sa & Sb) → (a = b ↔ (∀z)(z ∈ a ↔ z ∈ b))	1 ∀E
2	(4)	Sa & Sb	2 &E
1,2	(5)	a = b ↔ (∀z)(z ∈ a ↔ z ∈ b)	3,4 →E
2	(6)	~(∃z)z ∈ a & ~(∃z)z ∈ b	2 &E
2	(7)	~(∃z)z ∈ a	6 &E
2	(8)	~(∃z)z ∈ b	6 &E
2	(9)	(∀z)z ∉ a	7 SI (QS)
2	(10)	(∀z)z ∉ b	8 SI (QS)
2	(11)	c ∉ a	9 ∀E
2	(12)	c ∉ b	10 ∀E
2	(13)	c ∉ a & c ∉ b	11,12 &I
2	(14)	(c ∈ a & c ∈ b) ∨ (c ∉ a & c ∉ b)	13 ∨I
2	(15)	c ∈ a ↔ c ∈ b	14 SI ((16), page 119)

2	(16)	$(\forall z)(z \in a \leftrightarrow z \in b)$	15 $\forall$I
1,2	(17)	$a = b$	5,16 SI
1	(18)	$((Sa \ \& \ Sb) \ \&$ $[{\sim}(\exists z)z \in a \ \& \ {\sim}(\exists z)z \in b]) \to a = b$	2,17 $\to$I
1	(19)	$(\forall y)[((Sa \ \& \ Sy) \ \&$ $[{\sim}(\exists z)z \in a \ \& \ {\sim}(\exists z)z \in y]) \to a = y]$	18 $\forall$I
1	(20)	$(\forall x)(\forall y)[((Sx \ \& \ Sy) \ \&$ $[{\sim}(\exists z)z \in x \ \& \ {\sim}(\exists z)z \in y]) \to x = y]$	19 $\forall$I ◆

Chapter 8, §5

(I.4) 'is similar in color to' is reflexive and symmetric, but not transitive. Consider a sequence of objects each of which is similar in color to objects adjacent to it. The last object in the sequence may not be similar in color to the first. (See further Chapter Eleven.)

(I.6) 'semantically entails' is single-premise reflexive and transitive, since for every p, $p \vDash p$ and for every $p, q, r, p \vDash q$ and $q \vDash r$ imply that $p \vDash r$ (suppose that $p \vDash q$ and $q \vDash r$, then if some assignment makes p true and r false, that assignment makes q false, since $q \vDash r$, but then $p \nvDash q$, contrary to hypothesis). But 'semantically entails' is not symmetric: A & B $\vDash$ A but A $\nvDash$ A & B.

(II.2) D = $\{\alpha, \beta\}$; Ext(R) = $\{\langle\alpha,\beta\rangle, \langle\beta,\alpha\rangle, \langle\alpha,\alpha\rangle, \langle\beta,\beta\rangle\}$. Then R is reflexive and transitive (note that '(Rab & Rba) $\to$ Raa' and '(Rba & Rab) $\to$ Rbb' are both true, and also totally connected, but R is not anti-symmetric: '((Rab & Rba) $\to$ a = b)' is false.

(II.6) D = $\{\alpha, \beta, \gamma\}$; Ext(R) = $\{\langle\alpha,\alpha\rangle, \langle\beta,\beta\rangle, \langle\gamma,\gamma\rangle, \langle\alpha,\beta\rangle, \langle a,\gamma\rangle\}$. Then R is reflexive and transitive, but not directed, since '(Rab & Rac) $\to$ ($\exists$w)(Rbw & Rcw)' is false. 'Rab & Rac' is true but '($\exists$w)(Rbw & Rcw)' is false, since 'Rba & Rca', 'Rbb & Rcb' and 'Rbc & Rcc' are all false.

(III.2)	1	(1)	$(\forall x)(\forall y)(\forall z)((Rxy \ \& \ Ryz) \to Rxz)$	Premise
	2	(2)	$(\forall x)(\forall y)(\forall z)((Rxy \ \& \ Ryz) \to {\sim}Rxz)$	Premise
	3	(3)	Rab	Assumption
	4	(4)	Rba	Assumption
	3,4	(5)	Rab & Rba	3,4 &I
	1	(6)	(Rab & Rba) $\to$ Raa	1 $\forall$E thrice
	1,3,4	(7)	Raa	6,5 $\to$E
	2	(8)	(Rab & Rba) $\to$ ${\sim}$Raa	2 $\forall$E thrice
	2,3,4	(9)	${\sim}$Raa	8,5 $\to$E
	1,2,3,4	(10)	$\curlywedge$	9,7 ${\sim}$E
	1,2,3	(11)	${\sim}$Rba	4,10 ${\sim}$I
	1,2	(12)	Rab $\to$ ${\sim}$Rba	3,11 $\to$I
	1,2	(13)	$(\forall y)(Ray \to {\sim}Rya)$	12 $\forall$I
	1,2	(14)	$(\forall x)(\forall y)(Rxy \to {\sim}Ryx)$	13 $\forall$I ◆

(IV)	1	(1)	$(\forall x)(\forall y)(\forall z)((Rxy \ \& \ Rxz) \rightarrow Ryz)$	Premise
	2	(2)	Rab	Assumption
	1	(3)	$(Rab \ \& \ Rab) \rightarrow Rbb$	1 $\forall$E thrice
	2	(4)	Rab & Rab	2,2 &I
	1,2	(5)	Rbb	3,4 $\rightarrow$E
	1	(6)	Rab $\rightarrow$ Rbb	2,5 $\rightarrow$I
	1	(7)	$(\forall y)(Ray \rightarrow Ryy)$	6 $\forall$I
	1	(8)	$(\forall x)(\forall y)(Rxy \rightarrow Ryy)$	7 $\forall$I ◆

Chapter 8, §8

(I.2) $(\exists X)(Xa \ \& \ \sim Xb)$.   (I.5) $(\forall x)(Xa \leftrightarrow Xb)$.

(II.2) Let D = {1,2,3,...,10}, Ext(F) = {1,2,3,4,5,6}, Ext(G) = {1,2}. Then 'Few Fs are G' is true because only two of the six Fs are G. But '($\mathcal{W}$x)(Fx $\rightarrow$ Gx)' is false since it has six true instances on this interpretation. Therefore '($\mathcal{W}$x)(Fx $\rightarrow$ Gx)' is not a correct symbolization of 'Few Fs are G'.

Chapter 9, §2

(6) $W = (w^*, u)$; w^*: A $\mapsto \perp$, B $\mapsto \perp$; u: A $\mapsto \top$, B $\mapsto \perp$. Then $w^*[A \rightarrow B] = \top$, hence $w^*[\diamond(A \rightarrow B)] = \top$. But $u[A] = \top$ so $w^*[\diamond A] = \top$, while $w^*[B] = \perp$ and $u[B] = \perp$, so $w^*[\diamond B] = \perp$. Therefore $w^*[\diamond A \rightarrow \diamond B] = \perp$.

(11) $W = (w^*, u)$; w^*: A $\mapsto \perp$, B $\mapsto \top$ (or $\perp$); u: A $\mapsto \top$, B $\mapsto \perp$. $u[A \rightarrow B] = \perp$, hence $w^*[\Box(A \rightarrow B) = \perp$. But $w^*[\Box A] = \perp$ since $w^*[A] = \perp$. Hence $w^*[\Box A \rightarrow \diamond B] = \top$.

Chapter 9, §3

(I.4) Trans$[\Box(A \rightarrow \diamond \lambda)] = (\forall w)$Trans$[A \rightarrow \diamond \lambda] = (\forall w)($Trans$[A] \rightarrow$ Trans$[\diamond \lambda]) = (\forall w)(A'w \rightarrow (\exists w)$Trans$[\lambda]) = (\forall w)(A'w \rightarrow (\exists w)\lambda)$.

(II.2) '$(\exists w)(\forall w)A'w \rightarrow A'w$' is the translation of the conditional '$\diamond \Box A \rightarrow A$'.

Chapter 9, §4

(6)	1	(1)	$\Box(A \ \& \ \diamond B)$	Premise
	1	(2)	A & $\diamond$B	1 $\Box$E
	1	(3)	A	2 &E
	1	(4)	$\diamond$B	2 &E
	5	(5)	B	Assumption
	1,5	(6)	A & B	3,5 &I
	1,5	(7)	$\diamond$(A & B)	6 $\diamond$I
	1	(8)	$\diamond$(A & B)	4,5,7 $\diamond$E ◆

(19)	1	(1)	$\Box(A \to \Box B)$	Premise
	2	(2)	$\Diamond A$	Assumption
	3	(3)	A	Assumption
	1	(4)	$A \to \Box B$	1 $\Box$E
	1,3	(5)	$\Box B$	4,3 $\to$E
	1,2	(6)	$\Box B$	2,3,5 $\Diamond$E
	1,2	(7)	B	6 $\Box$E
	1	(8)	$\Diamond A \to B$	2,7 $\to$I
		(9)	$\Box(\Diamond A \to B)$	8 $\Box$I ◆

Chapter 9, §5

(5) $W = \{w^*, u\}$, $w^*(D) = \{\alpha\}$, $u(D) = \{\alpha\}$ (or $\varnothing$), $w^*[F] = \{\alpha\}$, $w^*[G] = \varnothing$, $u[F] = \varnothing$, $u[G] = \{\alpha\}$. Then $w^*[\Box Fa] = \bot$, $w^*[\Box Ga] = \bot$, so $w^*[(\forall x)(\Box Fx \leftrightarrow \Box Gx)] = \top$. However, $w^*[Fa \leftrightarrow Ga] = \bot$ so $w^*[\Box(Fa \leftrightarrow Ga)] = \bot$ so $w^*[(\forall x)\Box(Fx \leftrightarrow Gx)] = \bot$.

(10) $W = \{w^*, u\}$, $w^*(D) = \{\alpha\}$, $u(D) = \{\beta\}$. $w^*[R] = \varnothing$, $u[R] = \langle \alpha, \beta \rangle$. Then $u[(\exists y)Ray] = \top$ so $w^*[\Diamond(\exists y)Ray] = \top$. $\alpha \in w^*(D)$, so $w^*[(\exists x)\Diamond(\exists y)Rxy] = \top$. However, $w^*[(\exists x)(\exists y)Rxy] = \bot$ and also $u[(\exists x)(\exists y)Rxy] = \bot$ ('$(\exists y)Ray$', though true at u, does not make '$(\exists x)(\exists y)Rxy$' true at u since $\alpha \notin u(D)$). So $w^*[\Diamond(\exists x)(\exists y)Rxy] = \bot$.

(16) $W = \{w^*, u\}$, $w^*(D) = \varnothing$, $u(D) = \{\alpha\}$, $w^*[F] = \{\alpha\}$ (or $\varnothing$), $u[F] = \{\alpha\}$. Then $w^*[(\forall x)\Box Ex] = \top$ since $w^*(D) = \varnothing$, and $w^*[(\exists x)\Diamond Fx] = \bot$ for the same reason. But $u[(\exists x)Fx] = \top$ so $w^*[\Diamond(\exists x)Fx] = \top$.

Chapter 9, §6

(5)	1	(1)	$(\forall x)\Box Ex$	Premise
	2	(2)	$(\exists x)\Diamond Fx$	Assumption
	3	(3)	$Ea \& \Diamond Fa$	Assumption
	3	(4)	Ea	3 &E
	1	(5)	$Ea \to \Box Ea$	1 $\forall$E
	1,3	(6)	$\Box Ea$	5,4,$\to$E
	7	(7)	$\Box Ea$	Assumption
	3	(8)	$\Diamond Fa$	3 &E
	9	(9)	Fa	Assumption
	7	(10)	Ea	7 $\Box$E
	7,9	(11)	$Ea \& Fa$	10, 9 &I
	7,9	(12)	$(\exists x)Fx$	11 $\exists$I
	7,9	(13)	$\Diamond(\exists x)Fx$	12 $\Diamond$I
	7,3	(14)	$\Diamond(\exists x)Fx$	8,9,13 $\Diamond$E
	3	(15)	$\Box Ea \to \Diamond(\exists x)Fx$	7,14 $\to$I
	1,3	(16)	$\Diamond(\exists x)Fx$	15,6 $\to$E
	1,2	(17)	$\Diamond(\exists x)Fx$	2,3,16 $\exists$E
	1	(18)	$(\exists x)\Diamond Fx \to \Diamond(\exists x)Fx$	2,17 $\to$I ◆

(12)	1	(1)	◇(∀x)□Ex	Premise
	2	(2)	□(∀x)Fx	Assumption
	3	(3)	(∀x)□Ex	Assumption
	3	(4)	Ea → □Ea	3 ∀E
	2	(5)	(∀x)Fx	2 □E
	2	(6)	Ea → Fa	5 ∀E
	7	(7)	□Ea	Assumption
	7	(8)	Ea	7 □E
	2,7	(9)	Fa	6,8 →E
	2,7	(10)	□Fa	9 □I
	2,7	(11)	Ea → □Fa	10 SI (PMI)
	2	(12)	□Ea → (Ea → □Fa)	7,11 →I
	2,3	(13)	Ea → (Ea → □Fa)	4,12 SI (Chain)
	2,3	(14)	Ea → □Fa	13 SI (NK)
	2,3	(15)	(∀x)□Fx	14 ∀I
	3	(16)	□(∀x)Fx → (∀x)□Fx	2,15 →I
	3	(17)	◇[□(∀x)Fx → (∀x)□Fx]	16 ◇I
	1	(18)	◇[□(∀x)Fx → (∀x)□Fx]	1,3,17 ◇E ◆

(19)	1	(1)	□(∀x)□Ex	Premise
	2	(2)	A	Premise
	3	(3)	◇(∀x)(Fx → □(A → ~Fx))	Premise
	4	(4)	◇A	Assumption
	1	(5)	□Ea	1 SI (Example 3)
	6	(6)	(∀x)(Fx → □(A → ~Fx))	Assumption
	7	(7)	□(A → ~Fa)	Assumption
	8	(8)	A	Assumption
	7	(9)	A → ~Fa	7 □E
	7,8	(10)	~Fa	9, 8 →E
	7,8	(11)	◇~Fa	10 ◇I
	4,7	(12)	◇~Fa	4,8,11 ◇E
	4	(13)	□(A → ~Fa) → ◇~Fa	7,12 →I
		(14)	Fa ∨ ~Fa	TI (LEM)
	15	(15)	Fa	Assumption
	6	(16)	Ea → (Fa → □(A → ~Fa))	6 ∀E
	1	(17)	Ea	5 □E
	1,6	(18)	Fa → □(A → ~Fa)	16,17 →E
	1,4,6	(19)	Fa → ◇~Fa	18,13 SI (Chain)
	1,4,6,15	(20)	◇~Fa	19,15 →E
	21	(21)	~Fa	Assumption
	21	(22)	◇~Fa	21 ◇I
	1,4,6	(23)	◇~Fa	14,15,20,21,22 ∨E
	1,3,4	(24)	◇~Fa	3,6,23 ◇E
	1,3,4	(25)	Ea & ◇~Fa	17,24 &I
	1,3,4	(26)	(∃x)◇~Fx	25 ∃I
	1,3,4	(27)	□(∃x)◇~Fx	26 □I
	1,3	(28)	◇A → □(∃x)◇~Fx	4,27 →I

	(29)	$A \to \Diamond A$	TI
1,3	(30)	$A \to \Box(\exists x)\Diamond{\sim}Fx$	29,28 SI (Chain)
1,2,3	(31)	$\Box(\exists x)\Diamond{\sim}Fx$	30,2 $\to$E ◆

Chapter 10, §2

(I.5) $S = \{\alpha,\beta,\gamma,\delta\}$, $\leqslant = \{\langle\alpha,\beta\rangle, \langle\beta,\gamma\rangle, \langle\alpha,\delta\rangle\}$, $warr(\alpha) = \varnothing$, $warr(\beta) = \{\text{'A'}\}$, $warr(\gamma)$ = $\{\text{'A'}, \text{'B'}\}$, $warr(\delta) = \varnothing$. Since $\alpha \leqslant \beta$, $\beta \Vdash A$ and $\beta \nVdash B$, we have $\alpha \nVdash A \to B$. However, $\alpha \Vdash {\sim}B \to {\sim}A$, since $\alpha \leqslant \delta$, $\delta \Vdash {\sim}B$, $\delta \Vdash {\sim}A$, and there is no other state at which '${\sim}B$' holds.

(I.12) $S = \{\alpha,\beta\}$, $\leqslant = \{\langle\alpha,\beta\rangle\}$, $warr(\alpha) = \varnothing$, $warr(\beta) = \{\text{'A'}\}$. Since $warr(\alpha)$ is empty, $\alpha \nVdash A$. Since $\beta \Vdash A$ and $\beta \nVdash B$, $\alpha \nVdash A \to B$. Hence $\alpha \nVdash A \vee (A \to B)$.

Chapter 10, §3

(6) $S = \{\alpha,\beta\}$, $\leqslant = \{\langle\alpha,\beta\rangle\}$, $dom(\alpha) = \{\mathbf{0}\}$, $dom(\beta) = \{\mathbf{0},\mathbf{2}\}$, $warr(\alpha) = \varnothing$, $warr(\beta) = \{\langle F,\mathbf{0}\rangle, \langle G,\mathbf{2}\rangle\}$. $\alpha \nVdash (\exists x)(Fx \to Gx)$ since $\alpha \nVdash Fa \to Ga$, since $\beta \Vdash Fa$ but $\beta \nVdash Ga$. However, since $\beta \Vdash (\exists x)Fx$ and $\beta \Vdash (\exists x)Gx$, we have $\alpha \Vdash (\exists x)Fx \to (\exists x)Gx$.

(9) $S = \{\alpha,\beta\}$, $\leqslant = \{\langle\alpha,\beta\rangle\}$, $dom(\alpha) = \{\mathbf{0}\}$, $dom(\beta) = \{\mathbf{0},\mathbf{2}\}$, $warr(\alpha) = \{F,\mathbf{0}\}$, $warr(\beta)$ = $\{F,\mathbf{0}\}$. $\beta \nVdash (\forall y)(F\mathbf{0} \to Fy)$ since $\beta \nVdash F\mathbf{0} \to F\mathbf{2}$. Hence $\alpha \nVdash (\exists x)(\forall y)(Fx \to Fy)$.

Bibliography

Boolos, G. and Jeffrey, R. *Computability and Logic*. 3d edition, Cambridge University Press, 1989.

Bull, R. and Segerberg, K. 'Basic Modal Logic'. In D. Gabbay and F. Guenthner, 1984, pp. 1-88.

Burgess, J. 'Basic Tense Logic'. In D. Gabbay and F. Guenthner, 1984, pp. 89-134.

Davidson, D. *Essays on Actions and Events*. Oxford University Press, 1980.

Dummett, M. 'The Philosophical Basis of Intuitionistic Logic'. In *Truth and Other Enigmas* by Michael Dummett. Harvard University Press 1978, pp. 215-247.

Forbes, G. 'But *a* Was Arbitrary'. *Philosophical Topics*, Fall 1993.

Gabbay, D. and Guenthner, F. (eds.). *The Handbook of Philosophical Logic*. Vol. 1, *Elements of Classical Logic;* Vol. 2, *Extensions of Classical Logic;* Vol. 3, *Alternatives to Classical Logic;* Vol. 4, *Topics in the Philosophy of Language*. D. Reidel, 1983, 1984, 1986, 1989.

Gentzen, G. 'Investigations into Logical Deduction'. In M. Szabo (ed.), *The Collected Papers of Gerhard Gentzen*. North Holland, 1969, pp. 68-131.

Grice, H. P., 'Logic and Conversation'. In A. P. Martinich (ed.), *The Philosophy of Language*. Oxford University Press, 1985, pp. 159-170.

Gustason, W. and Ulrich, D. *Elementary Symbolic Logic*. 2d ed., Waveland Press, 1989.

Harel, D. *Algorithmics*. Addison-Wesley, 1987.

Hintikka, J. "'is', Semantical Games and Semantical Relativity". *Journal of Philosophical Logic* 8 (1979): 433-68.

Hodges, W. 'Elements of Classical Logic'. In D. Gabbay and F. Guenthner 1983, pp. 1-131.

Jeffrey, R. *Formal Logic—Its Scope and Limits*, Second Edition, McGraw-Hill 1991.

Johnson-Laird, P. and Byrne, R. *Deduction*. Lawrence Erlbaum Associates, 1991.

Kandel, A. *Fuzzy Mathematical Techniques with Applications*. Addison Wesley, 1986.

Lemmon, E. J. *Beginning Logic*. Hackett, 1978.

Mendelson, E. *Introduction to Mathematical Logic*. Van Nostrand, 1979.

Partee, B., ter Meulen, A. and Wall, R. *Mathematical Methods in Linguistics*. Kluwer, 1990.

Prawitz, D. *Natural Deduction.* Almqvist & Wiksell, 1968.

Read, S. *Relevant Logic.* Basil Blackwell, 1988.

Russell, B. 'Lectures on Logical Atomism'. In *Logic and Knowledge*, R. C. Marsh (ed.), George Allen and Unwin, 1956.

Shapiro, S. *Foundations Without Foundationalism.* Oxford University Press, 1992.

Smullyan, R. *What Is the Name of This Book?* Simon & Schuster, 1978.

Strawson, P. F. *Logico-Linguistic Papers.* Methuen, 1971.

Sundholm, G. 'Systems of Deduction'. In D. Gabbay and F. Guenthner, 1984, pp. 133-188.

Van Benthem, J. and Doets, K. 'Higher-Order Logic'. In D. Gabbay and F. Guenthner, 1983, pp. 275-330.

Van Dalen, D. 'Intuitionistic Logic'. In D. Gabbay and F. Guenthner 1986, pp. 225-339.

Westerståhl, D. 'Quantifiers in Formal and Natural Languages'. In D. Gabbay and F. Guenthner 1989, pp. 1-131.

Index